Fitness Landscapes and the Origin of Species

MONOGRAPHS IN POPULATION BIOLOGY

EDITED BY SIMON A. LEVIN AND HENRY S. HORN

Completed series list follows Index

Fitness Landscapes and the Origin of Species

Sergey Gavrilets

PRINCETON UNIVERSITY PRESS
Princeton and Oxford
2004

Library of Congress Control Number 2003116848

ISBN: 0-691-11758-6   0-691-11983-X (paper)
British Library Cataloging-in-Publication Data is available

The publisher would like to acknowledge the author of this volume for
providing the camera-ready copy from which this book was printed

Printed on acid-free paper.

www.pupress.princeton.edu

Printed in the United States of America

10 9 8 7 6 5 4 3 2 1

To Galia

Contents

PART III
SPECIATION VIA THE JOINT ACTION OF DISRUPTIVE
NATURAL SELECTION AND NONRANDOM MATING

Preface

This book attempts to summarize and generalize in a systematic way what mathematical models tell us about the dynamics of speciation. Although the theoretical speciation research is 40 years old, there is not a single book or even a single journal review that has attempted to provide such a summary. The relevant theoretical literature is scattered across various journals. This has definitely contributed to the persistence of different controversies, common neglect of the previous work, and occasional confusion and propagation of unvalidated claims. This is definitely a good time to write such a book because currently there is a dramatic growth in the interest in speciation among biologists and an explosion in the amount of new mathematical theory. Given this growth, in a few years a comprehensive review of theory will hardly be feasible. Besides providing an overview of previous work, significant components of this book are my own previously unpublished results.

The intended audience of the book includes researchers, graduate students, and advanced undergraduates in life sciences, "complexity theory," mathematical biology, and applied mathematics. Although I do present and discuss a lot of different equations, the mathematical derivations are almost completely avoided. Therefore I believe that empirical biologists can handle the content of the book in spite of their fear of math, once they realize that the tiger is in the cage. Almost 90 figures illustrating different features of mathematical equations will help as well.

Although my review of the previous theoretical work is rather comprehensive, still, some existing models and papers have not been included intentionally. These are mostly nongenetic models utilizing mathematical approximations that I find difficult to justify biologically, as well as some models which, in my opinion, have limited applicability outside the specific biological situation they were meant to describe. I apologize if my interpretations were wrong.

A few words on what this book is not about. First, this book does not review empirical evidence relevant to speciation, although I do include a large number of brief discussions and references to the corresponding literature. In the last 20 years or so, at least ten excellent monographs providing sufficiently deep reviews of the empirical data were published. They should be consulted for more information. Second, statistical questions and hypotheses about speciation that can be tested using data are not a major focus of the book. Rather, the main purpose of models studied here is to provide insights into the speciation process, train our intuition, build a general framework for studying speciation, find key components in its dynamics, and identify crucial parameters that need to be measured. I believe that developing nontrivial statistical tests and hypotheses requires a sound mathematical foundation. Indeed, to come up with the idea of using the deviations from the Hardy-Weinberg proportions as a statistical test of random mating, one has to show first that random mating results in the Hardy-Weinberg proportions. In a similar way, to come up with the idea that differences in neutral genes can be used to reconstruct phylogenies, one has to figure out first that the rate of accumulation of neutral mutations depends only on the rate of mutation. Hopefully, the mathematical theory built in this book will serve as a foundation for developing statistical tests and hypotheses about speciation.

A number of my colleagues, friends, students, and postdocs have read various drafts of different chapters or the entire manuscript. They are Frans Jacobs, John Endler, Mike Saum, Dan Howard, Janko Gravner, Roger Butlin, John Thompson, Günter Wagner, Reinhard Bürger, Aysegul Birand, Tadashi Fukami, Takehiko Hayashi, Michael Kopp, Jason Leonard, and Scott Sylvester. I am really grateful to all of them for criticism, suggestions, questions, and corrections. I am indebted to Galia Gavrilets, Roger Nisbet, and Sam Elworthy for initiating the whole project. I also wish to thank the National Institutes of Health, the National Science Foundation, Université P. et M. Curie, École Normale Supérieur, the Nederlandse Organisatie voor Wetenschappelijk Onderzoek, the Italian program "Mathematical Methods and Models in Biology," and the Biotechnology and Biological Sciences Research Council (UK) for support of my research.

Mathematical symbols

A, B, a, b, A_i, B_i	alleles at different loci
x, y, z	quantitative trait values
t	generation number, time
η	spatial coordinate
N, N_t	population size
$\mathcal{L}, \mathcal{A}, \mathcal{Q}, \mathcal{D}$	number of loci, alleles, traits, and dimensionality of genotype (phenotype) space
$\mathcal{P}, \mathcal{P}_c, \mathcal{P}_d$	probability of being viable, the threshold value of $\mathcal{P}$, and probability of being viable at distance d from a viable genotype
$\mathcal{N}, \overline{\mathcal{N}}, \mathcal{N}_p$	landscape neutrality, average landscape neutrality, and population neutrality
ρ	correlation function of fitness landscape, correlation between male and female traits
l_c	characteristic length, correlation length
p, q, p_i, q_i	allele frequencies
D, D_{ij}, D_{ijk}	(normalized) linkage disequilibria
ϕ_i	gamete frequencies
$\Phi_i, \Phi'_j, \Phi_{Aa}$	genotype frequencies
$\phi_f(x), \phi_m(y), \phi(z)$	distributions of quantitative traits in a population
$\phi(x), \phi'(y)$	distributions of quantitative traits in a population
w, w_i, w_{ij}, w_A	fitness, fitness component, mating success, induced fitness
v, f	viability and fertility
s, s_1, s_2, s_m	selection coefficients
S, S_a	products of s or s_a and $N/2, N, 2N$ or $4N$
$s_x(y), s_{x_1, x_2}(y)$	invasion fitnesses
$D(x)$	fitness gradient

$\bar{a}$	the average of a generic variable a over the population
V_a	variance of a
σ_a	standard deviation of a
CV_a	coefficient of variation on a
$f(a)$	distribution or function of a
$\mathrm{corr}(a, b)$	correlation of a and b
Δa	difference in a between two generations or two populations
a_c	critical or characteristic value of a
$E(\Delta a)$	the expected change in a
$\mathrm{var}(\Delta a)$	the expected variance of a change in a
l_i, l_i'	indicator variables (equal to 0 and 1, or to ± 1)
μ, μ_{ij}, ν	the rates of mutation per locus, from allele i to allele j, and per gamete
$r, r_1, r_2, r_{ij,k}$	recombination rates
m, M	migration rates, migration matrix
δ	extinction rate
$u(u^+, u^-)$	probabilities of fixation or speciation
$\kappa(\kappa^+, \kappa^-), \kappa_a, \kappa_d$	probabilities of fixation of underdominant, advantageous, and deleterious alleles relative to that of neutral alleles
$\omega(\omega^+, \omega^-)$	rates of fixation, rates of stochastic transitions, rates of gene duplication
$\mathcal{I}$	dispersion index
T	average waiting time to stochastic transition, speciation, or beginning of radiation
τ	average duration of stochastic transition, speciation, or radiation
g	genotypic value of a quantitative trait
V_G	genetic variance of an additive quantitative trait
V_g	genic variance
V_{LD}	contribution of linkage disequilibrium to V_G
V_E	microenvironmental variance

θ	optimum trait value
a_i	contribution of locus i to trait
$\Gamma(\cdot), \Gamma(\cdot,\cdot), \Psi(\cdot)$	gamma function, incomplete gamma function, and digamma function
$\Pr(x = i)$	probability that $x = i$
$P_{ij}, \Pr(x \to y)$	probabilities of stochastic transitions from i to j and from x to y
R, κ	various ratios of parameters
d	genetic distance
d_i	genetic distance at the i-th locus
$\overline{d}$	disparity, i.e., average genetic distance in a clade (population)
d_w	average genetic distance within a subpopulation (cluster)
d_b	average genetic distance between subpopulations
d_f	average genetic distance to a founder
$d_{w,max}$	cluster diameter (i.e., maximum genetic distance in a cluster)
ε_k	probability of incompatibility of k alleles
$J, J_k, \overline{J}$	number of incompatibilities and expected number of incompatibilities
k	parameter measuring complexity of incompatibility
Q	number of incompatibilities necessary for speciation
K	(average) number of substitutions required for speciation
K_i	threshold divergence at the i-th level of a taxonomic classification
$\psi_{ij}, \psi(x,y)$	female preference function
$\overline{\psi}_i, \overline{\psi}(x)$	female mating rate
$P(i \times j)$	overall frequency of $i \times j$ mating
Π_k, Π'_k	probabilities of joining mating group k
α	probability of assortative mating

$\xi_{ij}, \xi(x,y)$	pairwise competition function
$\bar{\xi}_i, \bar{\xi}(x), C(x)$	strength of competition experienced by individual
$K(x)$	carrying capacity at trait value x
c	strength of competition
$\mathcal{S}, \mathcal{R}, \mathcal{T}$	diversity (i.e., number of clusters in a clade), range size, and turnover rate
n	number of demes, matings, or alleles
I, I_A, I_B	heterozygote and hybrid deficiency indexes
γ	gene flow factor
F_1	first generation of hybrids
B	time to spread through the population
β_i	probability of spreading through a hybrid zone
a_i, b_i, c_i, C_i	coefficients
Δ	deviation of optimum from midrange
σ^2	the variance in distance between parent and offspring
a_w	cline width
r, A, B	coefficients
γ, ρ	coefficients in Chapter 9
$P(i), P(j)$	parents of i
f_{obs}, f_{exp}	observed and expected frequencies
C	covariance

COMMON ABBREVIATIONS

1D   one-dimensional
2D   two-dimensional
AD   adaptive dynamics
BDM   Bateson-Dobzhansky-Muller
DS   disruptive selection
My   million years
NM   non-random mating
RI   reproductive isolation

CHAPTER ONE

Introduction

Between 1.4 and 1.8 million species have been described (Stork 1993). Current estimates are on the order of 10 million species, with some estimates going as high as 100 million species (Ehrlich and Wilson 1991; Hammond 1995; May 1990). According to the World Resources Institute,[1] there are about 750,000 described species of insects, 250,000 species of multicellular plants, 123,000 species of non-insect arthropods, 50,000 species of mollusks, 47,000 species of fungi, 31,000 species of protozoa, 27,000 species of algae, 19,000 species of fish, 12,200 species of flatworms, 12,000 species of roundworms, 12,000 species of earthworms, 9,000 species of birds, 9,000 species of Cnidaria (jellyfish, corals, and comb jellies), 6,300 species of reptiles, 6,100 species of echinoderms, 5,000 species of sponges, 4,800 species of bacteria and blue-green algae, 4,200 species of amphibians, and 4,000 species of mammals.

Paleontologists have argued that the extant species represent less than 1% of the number of extinct species (Raup 1991). This gives an "average" rate of speciation on the order of three new species per year (Sepkoski 1998). Of course, any average rates of speciation are misleading because speciation takes place simultaneously in many different geographic locations and its rates vary between different groups of organisms as well as in time. More precise data exist. For example, according to Sepkoski (1998), the rate of genus origination (i.e., the number of originations per standing diversity per 1 million years (1 My)) for marine animals in Cambrian fauna ranged from 0.06 (for inarticulates) to 0.13 (for trilobites). For Paleozoic fauna, the rates ranged from 0.03 in stenolaemates to 0.06 in trilobites to 0.11 in cephalopods,

---

[1] *http://www.wri.org/wri/biodiv/f01-key.html.*

whereas for modern fauna this rate is approximately 0.03. As a rule of thumb, species origination rates are one order of magnitude higher than the corresponding genus origination rate. Thus, for example, an average species of Paleozoic crinoids produced a new species in 1 My with a probability of roughly 60%.

The maximum speciation rates known are much higher. The following are some data on speciation in lakes and on islands compiled by McCune (1997). The Hawaiian Islands have existed for about 5.6 My. During this time a large number of endemic species have originated there: 250 species of crickets, 860 species of drosophilids, 47 species of honeycreepers, 100 species of spiders, and 40 species of plant bugs. Fourteen endemic species of finches are known on the Galapagos Islands, which have existed for 5 to 9 My. Other examples are for fishes in lakes. Five species of cyprinodontids originated in Lake Chichancanab in 8,000 years and 22 species of cyprinodontids in Lake Titicaca in 20,000 to 150,000 years. Speciation of cichlids in the Great African Lakes has been extremely rapid: 5 species in Lake Nabugabo in 4,000 years, about 400 species in Lake Malawi in 0.7 to 2.0 My, dozens of species in Lake Tanganyika in 1.2 to 12 My, 11 species in Lake Borombi Mbo in 1.0 to 1.1 My, and, arguably the most spectacular speciation event known, speciation of hundreds of cichlid species in Lake Victoria, which according to some evidence was dry less than 15,000 years ago. Although many of these estimates can be questioned on a number of grounds (e.g., Fryer 2001), the important point is that speciation can be very rapid.

Speciation is the process directly contributing to the diversity of life as exemplified above. The origin of species has fascinated both biologists and the general public since the publication of Darwin's famous treatise *On the Origin of Species* in 1859. In Darwin's time, not much was actually known about the origin of species. Major steps towards identifying crucial features and factors of this process were made only in the 1930s, culminating in the two landmark books: Dobzhansky's *Genetics and the Origin of Species* (1937) and Mayr's *Systematics and the Origin of Species* (1942). New data, ideas, and theories have been steadily accumulating since then. Speciation has traditionally been considered as one of the most important and intriguing processes of evolution. In spite of this consensus and significant advances in both empirical and theoretical studies of evolution, understanding speciation still remains a major challenge faced by evolutionary biology, even after almost 150 years since Darwin's book came out (Templeton 1989, 1994; Harrison 1991; Coyne 1992; Claridge et al. 1997; Turelli et al. 2001; Kirkpatrick and Ravigné 2002).

There are two major reasons for this situation. The first is that speciation is indeed a remarkably complex process, which is affected by many different factors (e.g., genetical, ecological, developmental, and environmental) interacting in interwoven, non-linear ways. As a consequence, in different organisms speciation occurs in different ways. The second is the ineffectiveness of direct experimental approaches because of the time scale involved. Experimental work necessarily concentrates on distinct parts of the process of speciation, intensifying and simplifying the factors under study (Rice and Hostert 1993; Templeton 1996; Ödeen and Florin 2000, 2002; Florin and Ödeen 2002) which often leaves a lot of room for differences in interpretation and speculation.

Both this complexity and the difficulties of experimental approaches imply that mathematical models have to play a very important role in speciation research. The ability of models to provide insights into the speciation process, to train our intuition, to provide a general framework for studying speciation, and to identify key components in its dynamics are invaluable. A suite of tools for modeling the dynamics of genetic divergence is provided by theoretical population genetics and ecology. These two theories form a foundation for the quantitative/mathematical study of speciation.

Models of speciation were slow to appear. In particular, none of the four greats who are usually viewed as the founders of modern theoretical population genetics (Fisher, Wright, Haldane, Kimura) expressed much interest in developing mathematical models explicitly dealing with speciation.[2] The founders were mostly interested in how populations change and adapt to their environment rather than in determining how new species branch off their parent species. In this they followed Darwin, whose work was largely devoted to establishing the fact of evolutionary change with time and to identifying natural selection as the major mechanism of evolutionary change.

---

[2]Among their work most closely related to speciation are Fisher's verbal models of parapatric speciation and of runaway evolution caused by sexual selection (1930), Wright's verbal theory of shifting balance (1931,1982) and his model of assortative mating (1921), studies of stochastic peak shifts (Haldane 1931; Wright 1941; Kimura 1985), and clines (Haldane 1948; Fisher 1950).

was characterized as balkanization[4] by Kirkpatrick and Ravigné (2002). Different ideas, data, and verbal theories relevant to the origin of species have been covered in depth in numerous monographs (e.g., Mayr 1963; Grant 1971; White 1978; Brooks and McLennan 1991; Ridley 1993; Futuyma 1998; Levin 2000; Schluter 2000; Mayr and Diamond 2001). In contrast, no comprehensive attempts have been made to summarize and generalize what mathematical models tell us about the dynamics of speciation.

The book's goals are at least three-fold. First, I will provide a systematic and evenhanded review of what has been done in the field over the last 40 years. I will consider a number of questions: Which models are available to explain speciation? What are the explicit and implicit assumptions of the models? What are their predictions regarding the dynamics of speciation? How robust are these conclusions under realistic biological settings? What are the most important model parameters?

With regard to model predictions, a major focus will be on the conditions for speciation, the waiting time to speciation, the duration of speciation, and on different characteristics of emerging species. I will consider how these features are expected to depend on different evolutionary factors and parameters, such as the rates of migration and mutation, the strength of selection for local adaptation, population size, spatial structure of the population, and the genetic architecture of RI. Knowledge of these dependencies would allow one to evaluate the plausibility of different mechanisms of speciation and would help in interpreting different empirical observations. I will concentrate mostly on *simple* models, allowing for an analytical investigation. The main reason for this focus is that simple mathematical models (rather than complex and/or numeric models) lead to the most lucid conclusions. The critical review of previous theoretical work will be augmented by my own recent results. I will show that in spite of the complexity of the processes leading to speciation, analytical approaches can be successful.

Second, I will provide a synthesis of this very fragmented work. I will use a unified framework based on the notion of fitness landscapes introduced by Sewall Wright in 1932, generalizing them to explore the consequences of the huge dimensionality of fitness landscapes corre-

---

[4]Balkanization is the term coined before World War I to describe fragmentation of a region into smaller, often hostile, political units. The term comes from the Balkan Peninsula of Europe.

sponding to biological systems. In this framework, the divergence of (sub)populations along fitness landscapes under the joint action of different evolutionary factors results in the evolution of RI and, ultimately, speciation.

Third, I will attempt to accomplish the first two goals in a way that biologists who lack special mathematical training will be able to comprehend or at least get a good feel of. I will mostly use standard methods of theoretical population genetics and ecology, with a new emphasis on the consequences of the multidimensionality of biologically realistic fitness landscapes. The common wisdom is that a picture is worth a thousand words. In the exact sciences, an equation is worth a thousand pictures. Equations and their interpretations are the most concrete results a theoretician can come up with. Therefore, equations and their interpretations are a major focus of this book. The equations to be discussed below represent a kind of a do-it-yourself kit that can be used by biologists to check or train their intuition about speciation by substituting specific numerical values of parameters and interpreting the results. Throughout the book I will completely avoid or only briefly outline the intermediate steps of mathematical derivations, concentrating on the "final product" instead. I apologize to more mathematically oriented readers who might find such an approach somewhat frustrating. My excuse is that in most cases the derivations are rather straightforward, albeit tedious. I intend to make the derivations available from the book's web page.

1.1 GENERAL STRUCTURE OF THE BOOK

The book has three parts, each with three chapters, followed by a final chapter. The first part, "Fitness Landscapes," discusses different types of fitness landscapes. Chapter 2 starts by introducing the notion of fitness landscapes and proceeds to describe different classical types of landscapes. In Chapter 3, I consider classical models of evolution on rugged fitness landscapes via stochastic transitions between fitness peaks. Analyses of these models have important implications for the theories of founder effect speciation and chromosomal speciation. In Chapter 4, I consider recent advances in the theory of fitness landscapes, concentrating on the consequences of the extremely high dimensionality of landscapes describing real biological organisms. The most important feature of multidimensional fitness landscapes is the existence of nearly neutral networks of genotypes with approximately the same fitness that extend throughout the genotype space. I provide a number of biological examples of these networks. Finally, I introduce the notion of holey landscapes and consider some features of evolutionary dynamics on these landscapes.

The second part, "The Bateson-Dobzhansky-Muller Model," deals with a class of simple models of speciation and diversification. In most models studied there, I assume that fitness landscapes are constant (which implies that there is no frequency dependence or temporal variation in selection) and that the levels of within-population genetic variation are very low. These assumptions allow one to neglect genetic details to a certain extent. Chapter 5 starts by describing a classical two-locus, two-allele model of RI, which provides the simplest example of holey fitness landscapes. Although this model, which I call the Bateson-Dobzhansky-Muller (BDM) model, is probably unrealistic (because usually more than two loci are involved in RI), it can be studied in much detail and, therefore, is very helpful in training intuition. I will use the BDM model to analyze the dynamics of allopatric and parapatric speciation. In Chapter 6, I proceed to build multilocus generalizations of the BDM model, which I use to study allopatric and parapatric speciation in a system of two demes. In Chapter 7, I extend the approach to the case of multiple demes, using one- and two-dimensional stepping-stone systems as the underlying model of spatial structure. I use both individual-based and deme-based models to study the dynamics of diversity and disparity, species turnover, and the genetic structure of clades undergoing diversification.

In the third part, "Speciation via the Joint Action of Disruptive Selection and Nonrandom Mating," I consider a number of different models in which genetic details cannot be neglected. These models require explicit description of genotype frequencies and allow for fitness landscapes that vary in space (that is, describe spatially heterogeneous selection) or vary with the genetic state of the population (that is, describe frequency-dependent selection). To understand how different factors interact, it always helps to have a pretty good understanding of how they work in separation. Chapter 8 reviews models of the maintenance of genetic variation under disruptive selection. Chapter 9 reviews and generalizes models of nonrandom mating and fertilization. Some of these models are applicable to the scenario of speciation via (runaway) sexual selection. In Chapter 10, I bring together the models for disruptive selection and nonrandom mating to study the dynamics of sympatric and parapatric speciation. This chapter describes a number of new analytical results on sympatric speciation.

In the final chapter, Chapter 11, I return to the different scenarios of speciation and summarize what mathematical models tell us about them. At the end of some chapters, I have boxes in which I expand on different theoretical issues, give pointers to other relevant information, and provide brief summaries of more technical material used in the book. These boxes are there for more theoretically oriented readers.

In the remainder of this introductory chapter, I outline some important notions and ideas to be used and explored throughout the book.

1.2   SOME BIOLOGICAL IDEAS AND NOTIONS

1.2.1   Species definition and the nature of reproductive isolation

Speciation is the process of multiplication of species. A number of different species definitions have been proposed, each with some advantages and disadvantages (e.g., Mallet 1995; Harrison 1998; Templeton 1998; Shaw 1998; de Queiroz 1998). Although the argument on the appropriate way(s) to define a species is still unsettled, ultimately speciation is a consequence of genetic divergence. Here, I will view the dynamics of speciation as the dynamics of genetic divergence between different populations (or between parts of the same population) resulting in substantial RI. Genetic divergence and RI are typically accompanied by other differences, such as morphological and behavioral. My approach is closely related to the *biological species concept*, according to which species are groups of actually or potentially interbreeding populations that are reproductively isolated from other such groups (Dobzhansky 1937; Mayr 1942). In contrast to the classical version of the biological species concept, I will not require RI to be complete and, thus, will allow for some gene flow.

Reproductive isolation is defined as reduction or prevention of gene flow between populations by some differences between them (e.g., Futuyma 1998). There are numerous mechanisms, which are called *isolating mechanisms* or *reproductive barriers*, that prevent gene flow between populations. Reproductive barriers can be classified into *prezy-*

gotic (i.e., acting before zygotes are formed) and *postzygotic* (i.e., acting after zygotes are formed). Prezygotic barriers include, for example, seasonal and habitat differences preventing potential mates from meeting each other, behavioral isolation preventing mating, mechanical isolation preventing transfer of male gametes during copulation, and gametic incompatibility preventing fertilization. Postzygotic barriers include embryo inviability, hybrid inviability and sterility, and hybrid breakdown (i.e., reduced viability or fertility in subsequent generations). In general, RI can be due to *genetic factors* (such as incompatibility of specific genes, specific combinations of genes, whole chromosomes, or specific traits) or *ecological factors* (such as lack of an appropriate ecological niche for hybrids or reduction in hybrid fitness due to strong competition with parental forms). In certain groups of organisms, RI can be due to *conditioning*, that is, a learning process in which an organism's behavior becomes dependent on the occurrence of a stimulus in its environment. For example, in some insects, adults get together on the host species from which they emerged to mate. Therefore, adults emerging on different species will be reproductively isolated. RI can also be controlled by a *culturally inherited trait* (such as song type in birds).

Theoretical population genetics has identified a number of factors controlling evolutionary dynamics, such as mutation, random genetic drift, recombination, and natural and sexual selection (Crow and Kimura 1970; Ewens 1979; Nagylaki 1992; Hartl and Clark 1997; Bürger 2000). These factors can be classified as *stochastic* or *deterministic*. The most important stochastic factors are random genetic drift, mutation, and recombination. The most important deterministic factors are natural selection (resulting from abiotic and/or biotic factors) and sexual selection. Stochastic factors are usually more important in small populations, whereas deterministic factors are usually more important in large populations. Because the environment is typically variable, natural selection can have a stochastic component. Experimental work has shown that both stochastic and deterministic factors can result in genetic divergence and RI (Rice and Hostert 1993; Templeton 1996).

1.2.2   Geographic modes of speciation

A straightforward approach for classifying different mechanisms and modes of speciation is according to the type and strength of the factors controlling or driving genetic divergence. In principle, any of the factors listed above can be used at any level of classification. However, in evolutionary biology the discussions of speciation mechanisms are traditionally framed in terms of a classification in which the primary division corresponds to the level of migration between the diverging populations (Mayr 1942, 1963). In this classification the three basic (geographic) modes of speciation are allopatric, parapatric, and sympatric (Smith 1955, 1965, 1969; Bush 1975; Endler 1977; Futuyma 1998). The traditional stress on the spatial structure of populations as the primary factor of classification rather than, say, on selection reflects both the fact that it is most easily observed (relative to the difficulties in inferring the type and/or strength of selection acting in natural populations) and the growing realization that spatial structure of populations is very important.

Allopatric speciation Allopatric speciation (from the Greek words *allos* meaning "other" and *patra* meaning "fatherland" or "country") is the origin of new species from geographically isolated populations. In the allopatric case, there is no migration of individuals (and, consequently, no gene exchange) between the diverging populations. The chain of events typically implied in this scenario is (i) the appearance of spatially isolated populations of the same species, followed by (ii) genetic divergence of these populations, which results in (iii) increasing RI (as a by-product of genetic divergence), which eventually leads to (iv) the emergence of one (or more) new species.

For each of these steps, different evolutionary factors and mechanisms can be of importance. There are three general mechanisms by which populations of the same species can become spatially isolated: *vicariance events, extinction of intermediate links in a chain of populations*, and *dispersal*. In the first case, a new extrinsic barrier such as a body of water or a mountain chain isolates parts of a formerly continuous population. In the second case, a continuous population becomes fragmented when some intermediate subpopulations go extinct and the remaining subpopulations do not quickly expand to fill the resulting gaps. In the third case, new isolated populations are formed when some individuals (or propagules) manage to overcome an existing barrier and succeed in colonizing a new territory. All these mechanisms are important in natural systems.

Allopatric speciation is widely regarded as the most common form of speciation (e.g. Mayr 1942, 1963; Brooks and McLennan 1991;

Ridley 1993; Rosenzweig 1995; Futuyma 1998). This view is strongly supported by a general pattern elevated by Jordan (1905) to the level of a "law of distribution": "Given any species in any region, the nearest related species is not likely to be found in the same region nor in a remote region, but in a neighboring district separated from the first by a barrier of some sort" (p. 547).

Sympatric speciation Sympatric speciation (from the Greek word *sym*, meaning "the same") is seen as the opposite of allopatric speciation. There are multiple definitions of sympatric speciation: "speciation without geographic isolation," "speciation within the dispersal area of the offspring of a single deme," "speciation within a single geographical area," "speciation within a single, initially randomly mating population," "speciation within the 'cruising range' of members of an existing species." These definitions, although intuitively appealing, are not precise enough for modeling purposes. Indeed, to use such a definition in a mathematical model one would need to additionally define the exact meaning of "without geographic isolation," "single geographic area," and "cruising range". I will define sympatric speciation as "the emergence of new species from a population where mating is random with respect to the birthplace of the mating partners." Note that, implicitly, mating is allowed to be nonrandom with respect to, for example, genotype, phenotype, and culturally inherited traits. This definition is actually implied in most mathematical models of sympatric speciation.

The chain of events typically implied in this scenario is (i) an increase in genetic variation within the population, which leads to (ii) the emergence of separate genetic clusters (subpopulations) within the population, which is followed by (iii) increasing genetic divergence of these clusters accompanied by increasing RI between then, which eventually results in (iv) the emergence of one (or more) new species. Strong disruptive selection counteracting the homogenizing effects of gene flow and promoting the evolution of assortative mating within the population is generally viewed as a prerequisite of sympatric speciation. Although this mode of speciation was originally preferred by Darwin (1859, Chapter 4), support for sympatric speciation was very weak until relatively recently due to the lack of convincing theoretical and empirical evidence and forceful arguments put against its plausibility by Mayr (1942, 1963).

Parapatric speciation The intermediate cases, when gene flow between diverging (sub)populations is neither zero nor the maximum

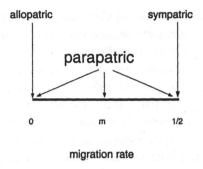

FIGURE 1.1. Geographic modes of speciation in a system of two demes exchanging a proportion m of its members each generation. (From Gavrilets (2003b), fig.1.)

possible, fall within the domain of parapatric speciation (from the Greek word *para* meaning "beside," "side-by-side," "next to"). Parapatric speciation is the process of species formation in the presence of some gene flow between diverging (sub)populations (Smith 1955, 1965, 1969; Bush 1975; Endler 1977). Parapatric speciation is associated with plants and animals of low vagility relative to the size of the species range. The chain of events as well as mechanisms and factors typically implied in the parapatric speciation scenario are similar to those of allopatric speciation, except that spatial isolation of the diverging (sub)populations is partial rather than complete.

To illustrate this classification of speciation modes, let us consider a system of two demes (subpopulations) that exchange a proportion m ($\leq 1/2$) of its members each generation. Assume that mating follows migration and is random within each deme. Then the allopatric case corresponds to $m = 0$, the sympatric case corresponds to $m = 1/2$, and the parapatric case corresponds to $0 < m < 1/2$ (see Figure 1.1). Both Figure 1.1 and biological intuition suggest that parapatric speciation is the most general (geographic) mode of speciation. Endler (1977) has already made the same point on the basis of empirical data. Moreover, the geographic structure of most species, which are usually composed of many local populations experiencing little genetic contact for long periods of time (Avise 2000), matches the one that is characteristic for parapatric speciation. In spite of this, parapatric speciation has received relatively little attention compared to the large number of empirical and theoretical studies devoted to allopatric and sympatric modes. I note that from the theoretical point of view, parapatric speciation is the most difficult scenario to analyze because one has to treat the dynamics of spatial distributions of genes explicitly.

I conclude this discussion with a few comments that were made repeatedly in the past (e.g., Smith 1955, 1969, Endler 1977) but still are worth repeating. The first is that a consistent use of terminology is very important. Sometimes the notion of sympatric speciation is used as an antonym of allopatric speciation and, thus, includes both parapatric speciation and sympatric speciation as defined above. This lays the ground for confusion. Even more misleading is the usage of the term *sympatric speciation* as a synonym of "the origin of species in a well-defined geographic area." For example, both in the empirical and theoretical literature, one can easily find statements about "sympatric speciation of Lake Victoria cichlids," which are only a little less misleading than something like "sympatric speciation of North American birds." One should also be absolutely clear that the definitions of the geographic modes of speciation imply nothing about the forces that drive genetic divergence leading to speciation. Any given evolutionary factor can play a role within each of the three geographic modes. For example, sympatric and parapatric speciation can be driven by mutation and random drift, and allopatric speciation can be driven by ecological factors. Models describing these examples will be discussed later in the book. Finally, the distinction among the three modes of speciation is sharp only in theoretical models. For most real-life situations, especially those that concern the distinction between parapatric and sympatric cases, there almost always exists room for differences in interpretation and for some controversy. Fortunately, this will be less of a problem for this book, which is mostly about models of speciation.

1.2.3   Some speciation scenarios and patterns

Next I briefly outline some scenarios of speciation that have received the most attention in the literature (for more discussion see Mayr 1963; White 1978; Brooks and McLennan 1991; Ridley 1993; Rosenzweig 1995; Futuyma 1998). The theory to be presented in this book will be relevant to most of them. These scenarios are not mutually exclusive.

Vicariant speciation   In this scenario, speciation follows the division of the range of a widespread population by an extrinsic barrier. It is tacitly assumed that the ranges of the resulting subpopulations are relatively large and comparable. Genetic divergence leading to speciation is thought to be driven both by divergent selection and stochastic factors. For example, the emergence of the Isthmus of Panama 3.0 to 3.5 My ago divided the ranges of many marine organisms into Atlantic and Pacific populations, some of which are now recognized as different species (e.g., Lessios 1998).

Peripatric speciation In this scenario, a new species emerges from a small, isolated subpopulation on the periphery of a larger ancestral population. According to Mayr (1954, 1963), peripheral populations are particularly likely to split off from the rest of the species because they experience less (or no) gene flow, have smaller sizes, and experience different selection regimes. This theory puts special emphasis on dispersal as the mechanism for the formation of spatial isolates. In a variant of this scenario known as *founder effect speciation*, a few individuals found a new population which rapidly grows in size. A new adaptive combination of genes is formed by random genetic drift during a short time interval, when the size of the expanding population is still small. This hypothesis was largely stimulated by studies of the Hawaiian *Drosophila* (Carson and Templeton 1984). In the Hawaiian archipelago, most species of *Drosophila* are endemic to a single island (Carson 1982). The evidence from geology, chromosomal inversion patterns, and prevailing winds suggests that the majority of speciation events occurred after colonization of a new island that had recently become habitable (Carson and Clague 1995; DeSalle 1995).

Centrifugal speciation Brown (1957) argued that new species often emerge from central populations (occupying more favorable areas) when they become isolated from peripheral populations (occupying less favorable areas) as a result of density fluctuations and range contraction affecting all or part of the species at one time. According to Brown, central populations are more likely to produce new species because they typically have higher genetic variation. Until very recently, the theory of centrifugal speciation was largely neglected by biologists. Brown's original paper cites a number of examples fitting his theory. An additional example is speciation of marine organisms in the Indo-West Pacific, where a relatively small East Indies area has apparently operated as a center of evolutionary novelty over the past 10 My (Briggs 1999a,b).

Punctuated equilibrium   According to the theory of punctuated equilibrium, the time interval during which the intermediate forms between an ancestral and a new species exist (and, thus, can be observed in the fossil record) is, as a rule, much shorter than the time interval during which the species exists with no major changes (Eldredge 1971; Eldredge and Gould 1972; Gould 2002). One of the best examples is provided by studies of tropical American Neogene cheilostome bryozoans *Metrarabdotos* and *Stylopoma* (Jackson and Cheetham 1999). In these genera, all long-ranging species (11 in each genus) persisted essentially morphologically unchanged for 2 to 16 My. New species appear abruptly in the fossil record, with morphological change occurring within the limits of stratigraphic resolution of sampling (≈150,000 years). The *turnover pulse* (Vrba 1985) and *coordinated stasis* (e.g., Brett et al. 1996) hypotheses state that not only single species but entire communities of different species living in the same geographic area undergo periods of stasis and periods of rapid turnover.

Chromosomal speciation   Closely related species often differ in the number, shape, and structure of chromosomes (White 1978; King 1993). Such differences often result in hybrid sterility or inviability. Chromosomal speciation is speciation that is caused (or initiated) by chromosome rearrangements that reduce hybrid fitness. For example, a major mode of speciation in plants (Grant 1991) is allopolyploidy, that is, the multiplication of the normal chromosome number as a result of hybridization. Spontaneous structural chromosome rearrangements are common in some animals. A classical example of chromosomal speciation apparently resulting from such rearrangements is the radiation of subterranean rodents, Spalacidae, in the Near East (Nevo 2001).

Speciation via hybridization   New species can also arise as a result of hybridization without a change in the chromosome number. In this case, hybridization of two diverged lineages creates new combinations of genes that give rise to a new, reproductively isolated species (e.g., Bullini 1994; Rieseberg 1995; Arnold 1997). An example of hybrid speciation is the origin of a wild sunflower species, *Helianthus anomalus*, which is a hybrid derivative of two widespread sunflowers, *H. annuus* and *H. petiolaris* (Rieseberg 1999).

Reinforcement   In this scenario, there are two populations of a species that have diverged genetically to such a degree that hybrids have reduced fitness. Upon their secondary contact, natural selection is expected to strengthen (and potentially make complete) prezygotic RI between the populations because this will decrease maladaptive hybridization. Since being developed and promoted by Dobzhansky (1940), reinforcement keeps generating a lot of interest among biologists (e.g., Butlin 1987, 1995; Howard 1993; Noor 1999). One expected consequence of reinforcement is increased prezygotic RI between sympatric populations relative to allopatric populations. The classic example of such a process is the divergence in songs of two frog species, *Litoria ewingi* and *L. verreauxi* in an area where the species are sympatric (Littlejohn 1965). In contrast, in allopatry the two species have very similar calls. Increased prezygotic RI between sympatric species pairs is a general pattern for *Drosophila* (Coyne and Orr 1989, 1997) and a wide range of other taxa from insects to fish and frogs (Hostert 1997).

Competitive speciation Competition for a limited resource among phenotypically similar members of the same population (or species) can be rather strong (Darwin 1859). In such a situation, natural selection can favor a splitting of the population into phenotypically diverged groups that specialize in different types of the resource and, consequently, experience less competition. This scenario was originally favored by Darwin (1959, Chapter 4), whose arguments were later extended by Rosenzweig (1978); Rosenzweig (1995) and Pimm (1979) (see also Gibbons 1979). Some empirical support for this scenario is provided by examples of *character displacement*, that is, morphological differences between closely related sympatric species (Brown and Wilson 1956). The classical example of character displacement are two species of birds of Eurasia, a large species of rock nuthatch, *Sitta tephronota*, and a smaller species, *S. neumayer*. These species exhibit profound differences in beak and body size in Iran, where they are sympatric, whereas individuals from allopatric populations look identical. An experimental illustration of character displacement in threespine stickleback fish within a single generation is provided by Schluter (1994).

Speciation along environmental gradients Environmental factors that control fitness of individuals are often found in gradients (e.g., of temperature or humidity). As a consequence, different parts of a widespread species may experience selection in opposing directions. If selection is sufficiently strong and migration is limited, genetic differentiation in the population can be rather substantial. This potentially may introduce strong post-zygotic RI as a by-product. Alternatively, because mating with immigrants that bring locally maladaptive genes decreases fitness of local populations, prezygotic RI may evolve in a way similar to the one implied by the reinforcement scenario (Endler 1977). An intermediate step of the speciation along environmental gradient can be the formation of a *hybrid zone*, that is, the area where hybrids between the two diverging groups are formed (e.g., Endler 1977; Harrison 1990; Barton and Gale 1993).

Speciation via host shift Factors that control fitness of individuals often vary substantially on a local scale. For example, for phytophagous insects, neighboring host plants of different species can represent very different microhabitats. Microhabitat-specific selective pressures accompanied by habitat selection and mating in the preferred habitat can result in speciation (e.g., Bush 1969, 1975; see Mayr 1947, 1963 for an earlier critique of this view). In this scenario, the role of geographic barriers reducing gene flow is played by *host fidelity*, that is, the tendency of an insect to reproduce on the same host species that it used in earlier life-history stages (Feder et al. 1994). The most studied example of a host shift that has resulted in significant RI is provided by the apple maggot fly *Rhagoletis pomonella* (Bush 1969; Feder 1998). Over the last 150 years or so, a widespread race of *R. pomonella* that uses cultivated apples as a host has emerged and genetically diverged from the ancestral species, which is a pest of hawthorns.

Speciation via runaway sexual selection If females happen to have a preference for a certain male trait and mate more often with males that possess that trait, then their offspring will inherit both the genes for the preference and the male trait. In this case, the female preference and the male trait can evolve in a self-reinforcing fashion, resulting in the evolution of exaggerated traits (Fisher 1930). In principle, this mechanism works with arbitrary initial preferences. Therefore, different populations may diverge with respect to both the male traits and female preferences. As a consequence, females from a population may have a reduced preference for males from a different population, that is, the two populations may become reproductively isolated (Lande 1981).

What are the relationships between these scenarios and the three geographic modes of speciation? Vicariant speciation is inherently allopatric and speciation along an environmental gradient is inherently parapatric, whereas competitive speciation is inherently sympatric. Peripatric, chromosomal, and centrifugal speciation can be treated within both the allopatric and parapatric frameworks. Reinforcement, speciation via host shift, and speciation via hybridization can be treated within both the parapatric and sympatric frameworks. Speciation via runaway sexual selection can occur within any geographic mode.

PART I
FITNESS LANDSCAPES

CHAPTER TWO

Fitness landscapes

Biological evolution is an extremely complex process influenced by a large number of genetic, ecological, environmental, developmental, and other factors. To understand a very complex phenomenon, it is very helpful to have a simple metaphor. During the last 70 years Wright's (1932) metaphor of "fitness landscapes," which are also known as "adaptive landscapes," "adaptive topographies," and "surfaces of selective value," has been a standard tool for visualizing biological evolution and speciation. Wright's metaphor is widely considered as one of his most important contributions to evolutionary biology (e.g., Provine 1986; Coyne et al. 1997; Arnold et al. 2001). At the same time, the metaphor has also become a subject of some controversy. As articulated by Provine (1986), there are actually two rather different versions of fitness landscapes which Wright himself used interchangeably, laying the ground for confusion about their exact meaning, dimensionality, and justification. Unfortunately, neither Provine's (1986) wonderful book nor recent review papers (e.g., Coyne et al. 1997; Fear and Price 1998; Arnold et al. 2001) adequately addressed this controversy. In this chapter, I attempt to clarify this issue. I also discuss and illustrate some classical fitness landscapes and introduce a generalization of fitness landscapes for the case of mating pairs of individuals that allows one to handle both natural and sexual selection within a single unified framework. Fitness landscapes considered in this chapter provide a general framework for treating speciation in subsequent chapters.

A key idea of evolutionary biology is that individuals in a population differ in fitness (because they have different genes and/or have experienced different environments). Differences in fitness that have genetic bases are the most important ones because it is changes in genes that make adaptation and innovations permanent. The relationship between genotype and fitness (direct or mediated via phenotype) is obviously of fundamental importance. Wright's metaphor of fitness landscapes provides a simple way to visualize this relationship. Implicitly, it also emphasizes the role of specific mechanisms and patterns in evolutionary dynamics. Before formally defining fitness landscapes, it is useful to start with a simple population genetics model.

2.1 WORKING EXAMPLE: ONE-LOCUS, TWO-ALLELE MODEL OF VIABILITY SELECTION

Let us consider a large randomly mating diploid population. Assume that generations are discrete and nonoverlapping. We will measure time in units of generations. This approach provides a realistic description of some species, such as annual plants and many spiders. For others it may still be a useful approximation that is widely used because of its mathematical simplicity.

I concentrate on a single diallelic locus. I will use bold letters to denote different alleles. In the diallelic case, alternative alleles at a locus will be designated using lowercase and uppercase letters without implying any dominance relation. With more alleles, I will use a subscript to designate different alleles. Let A and a be the alternative alleles at the locus under consideration, and let AA, Aa, and aa be the corresponding diploid genotypes. To complete the model, one needs to specify genotype fitnesses w_{AA}, w_{Aa}, and w_{aa}. Here, by a genotype's fitness I mean the genotype's *viability* defined as the probability to survive to the age of reproduction. In population genetics, viabilities are often normalized relative to the viability of a reference genotype, which can result in fitness values larger than one. Figure 2.1(a) visualizes the relationship between genotype and fitness in this model for a specific choice of fitness values resulting in two unequal "peaks" (at genotypes AA and aa) separated by a "valley" (at genotype Aa).

If the population size is very large and constant, the genetic structure of the population can be described in terms of the genotype frequencies. Under the Hardy-Weinberg equilibrium (which in this model is achieved in one generation, e.g., Nagylaki 1992), genotype frequencies in offspring can be expressed in terms of allele frequencies. Let p and $1 - p$ be the frequencies of alleles A and a. Then the frequencies of genotypes AA, Aa, and aa are p^2, $2p(1 - p)$, and $(1 - p)^2$, respectively. Now the genetic structure of the population can be described by

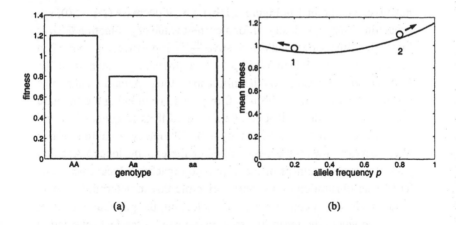

(a)   (b)

FIGURE 2.1. Visualization of the relationships between genotype and fitness in a one-locus, two-allele model. (a) Genotype fitnesses. (b) The mean fitness of the population $\overline{w}$ as a function of allele frequency p. Arrows show the direction of change in p for two different initial conditions.

a single variable p. With regard to fitness, the population state can be characterized by the mean fitness

$$\overline{w}(p) = w_{\text{AA}}\, p^2 + w_{\text{Aa}}\, 2p(1-p) + w_{\text{aa}}(1-p)^2, \qquad (2.1)$$

which depends on allele frequency p. Figure 2.1(b) visualizes the relationship between the mean fitness of the population $\overline{w}(p)$ and allele frequency p using the same numerical values of fitnesses as in Figure 2.1(a). Function $\overline{w}(p)$ has two unequal peaks (at $p = 0$ and $p = 1$) separated by a valley (at $p = 1/3$).

In the model under consideration, the function $\overline{w}(p)$ is very important because it controls the dynamics of allele frequency. Specifically, as shown by Wright (1935), the change in allele frequency, Δp, between two subsequent generations brought about by selection in a very large population can be written as

$$\Delta p = \frac{p(1-p)}{2\overline{w}(p)}\, \frac{d\overline{w}(p)}{dp}, \qquad (2.2)$$

where $\frac{d\overline{w}(p)}{dp}$ is the derivative of $\overline{w}$ with respect to p (see Box 2.1). In a polymorphic population (that is, in a population with $0 < p < 1$), the factor $\frac{p(1-p)}{2\overline{w}(p)}$ is always positive. Thus, if for a current value of p the function $\overline{w}(p)$ is decreasing (implying that the derivative $\frac{d\overline{w}(p)}{dp}$ is negative; e.g., point 1 in Figure 2.1(b)), p will decrease ($\Delta p < 0$), approaching 0 asymptotically. If for a current value of p, function $\overline{w}(p)$ is increasing (implying that the derivative $\frac{d\overline{w}(p)}{dp}$ is positive; e.g., point 2 in Figure 2.1(b)), p will increase ($\Delta p > 0$), approaching 1 asymptotically. In this model, depending on the initial state, the population will evolve towards a *fixation state* with $p = 0$ or $p = 1$. (Following the standard terminology, I will say that an allele is fixed if its frequency is 1 and that an allele is lost if its frequency is 0. Obviously, a fixation of one allele at a locus means the loss of all other alleles at the locus.) At either state, the average fitness of the population reaches a local maximum. This is an illustration of a general principle that in deterministic one-locus models with constant viability selection, the population evolves to a state where the mean fitness is maximized – the fact reflected in Fisher's fundamental theorem of natural selection (e.g., Nagylaki 1992).

If the population size is not very large, the dynamics of allele frequency will have a stochastic component due to random genetic drift (to be discussed in more detail later in this chapter). In this case, the population genetic structure is characterized by a probability distribution of allele frequency, say, $\phi(p)$. Under the Fisher-Wright binomial scheme for random genetic drift (see Section 2.4.3 below), the equilibrium distribution is proportional to a power of the mean fitness:

$$\phi(p) \sim \overline{w}(p)^{2N}, \tag{2.3}$$

where N is the population size (e.g., Wright 1969; Kimura 1964). This shows that the values of frequency p that are most probable (that is, result in the highest values of ϕ) are those maximizing the average fitness of the population $\overline{w}(p)$.

Figures 2.1(a) and 2.1(b) provide two alternative ways to visualize evolutionary dynamics. Using Figure 2.1(a), one can think of an individual as a point sitting on top of the bar corresponding to the individual's genotype (**AA**, **Aa**, or **aa**). The height of the bar specifies the individual's fitness. A population will be a group of points sitting at possibly different bars. As time progresses, the distribution of the points will change, concentrating more and more on one of the two homozygotes (**AA** or **aa**). In contrast, using Figure 2.1(b), one thinks of a population as a point on the curve $\overline{w} = \overline{w}(p)$. The point's abscissa characterizes the genetic structure of population (described by the allele frequency p). The point's ordinate specifies the mean fitness of the population. As time progresses, the point moves uphill, ending up on one of the two peaks (at $p = 0$ or $p = 1$).

Of the two relationships visualized in Figures 2.1(a) and 2.1(b), the first one, that is, the relationship between genotype and fitness, is simpler and more fundamental. It does not invoke any assumptions about the forces governing the evolutionary dynamics. The second relationship, that is, the one between the allele frequency p and the mean fitness $\overline{w}(p)$, is derived from the first one. Using the second construction implicitly invokes the assumption that evolution proceeds in the direction of increase in $\overline{w}$. In the example considered above, this assumption is perfectly justified. However, in other cases it might be not. A common feature of both relationships is the existence of multiple (here, two) local peaks with possibly different heights separated by valleys (here, one valley).

This simple model illustrates the differences between the two versions of fitness landscapes that Wright used interchangeably. In the first version, fitness landscape is a relationship between genotype and fitness (as depicted in Figure 2.1(a)). In the second version, fitness landscape is a relationship between the variables characterizing the genetic structure of a population and its mean fitness (as depicted in Figure 2.1(b)). The one-locus, two-allele model just considered is very simple, which allows one an easy and clear way to visualize both types of fitness landscapes. The following section considers more general and complex situations and illustrates the problems encountered in visualizing these relationships when there are more alleles and/or loci.

2.2 FITNESS LANDSCAPE AS FITNESS OF GENE COMBINATIONS

In Wright's 1932 original formulation, fitness landscapes represent fitnesses of gene combinations. To construct a fitness landscape, one first specifies a genotype space and then assigns fitness to each gene combination in the genotype space. *Genotype space* is defined as the set of all possible genotypes. (Wright himself referred to a "field of possible gene combinations" rather than to "genotype space.") Each genotype is described by a sequence of genes present at specific positions. An important notion used repeatedly in this book is that of *distance* in the genotype space. In what follows, I will extensively use a particular distance measure called the Hamming distance. Let each genotype be

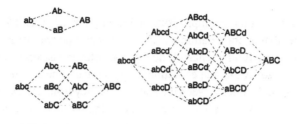

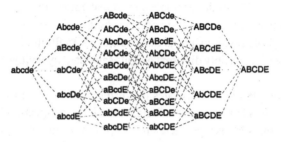

FIGURE 2.2. Genotype space in $\mathcal{L}$-locus, two-allele haploid systems, with $\mathcal{L} = 2, 3, 4$, and 5. Redrawn from Wright (1932).

represented by a sequence of $\mathcal{L}$ genes. Then the Hamming distance between two genotypes is

$$d = \sum_{i=1}^{\mathcal{L}} l_i, \qquad (2.4)$$

where $l_i = 0$ if the two sequences (i.e., genotypes) have the same element (i.e., allele) at the i-th position (i.e., locus), and $l_i = 1$ if the sequences have different elements at the i-th position ($i = 1, \ldots, \mathcal{L}$). That is, the Hamming distance between two sequences is equal to the number of elements that differ between the sequences. The range of possible values of d is from 0 to $\mathcal{L}$.

The genotype space corresponding to haploid organisms that differ with respect to $\mathcal{L}$ diallelic loci can be represented by the vertices of a $\mathcal{L}$-dimensional hypercube. For a biologist, a hypercube may sound like something extremely abstract and mathematical. However, using hypercubes for illustrative purposes goes back to Wright himself. Figure 2.2, redrawn from Wright's (1932) original figure, shows the hypercubes corresponding to $\mathcal{L} = 2, 3, 4$, and 5 loci. In this figure, the genotypes that are *one-step neighbors* (that is, are at distance $d = 1$) are connected by an edge, and different genotypes are arranged along the horizontal

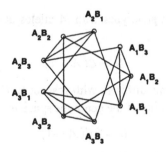

FIGURE 2.3. Genotype space in a two-locus, three-allele haploid system. Genotypes differing in a single gene (one-step neighbors) are connected by an edge.

direction according to the distance to a reference genotype represented by a chain of lowercase letters. With more than two alleles per locus, the genotype space can be represented by the vertices of a generalized hypercube (or an undirected graph). Figure 2.3 describes the genotype space corresponding to a two-locus, three-allele haploid system with alleles A_1, A_2, A_3 at the first locus and alleles B_1, B_2, B_3 at the second locus. In this figure, the pairs of genotypes that differ in a single gene are connected by an edge. Unfortunately, using the planar representations of hypercubes and generalized hypercubes such as those in Figures 2.2 and 2.3 for visualization purposes is useful only when the overall number of possible genotypes is small.

The genotype space corresponding to $\mathcal{L}$-locus diallelic haploid genotypes consists of $2^{\mathcal{L}}$ genotypes. Diploid organisms have twice as many genes as the haploids. Therefore, the genotype space corresponding to diploid organisms different with respect to $\mathcal{L}$ diallelic loci consists of $2^{2\mathcal{L}}$ genotypes. In general, with $\mathcal{A}$ alleles at each of $\mathcal{L}$ loci, there are $\mathcal{A}^{\mathcal{L}}$ possible haploid genotypes and $\mathcal{A}^{2\mathcal{L}}$ possible diploid genotypes.

The *dimensionality* of genotype space can be defined as the number of genotypes that can be obtained from a given genotype by changing single genes. In other words, this is the number of one-step neighbors of each genotype. This definition formalizes the ideas that the dimensionality of genotype space is both a measure of the number of possible directions for evolution and the number of independent coordinates that describe the position of a genotype in genotype space. If the individuals are haploid and the loci are diallelic ($\mathcal{A} = 2$), the dimensionality of genotype space, which I will denote as $\mathcal{D}$, is equal to the number of genes: $\mathcal{D} = \mathcal{L}$. In general, for haploid genotypes with $\mathcal{A}$ alleles at each of $\mathcal{L}$ loci, the dimensionality is

$$\mathcal{D} = \mathcal{L}(\mathcal{A} - 1). \qquad (2.5a)$$

For diploid organisms different with respect to $\mathcal{L}$ loci with $\mathcal{A}$ alleles each, the dimensionality of the genotype space is

$$\mathcal{D} = 2\mathcal{L}(\mathcal{A} - 1). \qquad (2.5b)$$

For example, the dimensionality of the genotype space corresponding to the two-locus, three-allele system depicted in Figure 2.3 is $2 \times (3 - 1) = 4$. With $\mathcal{L} = 100$ and $\mathcal{A} = 2$, there are approximately 1.27×10^{30} different haploid genotypes and approximately 1.61×10^{60} different diploid genotypes. In his original 1932 paper, Wright discussed an example of "a species with 1,000 loci each represented by 10 allellomorphs." In this case, the genotype space has $\mathcal{D} = 1,000 \times (10-1) = 9,000$ dimensions and consists of $10^{1,000}$ genotypes.

In general, the number of dimensions required for visualizing a fitness landscape is equal to the dimensionality of the corresponding genotype plus one extra dimension for fitness, that is, $\mathcal{L}(\mathcal{A} - 1) + 1$ dimensions in the haploid case, and $2\mathcal{L}(\mathcal{A} - 1) + 1$ dimensions in the diploid case. For example, with $\mathcal{L} = 100$ and $\mathcal{A} = 2$, the number of necessary dimensions is 101 and 201, respectively. Obviously, these numbers are much bigger than the two or three dimensions we are used to.

In some simple cases, visualization can be achieved with only two or three dimensions. For example, in the one-locus, two-allele diploid model ($\mathcal{L} = 1, \mathcal{A} = 2$) represented in Figure 2.1(a), we managed to use only one dimension for visualizing the genotype space rather than two $[2 \times 1 \times (2 - 1) = 2]$ because of the implicit assumption that heterozygotes **Aa** and **aA** have identical fitnesses. In biological terms, this assumption means the absence of *paternal/maternal effects* at this locus. Two examples of three-dimensional fitness landscapes are given in Figure 2.4. Figure 2.4(a) illustrates a fitness landscape implied in a one-locus, multi-allele diploid model with stepwise mutation used by Nei et al. (1983). (This model will be repeatedly used below.) Here, two dimensions (rather than $2 \times (5 - 1) = 8$ with $\mathcal{A} = 5$ alleles) are sufficient for visualizing the genotype space because we assumed no paternal/maternal effects and also explicitly assumed that alleles can be arranged in a natural order according to the stepwise pattern of mutations. Figure 2.4(b) illustrates a fitness landscape implied in a

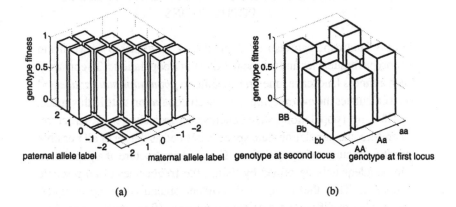

(a)   (b)

FIGURE 2.4. (a) Fitness landscape in Nei et al. (1983) one-locus, multi-allele model of postzygotic RI. It is assumed that allele A_i can mutate only to alleles A_{i+1} and A_{i-1} and that any two alleles separated by more than one mutational steps are incompatible (in the sense that the corresponding genotype has zero fitness). (b) Fitness landscape in a two-locus, two-allele model with $w_{11} = w_{13} = w_{31} = w_{33} = 1, w_{12} = w_{21} = w_{23} = w_{32} = 0.8, w_{22} = 0.6$.

two-locus, two-allele diploid model of viability selection with alleles **A** and **a** at the first locus and alleles **B** and **b** at the second locus (with $\mathcal{L} = \mathcal{A} = 2$). Models of this type, which will be extensively used below, are usually specified by a 3×3 fitness matrix:

	BB	**Bb**	**bb**
AA	w_{11}	w_{12}	w_{13}
Aa	w_{21}	w_{22}	w_{23}
aa	w_{31}	w_{32}	w_{33}

where, for example, w_{11} is fitness of genotype **AABB** and w_{12} is fitness of genotype **AABb** (e.g., Nagylaki 1992; Bürger 2000). In Figure 2.4(b), we used only two dimensions for the genotype space instead of four ($= 2 \times 2 \times (2 - 1)$) because of two implicit assumptions. The first assumption was that the paternal/maternal effects are absent. The second assumption was that fitnesses of double heterozygotes **AaBb** formed by gametes **AB** and **ab** and by gametes **aB** and **Ab** are identical. This assumption of the absence of *cis/trans* effects is often justified biologically.

2.3   FITNESS LANDSCAPE AS THE MEAN FITNESS OF POPULATIONS

Fitness landscape can also be defined as a relationship between the genotype frequencies and the mean fitness of a population.[1] To construct such a fitness landscape, one first specifies a *population state space* and then calculates the *average fitness* for each population state in this space.

Here, by a population's state I exclusively mean its genetic structure. Therefore, the population state space is defined as the set of all possible genetic structures of a population. The genetic structure of a population can be adequately described by listing the frequencies of all possible genotypes. Thus, the dimensionality of the population state space equals the number of different genotypes minus one. (Because the sum of all genotype frequencies is one, the frequency of one of them can be found if the frequencies of all other genotypes are known.) In general, with $\mathcal{L}$ diploid loci with $\mathcal{A}$ alleles each, the dimensionality of the population state space

$$\mathcal{D} = \mathcal{A}^{2\mathcal{L}} - 1. \tag{2.6a}$$

The genetic structure of a randomly mating population at a Hardy-Weinberg equilibrium under constant viability selection with no maternal/paternal or cis/trans effects can be described by the set of all possible haplotype frequencies (e.g., Nagylaki 1992), the number of which is $\mathcal{A}^{\mathcal{L}}$. (Here, *haplotype* will mean the set of genes in a gamete.) Thus, the dimensionality of the corresponding population state space is

$$\mathcal{D} = \mathcal{A}^{\mathcal{L}} - 1. \tag{2.6b}$$

For example, in the one-locus, two-allele model illustrated in Figure 2.1, there are only two different haplotypes (alleles). Therefore, a single variable (e.g., the frequency of an allele) is sufficient to describe the genetic structure of a population. In general, the dimensionality of the population state space is enormous. For example, with $\mathcal{L} = 100$ and $\mathcal{A} = 2$, equations (2.6a) and (2.6b) give 1.61×10^{60} and 1.27×10^{30}, respectively.

Assuming that haplotype frequencies are sufficient to describe a population's genetic structure, an adequate visualization of a fitness land-

---

[1] Some researchers use the term *adaptive landscapes* for these landscapes while reserving the term *fitness landscapes* for the landscapes in the first version.

scape requires $\mathcal{A}^{\mathcal{L}}$ dimensions. For example, in standard two-locus, two-allele diploid models of constant viability selection (such as the one specified by the fitness matrix on page 29), there are four different haplotypes, say, **AB, Ab, aB**, and **ab**. Let their frequencies be ϕ_1, ϕ_2, ϕ_3, and ϕ_4. The frequencies of any three haplotypes are sufficient to describe the genetic structure of such a population. Alternatively, one can use the frequencies of allele **A**, $p_1 = \phi_1 + \phi_2$, and allele **B**, $p_2 = \phi_1 + \phi_3$, and linkage disequilibrium

$$D = \phi_1\phi_4 - \phi_2\phi_3, \qquad (2.7)$$

which is a measure of nonrandom association of alleles in gametes. Thus, one needs four dimensions (three for the population genetic structure state plus one for fitness) to visualize fitness landscapes in these models. However, if fitnesses are additive, so that the genotype fitnesses are specified by a fitness matrix

	BB	**Bb**	**bb**
AA	$a_1 + b_1$	$a_1 + b_2$	$a_1 + b_3$
Aa	$a_2 + b_1$	$a_2 + b_2$	$a_2 + b_3$
aa	$a_3 + b_1$	$a_3 + b_2$	$a_3 + b_3$

where $a_i, b_i \geq 0$ are the corresponding contributions to fitness, the mean fitness is

$$\overline{w} = a_1 p_1^2 + 2a_2 p_1(1 - p_1) + a_3(1 - p_1)^2 + \\ b_1 p_2^2 + 2b_2 p_2(1 - p_2) + b_3(1 - p_2)^2.$$

Because $\overline{w}$ does not depend on D, three dimensions will be sufficient (see Figure 2.5). In a similar way, a fitness landscape corresponding to a $\mathcal{L}$-locus, two-allele model with additive fitnesses can be specified using $\mathcal{L}$ dimensions rather than $2^{\mathcal{L}}$. The same reduction can be achieved by assuming complete linkage equilibrium, so that the population genetic structure is adequately described by $\mathcal{L}$ allele frequencies.

I note that this version of fitness landscapes is often defined as a relationship between the allele (or gene) frequencies in a population and its mean fitness (e.g., Coyne et al. 1997; Fear and Price 1998; Arnold et al. 2001). Because allele frequencies in general are not sufficient for specifying the mean fitness of populations (unless fitness is additive or selection is very weak relative to recombination, so that linkage disequilibrium can be completely neglected), this definition is, strictly

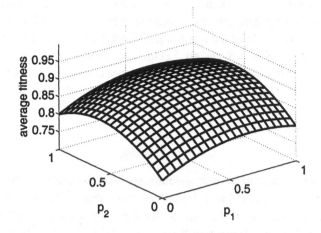

FIGURE 2.5. Fitness landscape as the average fitness of the populations in a two-locus, two-allele model with additive fitnesses. $a_1 = 0.45, a_2 = 0.50, a_3 = 0.40, b_1 = 0.40, b_2 = 0.50,$ and $b_3 = 0.35$.

speaking, meaningless in multilocus systems. Another common misconception is that the two versions of fitness landscapes are "wholly incompatible" (e.g., Provine 1986). Yet, it is straightforward to find mathematically the average fitness of the population $\overline{w}$ if one knows the individual fitnesses. That is, the first version can be transformed into the second version in a straightforward manner. However, knowing $\overline{w}$ indeed does not allow one to find individual fitnesses.

Evolutionary dynamics of populations on fitness landscapes is a very complex process. In general, all relevant evolutionary factors (e.g., natural and sexual selection, random genetic drift, mutation, spatial structure and migration, environmental variability) and their interactions are expected to play important roles. Excluding some special cases (e.g., one-locus models of viability selection), the features and patterns of evolutionary dynamics cannot be captured or predicted on the basis of any single characteristic such as the mean fitness. Indeed, it is well known that the mean fitness of the population does not necessarily increase and there are no general multilocus analogs of Wright's equations (2.2) and (2.3) that do not rely on rather strong approximations (such as those in Barton and Turelli 1991; Zhivotovsky and Gavrilets 1992; Turelli and Barton 1994). Therefore, this second version of fitness landscapes is not particularly useful in a multilocus context and for modeling speciation. The dimensionality of fitness landscapes in this version is very high and increases exponentially rather than linearly with the number of loci. Moreover, even with a relatively small number of loci, the number of possible genotypes is much larger than any reasonable population size. This implies that most genotype frequencies will be equal to zero. In this case, describing populations by listing genotype frequencies becomes equivalent to listing all genotypes present, which is exactly what is done in the first version of fitness landscapes. Finally, the process of speciation, during which a population splits into two different species, is impossible to describe in a framework where a population is the smallest unit. Finer resolution is necessary for describing the splitting of populations. In what follows I will primarily use the first version of fitness landscapes, treating fitness landscapes as fitnesses of gene combinations.

2.4   THE METAPHOR OF FITNESS LANDSCAPES

Perhaps our ultimate understanding of scientific topics is measured in terms of our ability to generate metaphorical pictures of what is going on. Maybe understanding *is* coming with metaphorical pictures.

Bak (1996)

The relationships between multilocus genotype and fitness (or any phenotypic characteristic) for real biological organisms are, in general, unknown. Only recently have direct studies of specific landscapes such as RNA and protein landscapes (to be discussed in Chapter 4) started to appear. However, some general features of fitness landscapes can be identified using available data, biological intuition, and mathematical reasoning. Having somehow identified these features, it may be helpful to have a way to visualize them graphically. Note that because of the huge discrepancy between the number of dimensions necessary to define a fitness landscape and the number of dimensions available for viewing, fitness landscapes cannot be graphically described in their entirety, even if they were known. One needs a simplified intuitive method. Such a method was proposed by Wright (1932), who used three-dimensional geographic landscapes as a metaphor for multidimensional relationships between genotype and fitness.

One has to realize that there is a clear distinction between a *fitness landscape* as defined in the previous sections and the *metaphor of fitness landscape* by which one means a two- or three-dimensional visualization of certain features of multidimensional fitness landscapes (Wright 1988). The former is defined in a precise mathematical sense. The latter is necessarily a simplification that emphasizes only those specific features of fitness landscapes thought to be most important while neglecting many other features. The requirements for metaphorical pictures are much less strict than for exact mathematical constructions. For example, in the precise mathematical sense, both the genotype space and the fitness landscape in the first version are discrete. However, in a metaphorical picture, it may be more convenient to visualize them as continuous. If one uses a graph to represent some exact dependencies, then all the axes and the corresponding scales have to be defined explicitly. However, this is not required for metaphorical pictures. Below, I will use the horizontal plane as a metaphor for the genotype space without giving any special separate meaning to the two horizontal axes or specifying the scale of measurement. The distance between two points in this plane will represent the degree of genetic dissimilarity between the corresponding genotypes. I will use the vertical axis for fitness, again without specifying the scale. Note that some of Provine's (1986) criticism of Wright's concept stems from Provine's attempt to judge a metaphor using rigorous mathematical criteria (Wright 1988).

In the following subsections, I will consider the three classical types of fitness landscapes, illustrating some of their applications with examples. I start with Wright's original concept of "rugged" landscapes.

2.4.1   Wright's rugged fitness landscapes

Wright's vision of a typical fitness landscape was that of a "rugged" surface having many isolated local "fitness peaks" of different height separated by "fitness valleys" of different depth (see Figure 2.6; note that Wright himself used two- rather than three-dimensional pictures to illustrate his idea). Fitness peaks represent high-fitness combinations

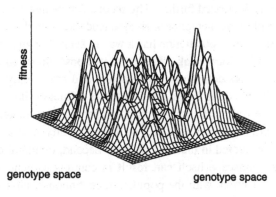

FIGURE 2.6. A rugged fitness landscape.

of genes ("coadapted gene complexes"). Wright reasoned that interactions of the effects of different loci on fitness coming from pleiotropy and epistasis will make the existence of multiple fitness peaks unavoidable. Different peaks can be viewed as alternative solutions to the problem of survival, which all biological organisms face. Fitness peaks that are sufficiently far away from each other in the genotype space may be thought of as corresponding to different species (real or potential). Clusters of peaks can be thought of as different species within the same genera. Fitness valleys between peaks represent low-fitness combinations of genes. These include both genotypes with deleterious genes (e.g., resulting from deleterious mutations) and genotypes with incompatible genes (e.g., resulting from hybridization).

Fitness peaks are important because of the expectation that natural selection will drive populations towards them. Within the framework of fitness landscapes, adaptive evolution is considered as "hill climbing." However, as soon as the population reaches a neighborhood of a local peak, any movement away from it is prevented by selection. It is important to realize that the peak the population has reached does not necessarily have the highest fitness. On the contrary, it is much more plausible that this peak has an intermediate height and that (much) higher fitness peaks exist nearby. Without some additional forces, a population evolving on a rugged landscape will stop changing after a relatively short transient time (see Box 2.2).

This conclusion leads to two important questions. The first is how fitness can be increased further. The second is how new species can be formed. Within the metaphor of rugged landscapes, both processes are impossible unless a population has a way to keep changing genetically after reaching a local peak. There are two possible solutions. First, additional factors acting against selection and overcoming it at least occasionally can drive the population across a fitness valley. The factor that has received most attention in this regard is random genetic drift. The effects of random genetic drift on the probability of escaping a local peak are considered in the next chapter. Second, temporal changes in the fitness landscape itself can result in continuous genetic changes driven by selection, with the population continuously climbing uphill and chasing a fitness peak that continuously moves away, as implied, for example, in the "Red Queen" scenario (Van Valen 1973).

The value of Wright's metaphor is that it attempts to explain something very complex (multidimensional fitness landscapes and evolutionary dynamics) using something that everybody has a close knowledge of – geographic landscapes. The view of rugged landscapes explicitly emphasizes the existence of different alternative solutions (alternative fitness peaks) to the problem of survival. In most three-dimensional geographic landscapes, peaks are isolated and there is no way to get from one peak to another without first descending to some valley. The metaphor of rugged fitness landscapes imposes a belief that the same is true in multidimensional fitness landscapes. The last point will be the main subject of Chapter 4.

2.4.2   Fisher's single-peak fitness landscapes

In contrast to Wright, Fisher (see Provine 1986, pp. 274-275; Ridley 1993, pp. 206-207) suggested that as the number of dimensions in a fitness landscape increases, local peaks in lower dimensions will tend to become saddle points in higher dimensions. In this case, according to Fisher, natural selection will be able to move the population to the global peak without any need for genetic drift or other factors. A typical fitness landscape implied by Fisher's views has a single peak (see Figure 2.7). This view is based on a belief that (i) there is one perfect combination of genes (rather than a series of more or less equivalent alternative combinations), and that (ii) this gene combination (fitness peak) can

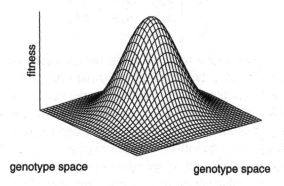

genotype space   genotype space

FIGURE 2.7. A single-peak fitness landscape.

be found by selection without the need for any additional factors such as genetic drift. It also implies that large populations are the major source of evolutionary innovations because they are more responsive to selection than small populations. Recently, these ideas were reiterated by Gordon (1994) and Whitlock et al. (1995).

However, Fisher's criticism is not warranted: the peaks that get transformed to saddle points by increasing the dimensionality of genotype space are well outnumbered by new local peaks brought about by the same process (Kauffman and Levin 1987; see also Box 2.2). This means that a typical fitness landscape is filled with local peaks, and finding the global peak by selection only is, in general, impossible.

Still, single-peak landscapes do arise in some simple models, including the two classical models of multilocus selection: additive and multiplicative. In the *additive fitness regime*, the fitness, w, of an organism is found by summing up the contributions w_i of $\mathcal{L}$ individual loci: $w = w_1 + w_2 + \cdots + w_{\mathcal{L}}$. The additive model may be a reasonable approximation if the contributions of individual loci to fitness (or another trait under consideration) are small. An example of additive fitnesses in the two-locus, two-allele case ($\mathcal{L} = 2$) was considered on page 31. In the *multiplicative fitness regime*, the fitness, w, of an organism is found by multiplying the contributions w_i of individual loci: $w = w_1 w_2 \ldots w_{\mathcal{L}}$. The following table gives an example of a fitness matrix arising in a multiplicative model with two diallelic loci:

	BB	**Bb**	**bb**
AA	$a_1 b_1$	$a_1 b_2$	$a_1 b_3$
Aa	$a_2 b_1$	$a_2 b_2$	$a_2 b_3$
aa	$a_3 b_1$	$a_3 b_2$	$a_3 b_3$

Here $a_i, b_i \geq 0$ are the corresponding contributions to fitness. The multiplicative model may be a reasonable approximation if the individual loci contribute to fitness (or another trait under consideration) at different time moments. In both the additive and multiplicative models, there is a single fitness peak. In the two-locus, two-allele multiplicative case, this is a genotype with fitness equal to $\max(a_i) \times \max(b_i)$. Here $\max(a_i)$ is the largest of the three values a_1, a_2, and a_3 and $\max(b_i)$ is the largest of the three values b_1, b_2, and b_3. Yet another important class of single-peak landscapes arises in models where fitness depends on the number of "deleterious" alleles (e.g., Kimura and Maruyama 1966; Kondrashov 1987).

Although situations with a single global fitness peak that can be reached by selection alone are hardly common, the single-peak metaphor still represents a very useful model for studying specific biological questions, including the levels and structure of genetic variation maintained by mutation (e.g., Schuster and Swetina 1982; Eigen et al. 1989; Higgs 1994) and the dynamics of adaptation in a neighborhood of a local fitness peak (e.g., Fisher 1930; Kimura 1983; Orr 1998).

2.4.3   Kimura's flat fitness landscapes

The major claim of the neutral theory (e.g., Kimura 1983) is that most evolutionary changes at the molecular level are neutral (i.e., do not result in changes in fitness). A typical fitness landscape implied by this view is flat (see Figure 2.8). There is extensive theoretical literature on the evolutionary dynamics of selectively neutral mutations by random drift (e.g., Crow and Kimura 1970; Ewens 1979; Kimura 1983; Gillespie 1991). Box 2.3 briefly summarizes some of the theoretical results that will be relevant here.

The neutral theory explicitly emphasizes the possibility of extensive genetic divergence by stochastic factors in the absence of deterministic forces of selection. When the population size N is finite, the genotype frequency change between subsequent generations will have a stochastic component coming from random variation in the number of offspring

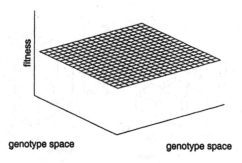

FIGURE 2.8. A flat fitness landscape.

between individuals and from the inherent randomness in the way that
alleles present in parents are sampled to pass to offspring. This stochasticity is commonly referred to as *random genetic drift*. There are different ways to describe random genetic drift mathematically (e.g., Ewens
1979). The most common way is based on the so-called Fisher-Wright
(or Wright-Fisher) binomial scheme. In the one-locus diploid case, let
p be the frequency of an allele in adults. In the Fisher-Wright binomial
scheme, the number of these alleles in offspring is treated as randomly
drawn from the binomial distribution with parameters p (probability of
success) and $2N$ (number of tries). Here, $2N$ is the overall number of
alleles in the (diploid) population. The underlying assumption is that
each adult produces a very large (effectively infinite) number of gametes of which $2N$ gametes are randomly sampled to form N diploid
offsprings. Under the Fisher-Wright scheme, the expected change in
p between two subsequent generations is equal to zero, $E(\Delta p) = 0$,
whereas the expected variance in the change in p per generation is

$$\text{var}(\Delta p) = \frac{p(1-p)}{2N}. \qquad (2.8)$$

The expected variance is maximized when N is small and p is close to
$1/2$. Genetic drift is generally considered as a major factor of evolution
(e.g., Kimura 1983; Gillespie 1991; Futuyma 1998). In the absence of
other factors, genetic drift results in the loss of genetic variation. Because genes get fixed and lost stochastically, different isolated populations that initially are polymorphic and have identical genetic structures
will diverge genetically.

Another stochastic factor, which is especially important in multilocus systems, is *mutational order*, that is, the random order at which new

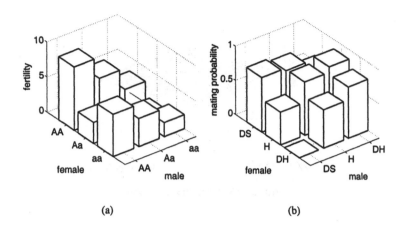

FIGURE 2.9. Fitness landscapes for mating pairs. (a) Fitness landscape in a one-locus, two-allele multiplicative fertility model. The corresponding male and female fertilities are $a_1 = 3, a_2 = 1, a_3 = 2$, and $b_1 = 3, b_2 = 2, b_3 = 1$ (see text). (b) Fitness landscape describing the probabilities of mating in no-choice tests involving *Drosophila silvestris* (DS), *D. heteroneura* (DH), and their F_1 hybrids (H) (data from Gavrilets and Boake 1998, Table 1).

genes are introduced in the population by mutation (Mani and Clarke 1990). Because the number of loci that can mutate is very large and the probability of each specific mutation is very small, different mutations will appear and accumulate in different local populations, leading to (stochastic) genetic divergence of these populations even under identical conditions. Mutational order has not received as much attention as it deserves. For example, only recently it became a subject of experimental work with viruses (Burch and Chao 1999) and bacteria (Finkel and Kolter 1999). Mutational order will be an important factor in several speciation models considered later in this book.

An important feature of evolution on flat fitness landscapes is that its rate does not depend on the population size (see Box 2.3).

2.5 FITNESS LANDSCAPES FOR MATING PAIRS

In the discussions of fitness landscapes, fitness is traditionally interpreted as a characteristic of an individual. In most cases, it is viability (i.e., the probability to survive to the age of reproduction). However, in many situations it is more appropriate to consider fitness as a characteristic of a mating pair of individuals. For example, fertility (i.e., the average number of offspring) often depends on whether maternal and paternal genes (or traits) match. The concept of fitness landscapes can be applied to mating pairs by redefining the genotype space and fitness. Now the genotype space has to be defined for combinations of genes each consisting of a male genotype and a female genotype, and fitness is assigned to a pair of genotypes. For example, in a classical model of multiplicative fertility selection (Bodmer 1965), the fertility f_{ij} of a mating pair is the product of the fertility effects in each sex: $f_{ij} = a_i b_j$, where a_i and b_j are fertility effects of male genotype i and female genotype j, respectively. Figure 2.9(a) shows a fitness landscape arising in a one-locus, two-allele version of this model.

Another component of fitness playing an extremely important role in discussions of sexual selection is the probability of mating between a male and a female given that they have encountered each other. This probability often depends on the degree of matching between the male's genes (or traits) and female preferences (e.g., Tregenza and Wedell 2000). For example, in Gavrilets and Boake's (1998) model of sexual selection, prezygotic RI was characterized in terms of a 3×3 matrix of the probabilities of mating. Figure 2.9(b) shows a fitness landscape for mating probabilities using the data from no-choice tests involving *Drosophila silvestris*, *D. heteroneura*, and their F_1 hybrids (Gavrilets and Boake 1998, Table 1).

2.6 FITNESS LANDSCAPES FOR QUANTITATIVE TRAITS

In the quantitative genetics framework, individuals are characterized by continuously varying traits rather than by discrete sequences of genes. Fitness landscapes for quantitative characters can be defined in two ways similar to those for discrete genes (Simpson 1953; Lande 1976).

2.6.1 Fitness landscape as fitness of trait combinations

In the first version, a fitness landscape is a relationship between a set of quantitative characters that an individual has and its fitness. Here, the role of the genotype space is played by the *phenotype space*, de-

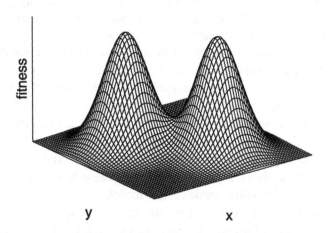

FIGURE 2.10. A fitness landscape with two quantitative characters x and y.

fined as the set of all possible phenotypes. The dimensionality $\mathcal{D}$ of the phenotype space is equal to the number $\mathcal{Q}$ of quantitative characters considered. In this version, a fitness landscape defines fitnesses of trait combinations. With quantitative characters, fitness landscapes are continuous surfaces rather than sets of distinct points. The dimensionality of the space in which such a fitness landscape is defined is $\mathcal{Q}+1$. Figure 2.10 illustrates fitness landscapes with two traits ($\mathcal{Q}=2$) in three dimensions. This figure shows two peaks separated by a valley.

2.6.2 Fitness landscape as the mean fitness of populations

In the second version, a fitness landscape defines the average fitness $\overline{w}$ of a population as a function of its genetic structure. The latter is usually described in terms of the moments of the phenotypic (or genotypic) distribution of the quantitative traits in the population. Thus, the dimensionality $\mathcal{D}$ of the population genetic structure space is equal to the number of phenotypic moments affecting the average fitness.

For example, let us consider a standard model of a quantitative trait where the trait value z is given by a sum of the genotypic (i.e., controlled by genotype) contribution g and a random microenvironmental (i.e., controlled by environment) effect e:

$$z = g + e \qquad (2.9)$$

(e.g., Lynch and Walsh 1998). Assume that the distribution of g in the population is Gaussian with mean $\bar{g}$ and genotypic variance V_G. Let e be independent of g and have a normal distribution with zero mean and constant variance V_e. Notice that under these assumptions the average trait value $\bar{z}$ is equal to $\bar{g}$. Let fitness $w(z)$ be given by a Gaussian function of the trait value:

$$w(z) = \exp\left(-\frac{z^2}{2V_s}\right), \qquad (2.10)$$

with $V_s > 0$ being a parameter characterizing the strength of selection. Such a fitness function is often used to model *stabilizing selection* with optimum zero. Small values of V_s imply that fitness decreases rapidly with the deviation from the optimum. This corresponds to strong selection. Large values of V_s imply that fitness decreases slowly with the deviation from the optimum. This corresponds to weak selection. Then the average fitness of genotype g is $w(g) = \int w(z)\phi(z|g)dz$, where $\phi(z|g)$ is the probability distribution of the trait value z given the genotypic contribution g. Using the assumptions about normality, one can evaluate the integral and show that

$$w(g) = \sqrt{\frac{V_s}{V_s + V_e}} \, \exp\left(-\frac{g^2}{2(V_s + V_e)}\right),$$

and that the mean fitness of the population is

$$\bar{w} = \sqrt{\frac{V_s}{V_G + V_s + V_e}} \, \exp\left(-\frac{\bar{g}^2}{2(V_G + V_s + V_e)}\right). \qquad (2.11)$$

Fitness $\bar{w}$ depends on both the average genotypic value $\bar{g}$ and genotypic variance V_G. In this case, a fitness landscape can be described in three dimensions (see Figure 2.11). I note that this version of fitness landscapes is often defined as a relationship between the mean fitness of the population and the mean trait values (e.g., Fear and Price 1998; Schluter 2000; Arnold et al. 2001). Because the mean trait values in general are not sufficient for specifying the mean fitness of the population (unless fitness is a linear function of trait values or selection is very weak), this definition is, strictly speaking, meaningless.

The average fitness of the population controls the change in the mean trait value per generation brought about by selection. This change is

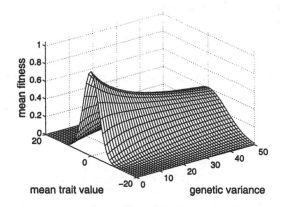

FIGURE 2.11. Fitness landscape as the mean fitness of a population, given by equation (2.11), in the case of a single quantitative character with $V_s = 10$ and $V_e = 1$.

approximated by an analog of Wright's equation (2.2), for which there are two alternative forms:

$$\Delta \bar{g} = V_G \frac{\partial \ln \bar{w}}{\partial \bar{g}} \qquad (2.12a)$$

and

$$\Delta \bar{g} = V_G \frac{d \ln w(g)}{dg} \Big|_{g=\bar{g}} . \qquad (2.12b)$$

Equation (2.12a) is justified if the distribution of g in the population is normal (Lande 1976). Equation (2.12b) does not use the normal approximation for the distribution of g, but requires some additional assumptions including the assumption that selection is weak (e.g., Iwasa et al. 1991; Gavrilets 1997a). If variance V_G is (approximately) constant, the population evolves in a gradient-type fashion by approaching a local peak in $\bar{w}$ or in $w(g)$ at a rate directly proportional to V_G. This behavior is analogous to that of the one-locus, two-allele model considered at the beginning of this chapter. Note that in discussing this version of fitness landscapes for quantitative characters (e.g., Fear and Price 1998; Arnold et al. 2001), it is customary to assume that only the average trait values change, whereas variances and other moments are constant. In some situations this assumption can be justified if selection is weak and the time interval considered is relatively short. In this case, the changes in the phenotypic distribution can be adequately described by the changes in the average trait values, and the dimensionality of the population state space is equal to the number Q of traits considered. This is the situation when this version of fitness landscapes is especially useful. However, in general, selection changes both the mean trait values and the corresponding variances (and covariances).

When the assumption about the constancy of variances cannot be justified, these fitness landscapes cannot be used for predicting evolutionary dynamics. However, they still can be very useful by providing descriptive statistics characterizing the effects of short-term natural selection on the phenotypic distribution. The most commonly used statistic is the selection gradient, $\frac{\partial \ln \bar{w}}{\partial g}$, measuring the strength of directional selection on the trait (e.g., Lande and Arnold 1983; Arnold and Wade 1984a,b; Endler 1986; Fear and Price 1998; Arnold et al. 2001; Hoekstra et al. 2001; Kingsolver et al. 2001). Because, in general, the mean fitness of the population depends on other moments besides the mean(s), the selection gradient and similar measures are inherently local. That is, changing the population genetic structure will result in new values of the selection gradient even if the fitness function itself has not changed. Therefore, one has to exercise caution in extrapolating results based on estimates of selection gradients.

2.6.3   Fitness landscapes for mating pairs

Typically, what one means by fitness in quantitative genetic models is individual viability. However, the notion of fitness landscapes can be applied to other types of fitness components as well. For example, a crucial component of many sexual selection models operating in terms of quantitative characters is a *preference function* ψ controlling the probability of mating between females and males with specific phenotypes. For example, in classical models of sexual selection proposed by Sved (1981a,b) and Lande (1981),

$$\psi(x, y) = \exp\left(-\frac{(x-y)^2}{2V_\psi}\right), \qquad (2.13)$$

where x and y are a female's and a male's trait values, and V_ψ is a positive parameter measuring mating tolerance. Small values of V_ψ imply that only males and females with very similar trait values have a chance to mate, whereas very large values of V_ψ imply that mating is largely

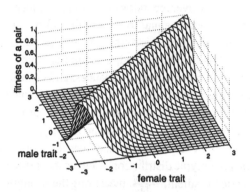

FIGURE 2.12. Fitness landscape for mating pairs defined by function (2.13) with $V_\psi = 0.5$.

indiscriminate. Preference functions can be interpreted as fitness functions of potential mating pairs. Figure 2.12 shows a fitness landscape corresponding to preference function (2.13). In this model, most possible mating pairs x and y have low fitness and a small proportion of mating pairs with high fitness (i.e., mating pairs with $x \approx y$) form a continuous ridge in the phenotype space.

2.7   GENERAL COMMENT ON FITNESS LANDSCAPES

The fitness landscapes considered in this chapter are static in the sense that they describe fitness components that do not change in time. Such landscapes are not appropriate for describing frequency-dependent selection and time-inhomogeneous selection. Of course, effects of many phenotypic characters (and, consequently, genes controlling them) on fitness change with changes in the biotic and abiotic environment faced by the organisms. For this reason, assuming the constancy of fitness landscapes corresponding to such genes and traits is a very strong and unrealistic assumption. However, the effects of genes underlying species differences and RI appear to be less affected by environment. Therefore, our general approach will be to use static landscape for describing species differences, but also to allow in some cases for nonstatic fitness components (e.g., resulting from frequency-dependent selection) to affect (or even drive) population divergence on a static landscape.

2.8 SUMMARY

• There are two versions of fitness landscapes for genotypes. In the first version, a fitness landscape is a relationship between possible combinations of genes and fitness (or a fitness component). A genotype space is the set of all possible genotypes. A fitness landscape can be formally defined as a map that assigns a fitness value w to each specific combination of genes in the genotype space.

• In the second version, a fitness landscape is a relationship between possible genetic structures of a population and its average fitness. The population genetic structure space is the set of all possible genetic structures of the population. The population genetic state is completely characterized by the frequencies of all possible genotypes. A fitness landscape can be formally defined as a map that assigns an average fitness value $\overline{w}$ to each population state in the population genetic structure space.

• Different fitness components are described by different fitness landscapes. Combining them in a single fitness landscape may not be easy.

• Wright's metaphor of fitness landscapes uses analogies with geographic landscapes to help visualize complex and generally unknown relationships between genotype (or genetic structure of a population) and fitness. Visualizing a *multidimensional* fitness landscape in two or three dimensions necessarily requires emphasizing only some aspects and features of the landscape.

• The view of *rugged* fitness landscapes emphasizes the existence of multiple high-fitness combinations of genes. Using analogies with three-dimensional geographic landscapes implicitly imposes the belief that evolution towards distinct high-fitness states necessarily requires overcoming selection and crossing fitness valleys.

• The view of *single-peak* landscapes emphasizes the potential for optimization by selection. At the core of this view are beliefs that there is a single best combination of genes and that selection alone is sufficient for finding this peak. The single-peak portrayal of fitness landscapes is appropriate only as a local approximation but not as a general picture.

• The view of *flat* landscapes emphasizes the possibility of extensive genetic divergence by mutation and random genetic drift. Chance and contingency play a very important role in the evolution on flat landscapes.

• Traditionally, fitness landscapes have been used only for describing viability selection. However, by appropriately redefining genotype space and defining the fitness of mating pairs, the notion of fitness landscapes can be directly expanded to other components of fitness to include fertility selection, sexual selection, and prezygotic RI.

• There are two versions of fitness landscapes for quantitative characters. In the first version, a fitness landscape is a relationship between a set of quantitative characters and a measure of fitness. This relationship can be treated as a map from a phenotype space (defined as the set of all possible combinations of quantitative traits) to fitness.

• In the second version, a fitness landscape is a map from a population state space to the average fitness of the population. The population state space is characterized by a set of different phenotypic moments (e.g., means, variance, and covariances). In this fitness landscape, the dimensionality of the population state space is equal to the number of the phenotypic moments affecting the average fitness of the population. This version of landscapes provides a set of useful descriptive statistics for evaluating effects of short-term natural and sexual selection on populations.

2.9   CONCLUSIONS

Scientific metaphors are similar to simple mathematical models in that they necessarily neglect some features of the phenomenon under consideration while (over)emphasizing others. It is important to realize that both the metaphors and models that we use not only help us solve specific (biological) problems but also affect the way we consider specific problems to be of importance. Fitness landscapes provide a very useful metaphor for thinking about a number of important evolutionary processes. The concept of fitness landscapes also provides a tool for bringing different forms of reproductive isolation within a unifying theoretical framework.

BOX 2.1

DYNAMICS OF ALLELE FREQUENCIES IN ONE-LOCUS, MULTIALLELE
POPULATIONS

Consider a randomly mating diploid population experiencing constant viability selection. Let p_i be the frequency of allele A_i and w_{ij} the viability (i.e., the probability of surviving to the age of reproduction) of a diploid organism $A_i A_j$. Assume that generations are distinct and nonoverlapping. Then the frequency of A_i in the next generation is given by the standard equation

$$p_i' = \frac{w_i}{\overline{w}} p_i \tag{1}$$

(Crow and Kimura 1970; Ewens 1979; Nagylaki 1992; Bürger 2000). Here, $w_i = \sum_j w_{ij} p_j$ is the induced fitness of allele A_i, and $\overline{w} = \sum_i w_i p_i = \sum_{ij} w_{ij} p_i p_j$ is the average fitness of the population.

Assume that there are only two alleles A and a at frequencies p and $q = 1 - p$, respectively. Let w_{AA}, w_{Aa}, and w_{aa} be the fitnesses of the three diploid genotypes. Then the above equation for the frequency p of allele A can be rewritten as

$$\Delta p = \frac{w_A - w_a}{\overline{w}} pq, \tag{2}$$

where $w_A = w_{AA}p + w_{Aa}q$ and $w_a = w_{Aa}p + w_{aa}q$ are the corresponding induced fitnesses, and $\overline{w} = w_{AA}p^2 + w_{Aa}2pq + w_{aa}q^2$ is the mean fitness of the population. Straightforward differentiation shows that the equality $\frac{d\overline{w}}{dp} = 2(w_A - w_a)$ is true. This shows the equivalence of equations (2) and (2.2). Note that Wright's equation (2.2) can be generalized for a multiallele case (Barton and Turelli 1987).

Assume that individuals surviving selection produce offspring that may differ from its parents as a result of mutations. Let μ_{ji} be the probability that genotype j mutates to genotype i ($i \neq j$), and $\mu_{ii} = 1 - \sum_{j \neq i} \mu_{ij}$ the probability that allele i is not changed by mutation. Then the genotype frequencies in offspring (that is, at the beginning of the new generation) are

$$p_i'' = \sum_j \mu_{ji} p_j'. \tag{3}$$

BOX 2.2

HILL CLIMBING ON A RUGGED FITNESS LANDSCAPE

Kauffman and Levin (1987; see also Kauffman 1993) introduced the following model. There are $\mathcal{L}$ haploid diallelic loci. The fitness value of each genotype is drawn independently and at random from a certain distribution. Let us consider adaptive walks on such landscapes. The walk starts on a randomly chosen genotype. At each time step, the walk samples one of the $\mathcal{L}$ one-step neighbors. If the neighbor has higher fitness, the walk moves there. Otherwise no change happens. The walk stops when it reaches a local fitness peak, so that all $\mathcal{L}$ one-step neighbors have smaller fitness.

Here, the walk can be thought of as representing the modal or average genotype in the population. A move to a fitter one-step neighbor corresponds to an advantageous mutation being fixed in a specific locus. The absence of a move after sampling a one-step neighbor corresponds to a deleterious mutation being rejected by selection. Obviously, modeling biological evolution using such a random walk makes several simplifications. Specifically, it is implicitly assumed that (i) there is no genetic variation in the population, (ii) deleterious mutations are never fixed, (iii) advantageous mutations are never lost, and (iv) the probability of fixation does not depend on the effect of mutation. The following is a sample of Kauffman and Levin's results for this model.

• The expected number of local peaks is $H = 2^{\mathcal{L}}/(\mathcal{L}+1)$. For example, if $\mathcal{L} = 100$, then $H \approx 1.26 \times 10^{28}$, and if $\mathcal{L} = 1,000$, then $H \approx 10^{299}$. These numbers are very large.

• The expected fraction of fitter one-step neighbors dwindles by $1/2$ on each improvement step.

• The walks to local peaks are very short. The average number of steps till reaching a local maximum is approximately $d = \log_2(\mathcal{L} - 1)$. For example, if $\mathcal{L} = 100$, then $d \approx 6.63$, and if $\mathcal{L} = 1,000$, then $d \approx 9.96$.

• The expected time to reach a local peak (including the waiting time till a fitter neighbor is found) is proportional to $\mathcal{L}$.

• The ratio of accepted to tried mutations in the course of the walk scales as $\ln k/k$, where k is the number of tried mutations.

• From most starting points, a walk can climb to only an extremely small fraction of the local peaks. Any one local peak can be reached from only an extremely small fraction of starting points.

• Kauffman and Levin's results lead them to the following conjecture ("The Complexity Catastrophe"): As $\mathcal{L}$ increases, the heights of accessible peaks fall towards the mean fitness over the whole landscape. This represents a fundamental restraint facing adaptive evolution: increasing complexity of the system leads to severe constraints on the levels of optimization that can be attained by selection.

BOX 2.3
EVOLUTION ON FLAT LANDSCAPES

Random walk on a $\mathcal{L}$-dimensional hypercube   The walk starts on a randomly chosen vertex of the hypercube. At each time moment there is a small probability ν that the walk moves to one of the neighboring vertices. The move represents fixation of a neutral mutation in a locus.

Equilibrium distribution. Asymptotically the walk can be found at any vertex with an equal probability $1/2^{\mathcal{L}}$. The number of steps necessary to reach this distribution is $\approx \frac{1}{4}\mathcal{L}\log \mathcal{L}$ (Diaconis et al. 1990).

Transient dynamics. Let d_t be the Hamming distance from the location of the walk at time t to its initial state. Distance d_t is expected to change according to equation

$$d_{t+1} = (1 - 2\mu)d_t + \mathcal{L}\mu,$$

where $\mu = \nu/\mathcal{L}$ can be viewed as the probability of a neutral mutation. An approximate solution to the above equation can be written as

$$d_t \approx \frac{\mathcal{L}}{2}(1 - e^{-2\mu t}). \tag{1}$$

This shows that d_t asymptotically approaches $\mathcal{L}/2$, with the characteristic time being $1/(2\mu)$. The distance between two independent walks (or between two lineages accumulating mutations) changes as

$$d_t \approx \frac{\mathcal{L}}{2}(1 - e^{-4\mu t}). \tag{2}$$

Now the characteristic time is twice as small as that in the previous model because there are twice as many mutations per unit of time.

The index of dispersion. The number of steps, χ_t, taken by the walk per time interval t has a Poisson distribution with parameter t/ν. Therefore, both the mean, $\overline{\chi}_t$, and the variance, $\mathrm{var}(\chi_t)$, of the number of steps per time interval t are equal to t/ν. This implies that the index of dispersion, defined as $\mathcal{I} = \mathrm{var}(\chi_t)/\overline{\chi}_t$, is equal to 1 (Kimura 1983).

Coalescence   Consider k alleles in a haploid population of constant size N with discrete nonoverlapping generations. Using the Fisher-Wright binomial scheme, the probability P_k that they all have different parents is

$$P_k = (1 - \frac{1}{N})(1 - \frac{2}{N})\ldots(1 - \frac{k-1}{N}) \approx 1 - \frac{k(k-1)}{2N}.$$

Thus, the probability that these k individuals have k different ancestors t generations ago and $k - 1$ ancestor $t + 1$ generations ago is

$$P_k^t(1 - P_k) \approx \frac{k(k-1)}{2N}exp\left[-\frac{k(k-1)}{2N}\,t\right].$$

(Box 2.3 continued)

This is the probability of a *coalescence* in generation $t + 1$. The distribution derived above is geometric with parameter $\lambda = k(k - 1)/(2N)$. The average time until a coalescence is $1/\lambda$. Thus, any two alleles ($k = 2$) can be traced back to a common ancestor approximately N generations ago.

The probability and the rate of fixation of new alleles   Assume first that all N alleles in the population are different. If there is no selection, each allele has exactly the same probability $1/N$ of being fixed. If an allele is represented by a copies, then the probability that this allele is fixed is a/N, that is, equals its initial frequency. Let the probability ν of mutations per allele per generation be very small. In each generation on average there are $N\nu$ mutations, of which $N\nu \times 1/N$ are fixed. Thus, the average number of mutations fixed per generation is equal to the mutation rate ν and, thus, is independent of N. The time between two fixations has the exponential distribution with mean $1/\nu$ and variance $1/\nu^2$.

Genetic distances   Assume that each allele is represented by a very long sequence of linked sites, such as nucleotides. This is a so-called *infinite site model* (Kimura 1969). During the time to coalescence of any two sequences (equal to $\approx N$ generations), each lineage accumulates $\approx N\nu$ mutations. Mutations accumulated in different sequences are expected to be different because of mutational order (see page 39). Then the expected genetic distance between a pair of randomly chosen sequences is

$$\bar{d} = 2N\nu. \tag{3}$$

Cluster structure   Consider a finite population of N sequences subject to mutation. Let us say that two sequences belong to the same t-family if their last common ancestor existed less than tN generations ago. The average number of t-families in the population is

$$S = \sum_{k=1}^{\infty} (2k - 1) \exp\left[-\frac{k(k - 1)}{2} t \right] \approx \frac{2}{t} \tag{4}$$

(Derrida and Peliti 1991). The members of the same t-family differ on average in $d_w = 2\nu \times tN$ genes. Then equation (4) implies that the individual sequences constituting the population can be clustered into

$$S \approx \frac{2\bar{d}}{d_w} \tag{5}$$

genetic clusters such that the average distance between the members of the same cluster is d_w.

CHAPTER THREE

Steps toward speciation on rugged fitness landscapes

Most biologists accept that the view of rugged fitness landscapes is much more general than those of single-peak or flat landscapes. Although on a local scale (that is, within a small subset of the genotype space) a fitness landscape can be treated as approximately single-peak or flat, on a larger scale Wright's picture of rugged landscapes with many peaks and valleys appears much more plausible. Because RI between different species is typically very strong or absolute, the valleys must be very deep. Within this picture, the fundamentally important question is how a population can move from one fitness peak against selection across a valley of maladaptation and to another peak. In this chapter, I first consider three simple models describing the dynamics of stochastic transitions between fitness peaks driven by random genetic drift. Next, I look at some consequences of spatial subdivision and density fluctuations for stochastic transitions. I also discuss the implications of these factors for the two classical theories of adaptation and speciation driven by stochastic transitions: the shifting balance theory and founder effect speciation. At the end of the chapter, I briefly outline some theories of peak shifts driven by selection.

3.1 STOCHASTIC TRANSITIONS BETWEEN ISOLATED FITNESS PEAKS

The mechanism for escaping a local peak that has received the most attention is random genetic drift. Drift is always present if the population size is finite. Below I will consider some simple models concentrating on the following interrelated characteristics of stochastic transitions between fitness peaks: (i) the probability of transition u, (ii) the average waiting time to transition T and the rate of transitions ω (that is, the number of transitions per unit time), and (iii) the average actual duration of transition τ.

Boxes 3.1 to 3.4 summarize some relevant mathematical results. The methods described there assume that the effects of selection and random drift on allele frequencies are small and comparable in magnitude. There are a number of excellent books (e.g., Ewens 1979; Gardiner 1983; Freidlin and Wentzell 1998; Grasman and van Herwaarden 1999) that should be consulted for more information on the mathematics of stochastic transitions.

3.1.1 Fixation of an underdominant mutation

Let us consider a randomly mating diploid population with discrete and nonoverlapping generations experiencing viability selection. We will concentrate on a single diallelic locus with alleles A and a, assuming that heterozygotes Aa have smaller fitness than homozygotes AA and aa. Alternatively, the results to be presented below can be interpreted in terms of two different chromosomes A and a, with the heterokaryotype Aa being less fit than both homokaryotypes AA and aa. This type of underdominance can arise either from improper segregation or the production of aneuploid gametes (i.e., gametes with a chromosome number that is not an exact multiple of the normal diploid number). Therefore the results given in this subsection have direct implications for the chromosomal speciation scenario (see page 16).

Equal peaks   Following Lande (1979), let the fitnesses be $w_{AA} = 1, w_{Aa} = 1 - s$, and $w_{aa} = 1$ ($0 \leq s \leq 1$). The fitness landscape corresponding to this model is similar to the one in Figure 2.1(a), with the difference that the two peaks have equal height. Parameter s measures the depth of the fitness valley separating the peaks. Assume for a moment that the population size is very large, so that genetic drift can be neglected. Under random mating and constant viability selection, the genotype frequencies in offspring are in Hardy-Weinberg proportions (e.g., Nagylaki 1992). The average fitness of the population is given by equation (2.1) which, with our parameterization of fitness, simplifies to

$$\overline{w} = 1 - 2sp(1 - p),$$

where p is the frequency of allele **A**. Using formula (2.2) and assuming that selection is weak ($s \ll 1$), the change in p per generation is

$$\Delta p = sp(1 - p)(2p - 1). \tag{3.1}$$

The allele frequency does not change ($\Delta p = 0$) if $p = 0, 1/2$, or 1. These three values specify the *equilibria* of dynamic equation (3.1). It is easy to see that $\Delta p > 0$ if $p > 1/2$, and $\Delta p < 0$ if $p < 1/2$. Thus, a population that is slightly perturbed away from the equilibrium states $p = 0$ or $p = 1$ will return to the corresponding state. In contrast, a population perturbed from the state $p = 1/2$ will not return but will end up at $p = 0$ or $p = 1$, losing one allele or another. Using the terminology of dynamical systems theory (e.g., Hofbauer and Sigmund 1988; Glendinning 1994), equilibria $p = 0$ and $p = 1$ are *locally stable*, whereas the equilibrium $p = 1/2$ is *unstable*. These considerations show that in a very large population, allele **A**, when initially rare ($p < 1/2$), will never make it to fixation (that is, to state $p = 1$), but will ultimately be lost.

Things are different if the population size is not too large, so that random genetic drift plays a role. Now, stochastic fluctuations can drive the allele frequency from values $p < 1/2$ across the unstable equilibrium at $p = 1/2$ into the domain of attraction of the stable equilibrium at $p = 1$. Let the population size be constant and equal to N. Assume that initially allele **A** has a low frequency p. Most likely this allele will be lost, but there is a small probability $u(p)$ that it will be fixed in the population. The latter event will represent a *peak shift*, that is, a transition from one fitness peak to another across a valley of maladaptation.

The probability of fixation, $u(p)$, can be found using general equations presented in Box 3.1. This requires one to specify the expected change and the expected variance of change in p in one generation. The former is given by the right-hand side of equation (3.1). The latter is given by equation (2.8) on page 39. Using equation (2) in Box 3.1, one can show that if the initial frequency of allele **A** is very small ($p \ll 1$), the probability of fixation is $u(p) = \kappa_u p$, where

$$\kappa_u = \frac{2}{\sqrt{\pi}} \frac{e^{-S}\sqrt{S}}{\mathrm{erf}(\sqrt{S})} \approx \frac{2}{\sqrt{\pi}} e^{-S}\sqrt{S}. \tag{3.2}$$

Here $S = Ns$, $\mathrm{erf}(x) = (2/\sqrt{\pi}) \int_0^x \exp(-y^2)dy$ is the error function (e.g., Abramowitz and Stegan 1965), and the approximation is good if $S > 2$. Because the probability of fixation on a neutral allele is

TABLE 3.1. The probability, κ_u, of fixation of underdominant alleles relative to that of neutral alleles.

N	s	Ns	κ_u
100	0.05	5	1.7×10^{-2}
1,000	0.01	10	1.6×10^{-4}
200	0.10	20	1.0×10^{-8}
1,000	0.05	50	1.5×10^{-21}

equal to its initial frequency (see Box 2.3), the factor κ_u characterizes the probability of fixation of underdominant alleles relative to that of neutral alleles. Numerical examples given in Table 3.1 show that, unless the population size is very small and the valley separating the peaks is shallow, stochastic transitions between the peaks are very unlikely.

Assume now that there is a constant input of alleles **A**, say, by mutation. Because the probability of fixation is very small, most mutant alleles will be lost. How long does one have to wait until a mutant allele becomes fixed? Let μ be the probability that allele a mutates to allele **A** per generation. Each generation there are on average $2N\mu$ mutant alleles each of which is fixed with probability $u = \kappa_u \times 1/(2N)$. If we assume that the number of mutants per generation is small, the rate of fixation (i.e., the probability of fixation per generation) is $\omega = \mu\kappa_u$. Correspondingly, the average time to fixation is

$$T = \frac{1}{\omega} = \frac{1}{\mu} \frac{\sqrt{\pi}}{2} \frac{e^S}{\sqrt{S}}. \tag{3.3}$$

For neutral alleles, $\omega = \mu$ and the average time until fixation is $1/\mu$. For underdominant alleles, this time grows exponentially with the product S of the population size and the depth of fitness valley. (Note that although the term $1/\sqrt{S}$ does decrease with increasing S, it is of lesser importance than the exponential term.) For example, with $\mu = 10^{-5}$ and $S = 20$, the above formula gives $T \approx 10^{13}$ generations.

If the average number of mutants per generation $2N\mu$ is not very small, equation (3.3) will not be applicable. Box 3.2 describes a general method for computing the waiting time to stochastic transition T using the diffusion theory. Box 3.3 presents two simple approximations for T. The second approximation is appropriate here. Using it requires one to specify the expected variance of change var(Δp) and the expected change $E(\Delta p)$ in p per generation. The former is given by equation (2.8) on page 39, whereas the latter is

$$E(\Delta p) \approx sp(1-p)(2p-1) + \mu(1-p). \tag{3.4}$$

Here, the first term in the right-hand side, which describes the effect of selection, is the same as in equation (3.1). The second term in the right-hand side gives the rate at which alleles **A** are introduced by mutation. This rate is equal to the frequency, $1 - p$, of allele **a** times the probability, μ, of **a** mutating to **A** per generation. Equation (3.4), which describes the deterministic dynamics of p, has two locally stable equilibria: a mutation-selection balance equilibrium at $p \approx \mu/s$ and a fixation equilibrium at $p = 1$, which are separated by an unstable equilibrium at $p \approx 1/2$. We are interested in the waiting time until stochastic transition from a neighborhood of $p = 0$ to a neighborhood of $p = 1$. Because the mutation-selection balance equilibrium is very close to 0 (assuming that $\mu \ll s$), we need to use the gamma approximation as described in Box 3.3. Following the procedure outlined in that box, one finds that the average time until fixation is approximately

$$T = \frac{\sqrt{1 + 8\pi N\mu}}{\mu} \frac{\sqrt{\pi}}{2} \frac{e^S}{\sqrt{S}} \exp\left(-4N\mu\left[1 + \ln\left(\frac{s}{2\mu}\right)\right]\right). \tag{3.5}$$

This equation shows that although T decreases with the average number of mutants per generation $N\mu$, the effect is small (because $N\mu$ is typically not too large). Note that if the number of mutants per generation is very small ($8\pi N\mu \ll 1$), the above expression reduces to equation (3.3). If $8\pi N\mu \gg 1$, T becomes proportional to $1/\sqrt{\mu}$ rather than to $1/\mu$. However, in either case, with moderately large S the average waiting time to transition is very long.

Most mutant alleles introduced into the population will be quickly lost. Therefore, the major part of time T is spent waiting for an allele destined to be fixed to enter the population. One can also find the average time τ that such an allele will take to get fixed. Time τ characterizes the average actual duration of transition (see Figure 3.1). The gamma approximation described in Boxes 3.3 and 3.4 results in

$$\tau = \frac{1}{s} + \frac{2}{s} \ln\left(\sqrt{\frac{S}{2}}\right). \tag{3.6}$$

The term $\ln(\sqrt{S/2})$ changes very slowly with S. Therefore, τ is on the order of $1/s$ generations and thus is much shorter than the average waiting time to transition T. The ratio of τ and T can be viewed as a

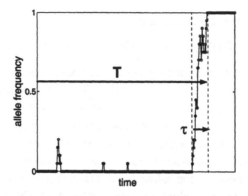

FIGURE 3.1. The dynamics of allele frequency in a stochastic simulation. The waiting time to fixation T and the duration of transition τ are explained with arrows. A few spikes observed initially represent mutations that were lost.

crude measure of the likelihood to observe the actual transition. The above results suggest that observing the transition is very unlikely.

Nonequal fitness peaks Following Walsh (1982), let us assume that the new peak has a selective advantage over the old peak. Let the fitness of genotype AA be $1 + \sigma$ rather than 1, used in the previous model. The fitness landscape corresponding to this model is similar to that presented in Figure 2.1(a). Using the same approach as before, the mean fitness of the population is

$$\overline{w} = 1 - 2sp(1-p) + \sigma p^2, \tag{3.7}$$

and the expected change in allele frequency per generation is

$$E(\Delta p) \approx sp(1-p)[p(2 + \beta) - 1], \tag{3.8}$$

where parameter $\beta = \sigma/s$ measures the advantage of the new fitness peak over the old fitness peak in units of the depth of fitness valley s. The only difference between this dynamic equation and equation (3.1) is that the unstable equilibrium is now at $p = 1/(2 + \beta)$ rather than at $p = 1/2$. That is, the unstable equilibrium is shifted towards the stable equilibrium at $p = 0$. One should expect that this will make a stochastic escape from a neighborhood of $p = 0$ easier.

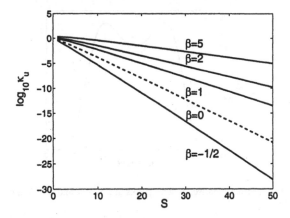

FIGURE 3.2. The relative rate of stochastic transitions κ_u in the underdominant model. The dashed curve corresponds to the case of equal fitnesses of homozygotes ($\beta = 0$). The curves above the dashed line correspond to the transitions to higher peaks. The curve below the dashed line corresponds to the transitions to a lower peak.

Proceeding as above, one finds that the relative probability of transition towards a higher peak is approximately

$$\kappa_u^+ = \frac{2}{\sqrt{\pi}} \frac{\exp\left(-\frac{S}{1+\beta/2}\right)\sqrt{S(1+\beta/2)}}{\frac{1}{2}\left[\operatorname{erf}\left(\sqrt{S}\frac{1+\beta}{\sqrt{1+\beta/2}}\right) + \operatorname{erf}\left(\sqrt{S}\frac{1}{\sqrt{1+\beta/2}}\right)\right]} \qquad (3.9)$$

(Walsh 1982). If $\beta = 0$, the above equation simplifies, as expected, to equation (3.2). Figure 3.2 illustrates the dependence of κ_u^+ on S and β. As expected, κ_u^+ increases relative to the case of equal peaks. However, one can see that unless the new peak is much higher than the old peak or S is small, the probability of stochastic transition across a valley of maladaptation is still very small.

If mutations are very rare ($N\mu \ll 1$), the waiting time till fixation of A can be approximated as $1/(\mu\kappa_u^+)$, whereas with a more frequent input of new mutations, the waiting time to fixation is approximately

$$T = \frac{\sqrt{1 + 8\pi N\mu}}{\mu} \frac{1}{\kappa_u^+}. \qquad (3.10)$$

In both cases, the relative rate of transitions κ_u^+ is given by equation (3.9). Under biologically reasonable conditions of large N and nonnegligible s, T is very large. One can also show that if a stochastic transition does happen, its actual duration τ is relatively short.

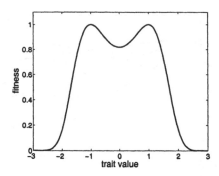

FIGURE 3.3. Fitness landscape with peaks at $z = -1$ and $z = 1$ and a valley at $z = 0$.

Stochastic transitions are also possible to a lower peak. The corresponding relative probability κ_u^- can be found by using a negative β. In Figure 3.2, the curve corresponding to the case $\beta = -1/2$ describes transitions to a lower peak when the genotype fitnesses are parameterized to be $1, 1 - s$, and $1 - s/2$.

Allowing for backward mutations (i.e., from **A** to **a**) will result in the population occasionally flipping back and forth between the two metastable[1] states in which one allele is maintained at low frequencies by a balance of selection, mutation, and random drift. The ratio of the rates of forward (i.e., towards a higher peak) and backward (i.e., towards a lower peak) transitions is $\kappa_u^+/\kappa_u^- \approx \exp(S\beta)$ (Slatkin 1981).

3.1.2   Peak shift in a quantitative character

Next, let us consider a diploid population of size N in which individuals differ with respect to a single additive quantitative trait z specified by equation (2.9) on page 43. Assume that the population is under viability selection defined by a fitness function $w(z)$ that has two peaks at z_0 and z_1 with a valley between them at z_v (see Figure 3.3). This model represents a quantitative trait analog of the model of underdominant selection considered in the previous subsection.

We wish to estimate the waiting time to stochastic transition of the mean trait value, $\bar{z}$, from a neighborhood of z_0 to a neighborhood of

---

[1]A state is metastable if it acts as a stable state for a relatively long period of time.

z_1. This time can be found using the normal approximation described in Box 3.3. Let V_G be the genetic variance of z in the population. Following standard practice, let us assume V_G remains constant during the whole process. In this model, the expected change, $E(\Delta \bar{z})$, in the mean trait value per generation is given by the right-hand side of equation (2.12b) on page 44. The variance of the expected change in $\bar{z}$ due to random genetic drift is

$$\operatorname{var}(\Delta \bar{z}) = \frac{V_G}{N} \qquad (3.11)$$

(Lande 1976; Bulmer 1980). This equation is analogous to equation (2.8), describing effects of random genetic drift on the frequency of alleles under the Fisher-Wright binomial scheme (see page 39). Using the formulas presented in Box 3.3, one finds that T can be written as

$$T = \frac{2\pi}{V_G} \left(-c_0 c_v\right)^{-1/2} \left[\frac{w(z_0)}{w(z_v)}\right]^{2N}, \qquad (3.12)$$

where $c_x = w^{-1} d^2 w(z)/dz^2$ is the curvature of the fitness function at $z = z_x$ (Barton and Charlesworth 1984; Lande 1985a).

For example, assume that fitness is given by function

$$w(z) = \exp[-s(1 - z^2)^2], \qquad (3.13)$$

where $s > 0$. This fitness function has two local peaks at $z_0 = -1$ and $z_1 = 1$ separated by a valley at $z_v = 0$ (see Figure 3.3). The height of the peaks is 1, the fitness at the bottom of the valley is $\exp(-s)$ (which is approximately $1 - s$ if s is small), and $(-c_0 c_v)^{-1/2} = \sqrt{2}/(8s)$. Substituting these values into equation (3.12), one finds that the average time to the peak shift is approximately

$$T = \frac{\sqrt{2}\pi}{4 V_G s} e^{2S}, \qquad (3.14a)$$

where, as before, $S = Ns$. Thus the waiting time to transition grows exponentially with S. Using the methods described in Box 3.4, one can also find the actual duration of transition τ:

$$\tau \approx \frac{\ln(8\sqrt{2}S)}{4 V_G s}. \qquad (3.14b)$$

Numerical examples given in Table 3.2 reaffirm the previous conclusion that stochastic transitions of even relatively small populations across even very shallow valleys of maladaptation are very improbable and take

TABLE 3.2. The average waiting time T and the average duration τ of peak shifts as predicted by equations (3.14). Genetic variance $V_G = 0.04$, which implies that the bottom of the valley is at five standard deviations from the peak. Both T and τ are measured in generations.

N	s	T	τ
100	0.01	2.1×10^4	516
500	0.01	6.1×10^7	2,522
100	0.05	1.2×10^7	504
500	0.05	2.9×10^{24}	706

an extremely long time. Table 3.2 also shows that the actual duration of transition is very short relative to the overall waiting time to transition (Lande 1985a).

Lande (1985a, 1986) and Newman et al. (1985) interpreted the results on the relative magnitude of T and τ as providing a theoretical justification for the punctuated equilibrium (see page 16). Within the framework used by these authors (and in this chapter), stochastic transitions are possible in a reasonable time only if the fitness valley separating the peaks is shallow. This implies that RI resulting from a single transition is very small. Potentially, strong or even complete RI (that is, speciation) can result from a series of peak shifts along a chain of intermediate fitness peaks, such that each individual transition is across a shallow valley but the cumulative effect of many peak shifts is large (Walsh 1982; to be considered in detail in Chapter 6). In this case, the above results actually imply that the population will spend a very long time at each of the intermediate fitness peaks. This would lead to a very long duration of speciation and plenty of possibilities for intermediate forms to leave a trace in the fossil record. Therefore, I doubt that these models really support the theory of punctuated equilibrium.

3.1.3   Fixation of compensatory mutations in a two-locus haploid population

Next, I discuss a model in which the transition to an alternative peak does not necessarily require the whole population to go through the valley. Let us consider a two-locus, two-allele haploid population with genotypes ab, Ab, aB, and AB having fitnesses (viabilities) $1, 1 - s, 1 - s$, and 1, respectively ($s \geq 0$). The fitness landscape corresponding to this

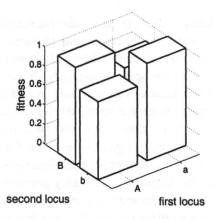

FIGURE 3.4. Fitness landscape in a two-locus, two-allele haploid model of compensatory mutations.

model is illustrated in Figure 3.4. Here, genotype **ab** has normal fitness and a pair of mutant genes **A** and **B** are individually deleterious, but in combination they restore the normal fitness. This model of *compensatory neutral mutations* was introduced by Kimura (1985). Kimura's model was in fact based on a diploid model introduced by Haldane (1931), who studied its deterministic dynamics. Compensatory mutations appear to be widespread in biological systems. For example, Haldane (1931) cites a paper by Gonzalez (1923), who showed that, in purple-eyed *Drosophila melanogaster*, arc wing and axillary speck (each controlled by a recessive gene) shortened life, but the two together lengthened it. For more examples, see Burch and Chao (1999) and Moore et al. (2000). Different generalizations of Kimura's model have been studied (e.g., Stephan 1996; Phillips 1996; Michalakis and Slatkin 1996; Iizuka and Takefu 1996; Higgs 1998; Innan and Stephan 2001) mostly through numerical simulations or numerical integration. Below, I present some simple analytical approximations for the waiting time to stochastic transition from one fitness peak to another that are valid for small recombination rates (see also Stephan 1996; Innan and Stephan 2001). To simplify presentation, I will assume that mutation rates from **a** to **A** and from **b** to **B** are the same and equal to μ and disregard the probability of backward mutations from **A** to **a** and from **B** to **b**. If mutation rates are small, allowing for backward mutations will not change the results (e.g., Higgs 1998).

Complete linkage First, let us assume that there is no recombination. Let ϕ_1, ϕ_v, and ϕ_4 be the frequencies of ancestral types **ab** (old peak), single deleterious mutants **aB** and **Ab** (valley genotypes), and double mutants **AB** (new peak), respectively. In general, this system is described by two dynamic variables, say, ϕ_1 and ϕ_v. However, as noted by Kimura, if selection is strong relative to mutation ($\mu \ll s$), then valley genotypes will remain at a very low frequency throughout the process. This frequency is given by equation 7 in Kimura (1985). Therefore, the population state can be approximately described by a single variable, say, the frequency ϕ_v of valley genotypes. Now one can use the theory of one-dimensional diffusion processes (see Box 3.1) to find the characteristics of the process. To simplify notation, let $p = \phi_v$. Kimura showed that the expected change in p is approximately

$$E(\Delta p) = \tilde{s}p(1 - p) + \tilde{\mu}(1 - p), \tag{3.15}$$

where $\tilde{s} \approx 2\mu$ and $\tilde{\mu} \approx 2\mu^2/s$ (Kimura 1985, equations 10 and 11). As before, the expected variance of the change in p is $\text{var}(\Delta p) = \frac{p(1-p)}{2N}$, where $2N$ is the number of genes in the population.

The waiting time to fixation of genotype **AB** can be found using the fact that equation (3.15) is equivalent to that describing the expected change in the frequency of an advantageous allele in a one-locus diploid population with additive fitnesses and one-directional mutation. Indeed, the first term in the right-hand side of equation (3.15) is the same as the one in equation (3) in Box 3.1. The second term in the right-hand side of equation (3.15) can be interpreted as a rate at which a deleterious allele, which has frequency $1 - p$, mutates to the advantageous allele if the rate of mutation is $\tilde{\mu}$. Now if there are $2N\tilde{\mu}$ mutants per generation, each of which is fixed with probability $u_a = 2\tilde{s}/[1 - \exp(-4N\tilde{s})]$ (see equation (4) in Box 3.1), then the average time to fixation can be approximated by the reciprocal of the average number of mutants fixed per generation, $2N\tilde{\mu} \times u_a$. Thus, returning to the original parameters μ and s, the average time to fixation is approximately

$$T = \frac{s}{2\mu^2} \frac{1 - \exp(-8N\mu)}{8N\mu}. \tag{3.16}$$

If the number of mutants per generation $2N\mu$ is small, the second ratio in the last formula is approximately 1, so that $T = s/(2\mu^2)$.

In contrast to the underdominant case, where the waiting time to the peak shift increases exponentially with the depths of fitness valley s (see equation (3.3)), here the dependence of T on s is linear. Table 3.3

TABLE 3.3. Average waiting time to fixation with $N = 1,000$ and $\mu = 0.0001$.

s	Underdominant mutation	Compensatory mutation
0.05	6.5×10^{25}	1.7×10^6
0.1	2.4×10^{47}	3.4×10^6

shows that peak shifts across relatively deep fitness valleys are much more probable in the compensatory mutation model with no recombination than in the underdominant model. The reason for this is that in the former case the population as a whole never has to go through the valley, whereas in the latter case the whole population has to cross the valley. The double mutants are not selected against once they have arisen. Therefore, in the compensatory mutation model with no recombination, the time scale of stochastic transitions is controlled by the process of double mutant formation (which takes, on average, s/μ^2 generations). In contrast, in the underdominant model, the time scale of transitions is controlled by the process of crossing the fitness valley.

Weak recombination   Next, let us assume that the recombination rate r is small relative to selection, but is much higher than the mutation rate ($\mu \ll r \ll s$). The effect of recombination is mainly to decrease the frequency of the new peak genotype **AB** through its crossing-over with the old peak genotype **ab**. The frequency ϕ_v of valley genotypes is given by equation 20 in Kimura (1985). To simplify notation, let p be the frequency of double mutants (i.e., $p = \phi_4$). The expected change in p becomes

$$E(\Delta p) = \tilde{s}p(1-p)[p(2+\tilde{\beta})-1)] + \tilde{\mu}(1-p), \qquad (3.17)$$

where $\tilde{\mu} \approx 2\mu^2/s$, $\tilde{s} \approx r$, and $\tilde{\beta} \approx 4\mu/r + 2r/s$ (Kimura 1985, equations 21 and 22). Once again, the waiting time until fixation can be found by noticing that equation (3.17) is similar to equation (3.8) from page 58 arising in the underdominant case with nonequal peaks, with recombination rate r playing the role of the depth of the fitness valley and a combined parameter $4\mu/r + 2r/s$ playing the role of the relative advantage β of the second peak. Note that our assumption $\mu \ll r \ll s$ implies that $\tilde{\beta} \ll 1$. Thus the waiting time to fixation is given by equation (3.3) from page 56, with Nr playing the role of the coefficient S:

$$T = \frac{s}{2\mu^2} \frac{\sqrt{\pi}}{2} \frac{e^{Nr} \operatorname{erf}(Nr)}{\sqrt{Nr}}. \qquad (3.18)$$

The last equation predicts an exponential increase in T with Nr. Thus, when the population size is large, even a small amount of recombination will significantly delay the peak shift. The reason for this is obvious: with recombination, the whole population has to pass through the fitness valley in a way similar to that in the underdominant case. An interesting implication of this analysis is that asexual populations should evolve more rapidly on rugged fitness landscapes than sexual populations, at least as far as the evolution by fixing compensatory mutations is concerned. This also suggests that in sexual populations, genes conferring RI via compensatory mutation mechanism should be more closely linked than other genes.

3.2   SOME CONSEQUENCES OF SPATIAL SUBDIVISION AND DENSITY FLUCTUATIONS

As the results of the previous section show, the probability of stochastic transitions quickly diminishes with increasing population size. Unless the fitness valley is very shallow, stochastic transitions with population sizes on the order of thousands are practically impossible. Population sizes of many species are much larger (e.g., Nei and Graur 1984). There are, however, two properties of natural populations that may keep the probability of stochastic transitions nonnegligible: populations are typically spatially subdivided, and the processes of extinction and colonization may be common. These properties can lead to a small *effective* population size (at a given spatial location or during some time interval) and make stochastic transitions plausible.

3.2.1   Spatial subdivision

Let us consider a population composed of n local demes of relatively small and equal size N receiving immigrants at a small rate m per deme per generation. We are interested in the case of underdominant selection acting on a single diallelic locus. Because N is small, deme extinction is a distinct possibility. Let the probability of extinction be δ per deme per generation. Assume that each local extinction is rapidly followed by recolonization from a single neighboring deme. In this model, the new fitness peak can spread by two mechanisms. First, immigrants with the new genotype can trigger a stochastic transition within the deme. Second, colonizers can bring the new peak genotypes into the deme previously occupied by organisms with the ancestral genotype. Both these mechanisms can also reestablish the ancestral genotype in the demes where it was lost.

Equal peaks Neglecting immigration, the rate of stochastic transitions to the alternative homozygous state caused by mutation within a deme is $\omega = \mu\kappa_u$ (see the paragraph before equation (3.3)). The rate Ω at which the entire population shifts to a new peak is equal to the number of demes in which the new peak is established each generation, $n\omega$, times the probability that a given one (or its descendants) will spread throughout the population. Because the demes are assumed to be equivalent, this probability is $1/n$. Thus the number of demes cancels and the transition rate (i.e., the probability of transition per unit of time) for the entire population is approximately the same as the transition rate for a single deme: $\Omega = \omega$ (Lande 1979). Therefore, the average waiting time to a peak shift is

$$T = \frac{1}{\Omega} = \frac{1}{\mu\kappa_u}. \tag{3.19}$$

If the size of local subpopulations is small, ω will be relatively large and T will be relatively small. Therefore, a strongly subdivided population has a much higher chance of shifting to a new peak than a single well-mixed population of the same size. For example, let $s = 0.05$ and $\mu = 10^{-5}$. Then the average waiting time to a peak shift in a single population of size $N = 1000$ is $T \approx 6.7 \times 10^{25}$ generations. In contrast, if the population is subdivided into $n = 20$ demes, each with $N = 50$ organisms, then $\kappa_u = 0.15$ and $T \approx 665{,}500$ generations.

How long the establishment of a new peak genotype in the whole population takes depends on the spatial arrangement of demes. Consider first the *circular stepping-stone system*, in which the demes are arranged in a circle and immigrants and colonizers are coming only from two adjacent demes. For example, one can think of a fish population inhabiting shallow areas along shores of a lake or of an animal population inhabiting a certain elevation range on a mountain. In this system, the average time to a peak shift in the entire population starting with a peak shift in a single deme is

$$\tau = \frac{n^2 - 1}{6(m\kappa_u + \delta)} \tag{3.20a}$$

(Slatkin 1981; Lande 1985a). In the *island model*, immigrants and colonizers can come from any deme. As the name suggests, here one can think of a population inhabiting an archipelago. In this model,

$$\tau = \frac{2(n-1)}{m\kappa_u + \delta} \qquad (3.20b)$$

(Slatkin 1981; Lande 1985a). For example, with the same parameter values as above and assuming that $m = \delta = 0.01$, $\tau \approx 6,700$ generations in the stepping-stone model and $\tau \approx 4,000$ generations in the island model. In both models, increasing the rate of local extinction δ can markedly accelerate the spread of the new genotype destined to be fixed.

Chromosomal speciation of blind subterranean mole rats, Spalacidae, in the Near East (Nevo 2001) appears to fit the scenario considered here. The *Spalax ehrenbergi* in Israel comprises four species (with chromosome numbers $2n = 52, 54, 58$, and 60). The population structure of *Spalax* is largely continuous across its central range, but at the periphery, small isolated or semi-isolated populations with up to $N = 100$ individuals occur. These could be the "cradles" where new chromosomal arrangements originate and then spread into ecologically vacant niches (Nevo 2001). Speciation in this case is peripatric.

Finally, I note that if there is no deme extinction, then the increase in the probability of fixation of an underdominant allele is noticeable only if migration is very small. For example, in the island model, the relative rate of fixation κ is given by equation (3.2) with S replaced by

$$S_e = N_T s \frac{Nm}{Nm+1}, \qquad (3.21)$$

where $N_T = Nn$ is the total population size (Cherry 2003). This shows that S_e is significantly smaller than $N_T s$ only if the expected number of immigrants per deme per generation, Nm, is very small.

Unequal peaks In this case, the relative probabilities of stochastic transitions within a deme towards higher (κ_u^+) and lower (κ_u^-) peaks are different. Using results for the classical Gambler's Ruin problem (e.g., Feller 1957), the transition rate towards the higher peak for the entire population can be approximated as

$$\Omega = \left(1 - \frac{\omega^-}{\omega^+}\right)\left[1 - \left(\frac{\omega^-}{\omega^+}\right)^n\right] \qquad (3.22)$$

(Slatkin 1981; Lande 1985a). Here $\omega^+ = m\kappa_u^+ + \delta$ and $\omega^- = m\kappa_u^- + \delta$ can be interpreted as the rates at which a deme shifts to a higher or lower peak as a result of stochastic transition triggered by immigrants or as a result of colonization.

The average waiting time until the entire population shifts to a higher peak given a peak shift in a single deme is $T = 1/\Omega$. The actual duration of transition, τ, can be found if there is no extinction (i.e., $\delta = 0$). In the circular stepping-stone system,

$$\tau = \frac{n}{2m\left(\kappa_u^+ - \kappa_u^-\right)},$$ (3.23a)

whereas in the island model,

$$\tau = \frac{\ln(n)}{m\left(\kappa_u^+ - \kappa_u^-\right)}$$ (3.23b)

(Slatkin 1981; Lande 1985a). With large n, τ in the stepping-stone model is much longer than that in the island model. In both cases, τ is much smaller than T, as before. The analysis above assumes that immigration rate is very small. Significant immigration of organisms with ancestral genotype can easily prevent stochastic transitions (Lande 1979; Barton and Rouhani 1991, 1993).

Shifting balance theory   The models studied above are related to the studies of Sewall Wright's shifting balance theory (Wright 1931, 1982), which he advanced as a general mechanism for a species to continuously increase its fitness by evolving from lower to higher peaks. Because this theory was thoroughly discussed by Coyne et al. (1997, 2000), with whose conclusions I largely agree (Gavrilets 1996), and because adaptation is not a major topic of this book, I will just give a very brief outline here. Wright considered a population to be spatially subdivided into a large number of small subpopulations exchanging migrants. Because local subpopulations are small and there are many of them, there is a nonnegligible probability of a stochastic peak shift in at least some of them. Wright divided the process of stochastic peak shift in a local subpopulation into two steps: stochastic movement by random genetic drift from a neighborhood of an old fitness peak into the domain of attraction of a new peak (phase I) and deterministic evolution towards the new peak once the subpopulation is within its domain of attraction (phase II). Wright reasoned that once a new adaptive combination of genes has been established in a subpopulation, it will take over the whole system as a result of higher emigration from the demes at the higher fitness peak (phase III). Wright's argument was mainly verbal. Recent formal analyses of different versions of the shifting-balance theory have led to the conclusion that although the mechanisms underlying this theory can, in principle, work, the conditions are rather strict.[2] The main problem is phase III – the spread of a new combination of genes from a local subpopulation to the whole system (Gavrilets 1996; Coyne et al. 1997, 2000). Mathematical analyses show that rather than spreading through the system after being established in a local subpopulation, the new combination of genes is much more likely to remain restricted to this subpopulation or to get swamped by immigrants from neighboring subpopulations with ancestral genotype.

Generally, for phase III to be successful, migration should be neither too strong nor too weak relative to selection, the new peak should be sufficiently higher than the old peak, and if there are several loci interacting epistatically, recombination rates have to be small. Elevated emigration resulting from higher population densities of the new peak populations facilitates phase III insignificantly unless there is a very strong asymmetry in the gene flow. For example, in the model of the spread of a new peak as a result of local extinction and colonization (see above), colonizers always arrived into an empty deme. Similarly, Peck and Welch (unpubl.) have shown that phase III of the shifting balance can be achieved relatively easily if deterioration of environmental conditions results in a significant reduction in population densities, so that migration is effectively one-directional, from higher fitness demes (which have relatively high density of individuals) into low fitness demes (which have very low density of individuals). A very important determinant of the ultimate outcome of competition between different peaks is the topological structure of the network of demes. Specifically, peak shifts in two-dimensional networks of demes are more difficult than in one-dimensional networks. In particular, increasing the dimensionality of the network from one to two dimensions increases the ease with which a higher peak can be swamped by migration. Moreover, spatial heterogeneity in selection regimes can severely impede phase III (Ohta 1972; Eldredge 2003; Gavrilets and Gibson 2002). Altogether these restrictions are consistent with Haldane's (1959) intuition; Haldane considered phase III as the weakest point of Wright's theory. Coyne et al. (1997, 2000) identified other weaknesses of Wright's theory, including the lack of convincing empirical support.

[2]For theoretical work arguing for the importance of shifting-balance in evolution, see Crow et al. (1990); Barton (1992); Kondrashov (1992); Phillips (1993); Moore and Tonsor (1994); Peck et al. (1998), and Wade and Goodnight (1998, 2000). The limitations of these approaches are discussed by Gavrilets (1996) and Coyne et al. (1997, 2000).

3.2.2 Stochastic transitions in a growing population

A different scenario that can potentially increase the plausibility of peak shifts is emphasized by the theory of founder effect speciation, which was mentioned in Chapter 1. Several versions of this theory have been proposed (Mayr 1942, 1954; Carson 1968; Carson and Templeton 1984; Templeton 1980; Kaneshiro 1980; see also Provine 1989 for a history of this theory). In founder effect speciation, the stochastic peak shift happens during a short time interval, when the size of the expanding population is still small. An inherent feature of the shifting balance that severely constrains this process is the necessity to spread the new adaptive combination of genes from a local deme to the rest of the population. During this stage (i.e., phase III) new combinations of genes have to compete with the old ones, which outnumber the former. Founder effect speciation avoids this difficulty by simply removing the necessity for the new combination to take over: a local subpopulation grows to become a new species without interacting with the ancestral one. The proponents of these theories proposed only verbal schemes without trying to formalize them. Later, formal analyses of founder effect speciation using analytical models and numerical simulation were undertaken by Lande (1980), Charlesworth and Smith (1982),Barton and Charlesworth (1984), Rouhani and Barton (1987), and Charlesworth and Rouhani (1988), and summarized in Barton (1989b). The general conclusion of these analyses is that a founder event cannot result in a sufficiently high degree of RI with a sufficiently high probability to be a reasonable explanation for speciation. To illustrate the reasons for this conclusion, I consider three different models.

Fixation of an underdominant allele in a growing population Assume again that there are three different diploid genotypes with fitnesses $w_{AA} = 1, w_{Aa} = 1 - s, w_{aa} = 1$ $(0 < s < 1)$. Initially the population has a single allele A, so that the starting frequency of this allele is $1/(2N_0)$, where N_0 is the size of the founder population. The population size increases deterministically with an exponential rate λ: $N_t = \exp(\lambda t)N_0$, where t is the generation number. To simplify numerical simulations, assume that stochastic transitions are possible only during the time interval that it takes for the population to reach some specified size N_{max}. The value N_{max} is considered to be sufficiently large, so that in populations with larger sizes, all stochastic effects on

TABLE 3.4. Probability of fixation of an underdominant mutation in an exponentially growing population (estimates based on 10^6 runs for each combination of parameters). T_s is the time to reach $N_{max} = 10,000$.

Initial size	Growth rate λ	Fixation probability		
		$s = 0.01$	$s = 0.05$	$s = 0.10$
$N_0 = 4$	$\lambda = 0$	0.122	0.114	0.101
	$\lambda = 0.05\ (T_s = 156)$	0.122	0.112	0.098
	$\lambda = 0.10\ (T_s = 78)$	0.117	0.105	0.089
	$\lambda = 0.20\ (T_s = 39)$	0.088	0.075	0.061
$N_0 = 8$	$\lambda = 0$	0.060	0.050	0.038
	$\lambda = 0.05\ (T_s = 143)$	0.057	0.042	0.029
	$\lambda = 0.10\ (T_s = 71)$	0.038	0.027	0.018
	$\lambda = 0.20\ (T_s = 36)$	0.012	0.008	0.005
$N_0 = 16$	$\lambda = 0$	0.028	0.019	0.010
	$\lambda = 0.05\ (T_s = 129)$	0.016	0.008	0.004
	$\lambda = 0.10\ (T_s = 64)$	0.005	0.002	0.001
	$\lambda = 0.20\ (T_s = 32)$	10^{-4}	10^{-4}	10^{-5}

allele frequencies effectively cease. The peak shift will be assumed to occur if the frequency of allele A is larger than 0.5 by the time the population size reaches N_{max}. In this case, the deterministic force of selection will eventually push the allele frequency towards 1. The results of numerical simulations of this model, based on the discrete Fisher-Wright binomial scheme allowing for selfing, are shown in Table 3.4. As expected, the probability of a peak shift increases with decreasing initial population size N_0, the rate of the population growth λ, and the depth of the fitness valley s. For example, if the founder population size is $N_0 = 4$ individuals (which is the best case scenario among those considered here), then the probability of fixation of a mutant allele is about 5% to 10%. This is a relatively high value.

However, the fact that fixation probability is high does not mean by itself that stochastic transitions are accomplished easily. One also has to consider the probability that a mutant allele is present in the founding population when its size is still very small (Charlesworth 1997). In principle, this mutant allele could be present among the $2N_0$ alleles carried by the founders or it could arise by mutation after the founding of the new population. Assuming mutation-selection balance, the frequency of the mutant allele in the ancestral population can be approximated as

μ/s, where μ is the probability of mutation. Therefore, the probability that a mutant allele present in the ancestral population makes it to the founder population is $\approx 2N_0\mu/s$. The probability that a new mutant arises in the founder population during the first τ generations while its size is still small is $\approx 2N_0\mu\tau$. For the parameter ranges used in Table 3.4, τ is on the order of 10 generations. In both cases, assuming that $\mu = 10^{-5}$ and $s = 0.1 - 0.01$, the probability that a new mutation makes it to a founder population is of order 10^{-3} or smaller. Therefore, the probability that a founder event results in a stochastic peak shift, which is equal to the product of the probability of fixation and the probability that a mutant allele is present, is of order 10^{-4} or smaller. Although this number is much higher than the probability of a stochastic transition in a stable population with a few hundred individuals, the absolute value of this probability is still very small. For the parameter values used above, it will take on average 10^4 or more founder events until a peak transition across a rather shallow (with $s = 0.01 - 0.1$) fitness valley occurs. Many more founder events will be necessary to accumulate significant RI. The above treatment did not consider the possibility of extinction of the population. Parameter values that favor the peak shift (i.e., small initial population size and slow population growth) also favor the extinction of the population. Therefore, if one accounts for the possibility of extinction, the probability of a successful peak shift will be even smaller than indicated by the analysis given here.

Transition in a quantitative character in a growing population
Next let us assume that individuals differ with respect to an additive quantitative character. First we completely neglect the effects of selection. Following a founding of a new population, a significant stochastic change in the mean trait value is most likely to occur when the population size N is still very small and genetic variance V_G is large. The variance of the change by drift in the mean trait value $\overline{z}$ per generation is given by equation (3.11). After t generations, the cumulative variance of the deviation of $\overline{z}$ from its initial value is

$$\text{var}(\overline{z}) = \sum_{i=1}^{t} \frac{V_{G,i}}{N_i}, \qquad (3.24)$$

where $V_{G,i}$ and N_i are the corresponding genetic variance and population size in generation i (Lande 1980). Random genetic drift is also expected to reduce genetic variance by the amount

$$\Delta V_{G,i} = V_{G,i+1} - V_{G,i} = -\frac{V_{G,i}}{2N_i} \tag{3.25}$$

per generation (e.g., Bulmer 1980). Combining equations (3.24) and (3.25) allows one to represent the cumulative variance of the deviation of $\bar{z}$ from the initial value over t generations as

$$\text{var}(\bar{z}) = 2 \sum_i (V_{G,i} - V_{G,i+1}) = 2(V_{G,0} - V_{G,t}) \tag{3.26}$$

(Rouhani and Barton 1987). The term $V_{G,0} - V_{G,t}$, which implicitly depends on N, is the total loss in additive genetic variance due to random drift in the founding population. Thus, in this model it is unlikely that random genetic drift will change the mean trait value by more than two standard deviations, $2\sqrt{\text{var}(\bar{z})}$, which is the same as $2.8\sqrt{V_{G,0} - V_{G,t}}$ (Rouhani and Barton 1987).

The derivations above neglected effects of selection. Rouhani and Barton (1987), who studied the case of a two-peak fitness landscape similar to that in Figure 3.3, argued that this is a reasonable approximation if the population growth rate is much higher than the depth of the fitness valley ($\lambda \gg s$). Charlesworth and Rouhani (1988) numerically studied founder effect speciation using the so-called infinitesimal model. The infinitesimal model assumes that the trait under consideration is controlled by a very large number of unlinked loci with very small additive effects, disregards changes in allele frequencies, and attributes changes in the phenotypic variance to the buildup of linkage disequilibrium (e.g., Bulmer 1980). Charlesworth and Rouhani (1988) showed that if the initial population size, the depth of the fitness valley, and the rate of population growth are all small, peak shifts inducing a change in the mean of two or more phenotypic standard deviations can occur with a frequency as high as 10%. However, in cases where peak shift probabilities are relatively high, the degree of RI is very small (less than 1%). Thus, in this model a single founder event seems to be unlikely to cause a very large change in the mean trait value or strong RI.

Fixation of a compensatory mutation in a growing population Slatkin (1996) briefly discussed a numerical study of stochastic transitions in a growing population that used a two-locus, two-allele haploid model with fitnesses $w_{ab} = 1, w_{Ab} = w_{aB} = 1 - s, w_{AB} = 1 + \sigma$ with $s, \sigma > 0$. The fitness landscape in this model is similar to that depicted in Figure 3.4, with the difference that the new peak (at genotype

AB) has a selective advantage σ over the old peak (at genotype aa). Slatkin's results show that if the population growth is fast, selective advantage of the new peak is large, and the initial frequency of AB is not too small, then peak shift occurs with very high probabilities (almost deterministically). The initial frequency of the new peak genotype AB does not have to be very high for a peak shift to occur, even if there is free recombination between the loci. For example, if $s = 0.1$ (i.e., the valley is shallow) and $\sigma = 0.5$, the initial frequency of AB has to exceed only 0.033 (i.e., at least one genotype AB per 30 founders). If $s = 0.5$ (i.e., the alley is deeper), the critical frequency of AB is 0.11 (i.e., at least one genotype AB per 9 founders).

How probable in this model is it to have at least a copy of AB out of N_0 founders? Assuming mutation-selection balance in the ancestral population, the frequency of each valley genotype Ab and aB will be μ/s. Mutation of these genotypes creates the new peak genotype AB, which will be present at an approximate frequency $2(\mu/s) \times \mu$. Thus, the probability that AB is present among N_0 founders is $\approx 2N_0\mu^2/s$. If $N_0 = 30$, $\mu = 10^{-5}$, and $s = 0.1$, this probability is $\approx 6 \times 10^{-8}$, which is very small. Once again, the implication is that a single founder event is unlikely to result in a peak shift.

3.3 PEAK SHIFTS BY SELECTION

Under certain conditions, transitions from one fitness peak to another can be accomplished by selection. This could happen if the fitness landscape has multiple peaks, but the mean fitness of the population has a single peak. For example, in models studied by Kirkpatrick (1982) and Whitlock (1995), individual fitness is a bimodal function of a quantitative trait value z (as in Figure 3.3), but the mean fitness of the population $\overline{w}$ changes from a bimodal to a unimodal form as genetic variance V_G increases. Now, if the variance of the quantitative trait becomes sufficiently large, the population can make a rapid deterministic transition from a lower peak to a higher peak, despite the presence of an intervening valley in the individual fitness function. The increase in the variance can be caused by a number of factors including a change in the external environment (which is specified by the fitness function), a change in the internal properties of the character (e.g., mutational or developmental), and random genetic drift. In the Price et al. (1993) model, a character

that has two alternative optima can be dragged from one optimum to
the other as a correlated response to selection on a second character. In
this model, high genetic correlation between the two characters makes
the peak shift easier. In the Pál and Miklós (1999) model, an increase in
the epigenetic variability creates a possibility to evolve around a fitness
valley by selection alone.

3.4 SUMMARY

• In both the one-locus, two-allele model of underdominant selection
and in the two-peak model of selection on an additive quantitative trait,
the probability of stochastic transitions between fitness peaks by random
genetic drift rapidly decreases with the population size and the depth
of the fitness valley. The waiting time until the stochastic peak shift is
very long. With a very small probability, stochastic transitions in these
models may happen. In this case, the actual duration of transition is very
short, being on the order of a few hundred to few thousand generations.
The very short duration of peak shifts is a rather general property of
transitions between different metastable equilibria of dynamical systems
driven by relatively weak stochastic factors. (Intuitively, if you need to
escape, you had better run very fast or you will not make it at all.)

• In the two-locus haploid model of compensatory mutations in
asexual populations and in sexual populations with very tight linkage,
stochastic transitions between the peaks are much more plausible than
in the models of underdominant selection and selection on a quantitative trait. The major difference between this model and the two other
models is that, in the former, the population as a whole never has to be
at the bottom of the adaptive valley. The stochastic transition is done
by just a few individuals whose offspring then take over the population.
However, the waiting time increases exponentially with the product of
the population size and the rate of recombination.

• Spatial subdivision can result in low effective population sizes,
which may make stochastic transitions in a local subpopulation more
plausible. However, the spread of a new fitness peak from a local subpopulation (deme) across the whole population (or species) is very difficult. This means that the mechanisms implied by Wright's shifting
balance theory are not very powerful in driving both adaptation and
speciation. Only if local densities in the new-peak demes are much

higher than the local densities in the old-peak demes can the spread of the new peak across the whole system be achieved relatively easily.

• Founding of new populations may imply that the population sizes are very small during relatively short time intervals. This may promote stochastic transitions between isolated fitness peaks. Although the relative probability of a peak shift in a small founder population is much larger than that in moderately large stable populations, the absolute value of this probability is very small. Many thousands (and more) founder events are required to result in a peak shift across even relatively shallow valleys.

• Stochastic peak shifts may be more likely in multideme populations when deme sizes and immigration rates are low, and there is a continuous extinction and recolonization process (rather than after a single colonization event). In this case, the rate of stochastic transitions in a population as a whole is equal to that in a single deme and, thus, can be relatively high if demes are relatively small.

• In certain situations, peak shifts can be accomplished by selection alone. No general theory of peak shifts driven by selection yet exists. The dynamics of peak shifts can be significantly affected by changes in fitness landscape induced by environmental changes. No general theory of these processes yet exists.

3.5   CONCLUSIONS

It is highly improbable that stochastic transitions requiring the majority of the population to pass through the bottom of a fitness valley are a major mechanism of genetic divergence (and speciation) of natural populations on a large scale. This is especially so if the population size is larger than a few hundred individuals and if the stochastic transition is to result in at least a moderate degree of RI. Evolution of strong RI and speciation are very unlikely outcomes of a single founder event if the fitness landscape has isolated peaks. If stochastic transition does happen, its duration is very short. This makes it extremely improbable to observe a trace of a single peak shift in the the form of intermediate genotypes (or phenotypes) in the fossil record.

BOX 3.1
DIFFUSION THEORY: THE PROBABILITY OF FIXATION

Let $a(x)$ and $b(x)$ be the expected change and expected variance of the change in the frequency x of an allele in one generation ($0 \leq x \leq 1$). The ultimate probability of fixation $u(p)$ of allele $\mathbf{A}$, given that its initial frequency is p, is

$$u(p) = \frac{\int_0^p F(x)dx}{\int_0^1 F(x)dx},$$ (1)

where

$$F(x) = \exp\left(-\int \frac{2a(x)}{b(x)}dx\right)$$ (2)

(e.g., Ewens 1979). For example, in the classical case of additive fitnesses, the fitnesses of genotypes $\mathbf{aa}$, $\mathbf{Aa}$, and $\mathbf{AA}$ are $1, 1 + s$, and $1 + 2s$, respectively. The expected change in p can be found using Wright's equation (2.2). Assuming weak selection (i.e., $s \ll 1$),

$$a(p) = sp(1 - p).$$ (3)

The variance of the change in the frequency p of allele $\mathbf{A}$ in one generation is given by equation (2.8) on page 39. Substituting these values into the general expression (2) and evaluating the integral, one finds that $F(p) = \exp(-4Nsp)$ and that the probability of fixation is

$$u(p) = \frac{1 - e^{-4Nsp}}{1 - e^{-4Ns}}$$

(e.g., Kimura 1983). If there is a single copy of the mutant allele initially, then $p = 1/(2N)$ and the probability of fixation is

$$u = \frac{1 - e^{-2s}}{1 - e^{-4Ns}} \approx \frac{2s}{1 - e^{-4Ns}}.$$ (4)

If s is positive (i.e., allele $\mathbf{A}$ is advantageous) and Ns is large (large population), $u \approx 2s$ (Haldane 1927). In large populations, selection will overwhelm drift once the advantageous allele is common. However, equation (4) shows that only a very small proportion $\approx 2s$ of advantageous mutations have a chance to become common. For a deleterious allele ($s < 0$),

$$u \approx \frac{2|s|}{e^{4N|s|} - 1}.$$ (5)

This equation implies that in small populations, the fixation of deleterious alleles can occur with a nonnegligible probability.

BOX 3.2
DIFFUSION THEORY: THE TIME TO FIXATION

The mean time $T(p)$ until the frequency x of an allele reaches the only absorbing state at $x = 1$, starting at $x = p$, is

$$T(p) = \int_0^1 t(x; p)dx, \tag{1}$$

where

$$t(x; p) = \frac{2}{b(x)F(x)} \int_p^1 F(y)dy, \ 0 \le x \le p, \tag{2}$$

$$t(x; p) = \frac{2}{b(x)F(x)} \int_x^1 F(y)dy, \ p \le x \le 1, \tag{3}$$

with F as in Box 3.1 (Ewens 1979).

Assume that the deterministic dynamics of x (that is, the dynamics expected if $b(x) = 0$) have two stable equilibria, at x_0 and x_1, separated by an unstable equilibrium at x_v.

Gaussian approximation   Approximating $F(x)$ by a Gaussian function $\alpha \exp(-\beta(x - \gamma)^2)$, one can approximate the average waiting time to transition from a neighborhood of x_0 to a neighborhood of x_1 as

$$T = 2\pi \sqrt{\frac{b(x_v)}{b(x_0)}} \frac{1}{\sqrt{|a'(x_0)|a'(x_v)}} \frac{F(x_v)}{F(x_0)} \tag{4}$$

(e.g., Gardiner 1983; Barton and Charlesworth 1984; Lande 1985a; Barton and Rouhani 1987).

Gamma approximation   If the equilibrium x_0 is close to a boundary of possible x values and the variance $b(x)$ vanishes at the boundary, then the normal approximation is not good because $F(x)$ will be strongly skewed. Approximating $F(x)$ by a function $x^{\alpha-1}\exp(-\beta x)$ yields

$$T = \sqrt{\frac{\pi b(x_v)}{a'(x_v)}} \frac{1}{a(0)} C\left[\frac{2a(0)}{b'(x_0)}\right] \frac{F(x_v)}{F(x_0)} \tag{5}$$

(Barton and Rouhani 1987), where $C[x] = \Gamma(x)x^{1-x}e^x\sqrt{1 + 2\pi x}$, with $\Gamma(\cdot)$ being the gamma function (e.g., Abramowitz and Stegan 1965; Gradshteyn and Ryzhik 1994).

BOX 3.3

DIFFUSION THEORY: THE DURATION OF TRANSITION

Most mutant alleles introduced in a population will be lost. The average waiting time $\tau(p)$ to fixation of a mutant allele destined to be fixed starting at frequency p is

$$\tau(p) = \int_p^1 \Phi(x)u(x)(1-u(x))dx + \frac{1-u(p)}{u(p)} \int_0^p \Phi(x)u(x)^2 dx, \quad (1)$$

where

$$\Phi(x) = \frac{2\int_0^1 F(x)dx}{b(x)F(x)} \quad (2)$$

(Kimura and Ohta 1969), and $u(p)$ is the probability of fixation.

Assume that the corresponding deterministic dynamics have two stable equilibria at $x = p_0$ and $x = p_1$, separated by an unstable equilibrium at $x = p_v$.

Gaussian approximation Approximating $F(x)$ by a Gaussian function and performing tedious but straightforward evaluations of the integrals in equation (1) leads to the following expression for the average actual duration of transition:

$$\tau(p_0) = \frac{1}{|a'(p_0)|} \ln\left[\sqrt{\frac{2|a'(p_0)|}{b(p_0)}}(p_v - p_0)\right]$$

$$+ \frac{1}{|a'(p_1)|} \ln\left[\sqrt{\frac{2|a'(p_1)|}{b(p_1)}}(p_1 - p_v)\right]$$

$$+ \frac{1}{a'(p_v)} \ln\left[\sqrt{\frac{2a'(p_v)}{b(p_v)}}(p_v - p_0)(p_1 - p_v)\right]. \quad (3)$$

This equation generalizes the one found by Lande (1985b).

Gamma approximation Approximating $F(x)$ by a gamma function leads to

$$\tau(p_0) = \frac{1}{|a'(p_0)|}(p_v - p_0) + \frac{1}{|a'(p_1)|}(p_1 - p_v)$$

$$+ \frac{1}{a'(p_v)} \ln\left[\sqrt{\frac{2a'(p_v)}{b(p_v)}}(p_v - p_0)(p_1 - p_v)\right]. \quad (4)$$

CHAPTER FOUR

Nearly neutral networks and holey fitness landscapes

As I already emphasized, an inherent assumption of Wright's picture of rugged fitness landscapes is that fitness peaks are isolated and that significant evolutionary divergence requires crossing deep valleys of maladaptation. This assumption appears both very natural and perfectly justified. Indeed, everybody knows from his or her own hiking experience that it is impossible to get from the top of one hill to another without having to descend to a kind of valley or a depression between them. Our intuition tells us that things will stay the same in landscapes with many more dimensions than the three we are so well familiar with, and that extended ridges or chains of very shallow valleys connecting high-fitness genotypes are improbable. If the fitness landscape is constant (in space and time), then stochastic fluctuations in the population genetic structure must be a major mechanism for crossing the valleys. However, the main conclusion of the previous chapter is that there is no satisfactory solution to the problem of stochastic transitions across even moderately deep valleys in static landscapes. Only if the valleys are very shallow, can stochastic transitions happen on a time scale short enough to be of biological significance. But both speciation and diversification seem to require crossing deep valleys in static landscapes describing RI. A logical conclusion is that speciation and diversification on rugged landscapes are impossible.

There must be some kind of an error or a weakness in the chain of arguments presented in the previous paragraph. Let us look at Wright's metaphor from a different perspective. The main reason why scientific metaphors are useful is that they help us understand very complex processes and phenomena using analogies with something simple and familiar. Wright's metaphor has been so successful exactly for this reason. It helps visualize evolutionary dynamics, which are an extremely complex process, using three-dimensional geographic landscapes. But are analogies coming from our experience with the three-dimensional world we live in actually good for understanding the properties of the genotype space, which has extremely high dimensionality?

The answer to this question, which I justify in this chapter, is no. The most important difference between three- and multidimensional landscapes is that although extended ridges connecting high-fitness genotypes are highly improbable in three dimensions, such fitness ridges are inevitable in real genotype spaces, which have extremely large dimensionality. This property leads to a view of fitness landscapes different from the one implied in Wright's picture and to different approaches for describing (and modeling) the processes of genetic diversification and speciation.

Below, I start by describing several simple models of fitness landscapes that illustrate the origin of ridges of high-fitness genotypes in multidimensional genotype spaces. Besides the term *ridges* I will use more formal notions of neutral and nearly neutral networks. A *neutral network* is defined as a contiguous set of sequences (or genotypes) possessing exactly the same fitness. Here, the word *contiguous* will mean that any two sequences of the set can be connected by a chain of one-step neighbors also belonging to the set. A *nearly neutral network* is defined as a contiguous set of sequences (genotypes) possessing approximately the same fitness. After considering the simple models, I will describe some properties of neutral networks of RNAs and proteins and discuss some additional evidence for the widespread existence of (nearly) neutral networks. Finally, I introduce a metaphor of holey landscapes and discuss some features of evolutionary dynamics on such landscapes.

4.1   SIMPLE MODELS

In this section, I consider six models: a model of a two-dimensional fitness landscape and five different models of multidimensional fitness landscapes. The first three models will describe uncorrelated landscapes, in which any genetic change (including substitutions at a single locus) results in an independent fitness value. The last three models will describe correlated landscapes where similar genotypes tend to have similar fitnesses.

The correlation patterns of fitness landscapes can be characterized by the correlation function ρ measuring the correlation of fitnesses of pairs of genotypes at the Hamming distance d from each other:

$$\rho(d) = \frac{\text{cov}[w(.), w(.)]_d}{\text{var}(w)} \tag{4.1}$$

(Eigen et al. 1989). Here, the term in the numerator is the covariance of fitnesses of two genotypes conditioned on them being at the Hamming distance d, and var(w) is the variance in fitness over the whole fitness landscape. Recall that the Hamming distance measures the number of differences between two sequences (see equation (2.4) on page 26). For uncorrelated landscapes, $\rho(d) = 0$ for $d \geq 1$. In contrast, for highly correlated landscapes, $\rho(d)$ decreases with d very slowly. Of the three models of correlated landscapes to be considered below, the first and the second will be characterized by linear and quadratic decay of ρ with distance d, whereas in the third model the rate of decay of ρ is controlled by a certain parameter.

A related useful measure of the correlation patterns of fitness landscapes is the correlation length l_c, which is defined as the inverse of the rate of decay of the correlation $\rho(d)$, with distance d in an exponential approximation of $\rho(d)$ for small d

$$\rho(d) \approx \exp(-d/l_c) \tag{4.2}$$

(e.g., Fontana et al. 1993). For genetic distances larger than l_c, fitnesses are effectively uncorrelated, whereas for genetic distances much smaller than l_c, fitnesses are very similar. For typical landscapes, the number (or density) of local peaks can be estimated from the correlation length l_c. Roughly, one expects one local peak within the volume of the genotype space corresponding to a radius that is determined by l_c (Stadler and Schnabl 1992; García-Pelayo and Stadler 1997). In general, a fitness landscape is rugged if it has many local peaks, if the length of adaptive walks (i.e., walks leading uphill) is short, and if the correlation of fitness values quickly decreases with genetic distance (i.e., if l_c is small).

4.1.1 Russian roulette model in two dimensions

Let us consider a two-dimensional lattice of square sites in which sites are independently painted black or white with probabilities $\mathcal{P}$ and $1 - \mathcal{P}$, respectively (see Figure 4.1). For each site, let its one-step neighbors be the four adjacent sites (directly above, below, on the left, and on the right). Let us say that two black sites are connected if there exists a sequence of black sites starting at one of them and going to another, such that subsequent sites in the sequence are neighbors. For any black site, let us define a *connected component* as the set of all black sites connected to the site under consideration. A simple numerical experiment shows that the number and the structure of connected components depend on the probability $\mathcal{P}$. For small values of $\mathcal{P}$ there are many connected components of small size (see Figure 4.1(a)). As $\mathcal{P}$ increases, the size of the largest connected component increases (see Figures 4.1a,b,c). As $\mathcal{P}$ exceeds a certain threshold $\mathcal{P}_c$, known as the *percolation threshold*, the largest connected component (known as the *giant component*) emerges, which extends (percolates) through the whole system and includes a significant proportion of black sites (see Figure 4.1d). In this model, describing a so-called site percolation on an infinite two-dimensional lattice, the percolation threshold is $\mathcal{P}_c \approx 0.593$ (e.g., Grimmett 1989).

The above model can be interpreted in simple genetic terms. Let us consider haploid individuals different with respect to $\mathcal{L} = 2$ genes, each with a very large number $\mathcal{A}$ of alleles. The alleles at the first locus will be denoted as A_i and the alleles at the second locus will be denoted as B_j. Assume a stepwise mutation pattern, as in Nei et al. (1983), allowing allele A_i only to mutate to alleles A_{i+1} and A_{i-1} and, in a similar way, allowing allele B_i only to mutate to alleles B_{i+1} and B_{i-1}. Because of these restrictions, the dimensionality of the genotype space reduces from $\mathcal{D} = \mathcal{A}^2$, as expected in the general case, to $\mathcal{D} = 2$, and the genotype space can be adequately represented by a two-dimensional lattice identical to that in Figure 4.1 (see also Figure 2.4 on page 29).

Following Gavrilets and Gravner (1997), assume that genotype fitnesses are generated randomly and independently and are only equal to 1 (viable genotype) or 0 (inviable genotype) with probabilities $\mathcal{P}$ and $1 - \mathcal{P}$. Here, one might think of the set of all possible genotypes playing one round of Russian roulette, with $\mathcal{P}$ being the probability to get a blank. The fitness landscape in this model is uncorrelated and any change in genotype no matter how large or small, results in a fitness value completely independent of the initial value.

Following our approach above, we will say that two viable genotypes are connected if there exists a sequence of viable genotypes that goes from one of them to another such that subsequent genotypes in the se-

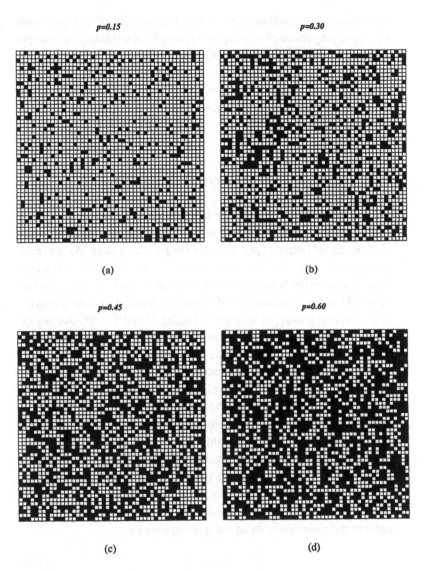

FIGURE 4.1. Percolation in two dimensions. (From Gavrilets (2003b), fig.4.)

quence are one-step neighbors. For any genotype, let us define a *neutral network* as a set of all viable genotypes connected to the genotype under consideration. The network is neutral in the sense that movement along it will not change fitness. Now the results from the first paragraph of this subsection immediately tell us that if probability $\mathcal{P} > 0.593$, then a positive proportion of viable genotypes will be connected in a single giant component (neutral network). In this model, the value of $\mathcal{P}$ that guarantees the existence of the giant component (i.e., the percolation threshold) is very large. However, as we will see in the next section, the percolation threshold dramatically decreases in genotype spaces of high dimensionality.

4.1.2   Russian roulette model on hypercubes

Assume next that there is a very large number of diallelic loci ($\mathcal{A} = 2, \mathcal{L} \to \infty$). Now the genotype space is mathematically equivalent to a $\mathcal{L}$-dimensional binary hypercube (see Figure 2.2 on page 26), and each genotype has $\mathcal{L}$ one-step neighbors. Let us assign fitnesses in exactly the same way as in the previous subsection, that is, fitnesses are generated randomly and independently and are only equal to 1 (viable genotype) or 0 (inviable genotype) with probabilities $\mathcal{P}$ and $1 - \mathcal{P}$. As before, viable genotypes will tend to form neutral networks. The following is a summary of the properties of these networks.[1]

If $\mathcal{P} > 1/2$, then all viable genotypes belong to a single neutral network, whereas if $\mathcal{P} < 1/2$, the probability of a single neutral network is close to 0 (Reidys 1997). Thus, the existence of a single neutral network requires an unrealistically high probability of being viable. For smaller values of $\mathcal{P}$, there are two qualitatively different regimes: subcritical, which takes place when $\mathcal{P} < \mathcal{P}_c$, and supercritical, which takes place when $\mathcal{P} > \mathcal{P}_c$, where $\mathcal{P}_c$ is the percolation threshold (Gavrilets and Gravner 1997). A very important, though counterintuitive, feature of this model is that the percolation threshold is approximately the reciprocal of the dimensionality of the genotype space:

$$\mathcal{P}_c \approx \frac{1}{\mathcal{L}}, \tag{4.3a}$$

and thus is very small. At the boundary of the two regimes, all properties of neutral networks undergo dramatic changes, a physical analogy

---

[1] Similar results are known for a model where each pair of vertices of a binary hypercube is connected by an edge with a certain probability $\mathcal{P}$, and the interest is in the properties of the networks of connected edges (Burtin 1977; Erdös and Spencer 1979; Ajtai et al. 1982; Bollobas 2001).

of which is a phase transition. In the subcritical regime, the largest neutral networks have order $\mathcal{L}$ different genotypes, and there are many of them. Typical members of a neutral network are connected by a single sequence of viable genotypes. Thus, there is a single possible evolutionary path. This path is straight in the sense that along it, substitution in a locus will typically happen only once. The supercritical regime takes place if the proportion of viable genotypes is small but not extremely small. In the supercritical regime, with probability approaching 1, there exists a neutral network of viable genotypes (a giant component) that includes a positive proportion of all viable genotypes. The giant component, which has order of $2^{\mathcal{L}}/\mathcal{L}$ genotypes, comes near every point of the genotype space in the sense that each point is within a few steps of the giant component. Typical members of the giant component are connected by many evolutionary paths, which are not extremely winding. A useful measure of the degree of a cluster's expansion is the *cluster diameter*, defined as the maximum Hamming distance between pairs of genotypes that belong to the cluster. In the supercritical regime, the diameter of the giant component is equal to the number of loci $\mathcal{L}$; the second largest neutral network has order $\mathcal{L}$ different genotypes.

Similar results hold if there are $\mathcal{A} > 2$ alleles, so that the genotype space can be represented by a generalized hypercube with dimensionality $\mathcal{D} = \mathcal{L}(\mathcal{A} - 1)$. In this case, all viable genotypes belong to a single network if $\mathcal{P} > 1 - 1/\sqrt[\mathcal{A}-1]{\mathcal{A}}$ (Reidys 1997), whereas the percolation threshold separating the subcritical and supercritical regimes is

$$\mathcal{P}_c \approx \frac{1}{\mathcal{D}} = \frac{1}{\mathcal{L}(\mathcal{A} - 1)} \qquad (4.3b)$$

(Gavrilets 1997c). Thus, as before, the existence of the giant component is guaranteed even with very small probabilities of being viable, provided the dimensionality of the genotype space is sufficiently high.

It is very easy to see why the percolation threshold is approximately $1/\mathcal{D}$. Because there are $\mathcal{L}$ loci with $\mathcal{A}$ alleles each, each genotype has $\mathcal{D} = \mathcal{L}(\mathcal{A} - 1)$ one-step neighbors. Pick up a viable genotype, say genotype 0 (see Figure 4.2). If at least one of its one-step neighbors, say genotype 1, is viable, there is a neutral network with at least two genotypes (0 and 1). Next, let us concentrate on genotype 1. It also has $\mathcal{D}$ one-step neighbors of which one is genotype 0. If at least one of the remaining $\mathcal{D} - 1$ one-step neighbors, say genotype 2, is viable, there is a neutral network with at least three genotypes (0, 1, and 2). The point

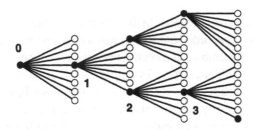

FIGURE 4.2. Branching on a hypercube. The figure presents a part of the genotype space for $\mathcal{D} = 8$.

is that if there is one viable genotype out of $\mathcal{D} - 1$ one-step neighbors, the size of the neutral network increases by one. It should be intuitively clear that if the dimensionality $\mathcal{D}$ is very large, then the probability of an infinite path starting at genotype 0 is approximately the same as the probability of survival of the branching process with $\mathcal{D} - 1$ successors where each successor is viable with probability $\mathcal{P}$ (see Figure 4.2). The branching process does not die out if the expected number of viable successors, which is $\mathcal{P}(\mathcal{D} - 1)$, is larger than one (e.g., Harris 1964; Karlin and Taylor 1975). Thus, a giant component representing a neutral network of viable genotypes expanding throughout the genotype space exists if $\mathcal{P} > 1/(\mathcal{D} - 1) \approx 1/\mathcal{D}$.

In the Russian roulette model, different genotypes will have different numbers of viable neighbors. The distribution of the number $\mathcal{N}$ of viable neighbors is binomial, which for large values of $\mathcal{D}$ can be approximated by a Poisson distribution:

$$\Pr(\mathcal{N} = i) = \binom{\mathcal{D}}{i} \mathcal{P}^i (1 - \mathcal{P})^{\mathcal{D}-i} \approx \exp(-\overline{\mathcal{N}}) \frac{\overline{\mathcal{N}}^i}{i!}, \qquad (4.4)$$

where $\overline{\mathcal{N}} = \mathcal{D}\mathcal{P}$ is the *average neutrality*, that is, the average number of neutral (here, viable) neighbors per viable genotype. This implies that the frequency of genotypes with at least $\mathcal{N}_c$ viable neighbors is

$$\Pr(\mathcal{N} > \mathcal{N}_c) \approx \sum_{i=\mathcal{N}_c}^{\mathcal{D}} \exp(-\overline{\mathcal{N}}) \frac{\overline{\mathcal{N}}^i}{i!} = 1 - \frac{\Gamma(\mathcal{N}_c, \overline{\mathcal{N}})}{(\mathcal{N}_c - 1)!}, \qquad (4.5)$$

where $\Gamma(.,.)$ is the digamma function (Gradshteyn and Ryzhik 1994). For example, if there are $\mathcal{L} = 100$ diallelic loci ($\mathcal{A} = 2$) and the probability of being viable is $\mathcal{P} = 0.05$, then each genotype has 5 viable neighbors on average. However, a proportion 1.6×10^{-8} of

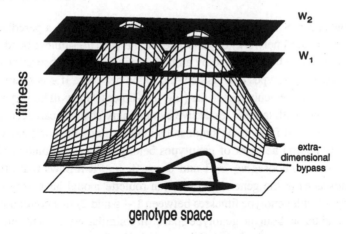

genotype space

FIGURE 4.3. Percolating fitness bands in the generalized Russian roulette model. The picture shows a landscape with two peaks and two clusters of genotypes that belong to a (w_1, w_2)-fitness band. Although in three dimensions these clusters appear disjointed, they are connected through some other dimensions if $w_1 - w_2 = \varepsilon > 1/\mathcal{D}$. This connection is marked as an *extra-dimensional bypass*. (From Gavrilets (1997a), fig.5.)

genotypes have at least 25 viable neighbors. Although this proportion is very small, it corresponds to about 2×10^{20} different genotypes.

4.1.3 Generalized Russian roulette model

The assumption that fitness can only take two values, 0 and 1, might be viewed as a serious limitation. To show that this is not so, let us consider the same genotype space as in the previous section, but now assume that genotype fitness, w, is a realization of a random variable having a uniform distribution between 0 and 1 (Gavrilets and Gravner 1997). The fitness landscape in this model is uncorrelated and is similar to the one discussed in Box 2.2.

Let us introduce threshold values w_1 and w_2, which differ by a small value ε. Let us say that a genotype belongs to the (w_1, w_2)-fitness band if its fitness w satisfies the conditions $w_1 < w \leq w_2$ (see Figure 4.3). Parameter ε can be viewed as the probability that a randomly chosen genotype belongs to the (w_1, w_2)-fitness band. Belonging to the (w_1, w_2)-fitness band is analogous to being viable in the previous model. Therefore, if the dimensionality of genotype space $\mathcal{D}$ is very large and $\varepsilon > 1/\mathcal{D}$, there exists a giant component (i.e., a percolating nearly neutral network) of genotypes in the (w_1, w_2)-fitness band. Its members can be connected by a chain of single-gene substitutions resulting in genotypes that also belong to the network. If ε is small, the fitnesses of the genotypes in the (w_1, w_2)-fitness band will be very similar. Thus, with large $\mathcal{D}$, extensive evolutionary changes can occur in a nearly neutral fashion via single substitutions along the corresponding nearly neutral network of genotypes belonging to a percolating cluster. If one chooses $w_2 = 1$ and $w_1 = 1 - \varepsilon$, it follows that fitness landscapes in the generalized Russian roulette model have very high ridges (with genotype fitnesses between $1 - \varepsilon$ and 1) that continuously extend throughout the genotype space. In a similar way, if one chooses $w_2 = \varepsilon$ and $w_1 = 0$, it follows that the landscapes have very deep gorges (with genotype fitnesses between 0 and ε) that also continuously extend throughout the genotype space. The maximum number of nonoverlapping (w_1, w_2)-fitness bands is $1/\varepsilon$, which, with ε just above the percolation threshold, is approximately $\mathcal{D}$. Thus, the maximum number of nonoverlapping percolating nearly neutral networks is equal to the dimensionality $\mathcal{D}$. It follows from the symmetry that each of the networks has approximately the same number of genotypes. As in the Russian roulette model, the *diameter* of each network, that is the maximum Hamming distance between its members, is equal to $\mathcal{L}$.

Finally, I note that the above conclusions apply not only for the uniform distribution of fitness values but for any random distribution of fitnesses, provided the overall frequency of genotypes that belong to a (w_1, w_2)-fitness band is larger than $1/\mathcal{D}$. That is, if $f(w)$ is the density function of the distribution of individual fitness values assigned randomly and independently, then genotypes within a (w_1, w_2)-fitness band form a percolating nearly neutral network if

$$\int_{w_1}^{w_2} f(w)dw > \frac{1}{\mathcal{D}}.$$

Once again, increasing dimensionality $\mathcal{D}$ makes percolation easier.

4.1.4 Multiplicative fitnesses

Let us now consider a correlated landscape. As before, assume that there are $\mathcal{L}$ diallelic loci. In a commonly used multiplicative fitness model (already discussed on page 37), alternative alleles are interpreted

as "advantageous" and "deleterious", and the fitness of an individual
with k deleterious alleles is chosen to be

$$w_k = (1 - s)^k, \tag{4.6}$$

with parameter $s > 0$ characterizing the strength of selection. Assuming
that s is small and that $\mathcal{L}$ is large, one can show that the correlation
function (see equation (4.1)) of this fitness landscape is

$$\rho(d) \approx 1 - 2\frac{d}{\mathcal{L}}. \tag{4.7}$$

Function $\rho(d)$ slowly decreases with d. The correlation length is equal
to half of the overall number of loci, $l_c = \mathcal{L}/2$, and is very large. The
fitness landscape in this model is highly correlated.

Here, the fitness landscape has a single peak at a genotype with
no deleterious alleles and $\mathcal{L} + 1$ different fitness levels $(1 - s)^k, (1 -
s)^{k-1}, \ldots, 1 - s, 1$. It is easy to see that any two genotypes at the same
fitness level can be connected by a chain of single-gene substitutions
leading not farther than the fitness level just below or just above the
original fitness level. For example, genotypes 01100 and 00011 with
two deleterious alleles 1 can be connected by evolutionary paths

$$01100\,(2) \rightarrow 01110\,(3) \rightarrow 00110\,(2) \rightarrow 00111\,(3) \rightarrow 00011\,(2)$$

or

$$01100\,(2) \rightarrow 00100\,(1) \rightarrow 00110\,(2) \rightarrow 00010\,(1) \rightarrow 00011\,(2),$$

where the numbers in the parentheses specify the number of deleterious
alleles present. Thus, the number of distinct nearly neutral networks
in this model is approximately $\mathcal{L}/2$. These networks can be imagined
as spherical shells in the genotype space at a constant Hamming distance from the optimum genotype $00 \ldots 0$. In contrast to the previous
model where different nearly neutral networks had similar sizes, here
different networks have different sizes. The diameter of the network
decreases from $\mathcal{L}$ for genotypes with an equal number of advantageous
and deleterious alleles to 0 for the most and least fit genotypes.

4.1.5 Stabilizing selection on an additive trait

Next, let us consider a model of an additive quantitative character z similar to the models discussed in Section 2.6. Assume that the genotypic

value g is determined by a sum of effects of $\mathcal{L}$ haploid diallelic loci,

$$g = \sum_{i=1}^{\mathcal{L}} a_i, \tag{4.8}$$

where a_i is the contribution of the i-th locus to the trait. Assume that environmental effects are absent ($e = 0$), so that the trait value is equal to the genotypic value ($z = g$).

Let us consider the case of stabilizing selection in which fitness $w(z)$ decreases with the deviation of the trait value from an optimum value θ. For simplicity, assume that all genes have equal effects: $a_i = 0$ or $a_i = 1$ depending on whether a certain allele is present or absent at the i-th locus. Assume also that optimum θ is at the midvalue of the trait range ($\theta = \mathcal{L}/2$) and that the fitness function is symmetric about the optimum, so that $w(z - \theta) = w(\theta - z)$ for all trait values z. These assumptions result in a fitness landscape with $\mathcal{L}/2$ different fitness values $w(0), w(1), w(2), \ldots, w(\theta)$ corresponding to different deviations of z from the optimum. As in the multiplicative fitness model, any two genotypes at the same fitness level can be connected by a chain of single substitutions leading not farther than the previous or the next fitness level. Thus, the number of distinct nearly neutral networks is $\approx \mathcal{L}/4$. In contrast to the multiplicative model, where the fittest nearly neutral network has the smallest diameter, in the present model such a network has the largest diameter, which is equal to $\mathcal{L}$. This property implies that extensive nearly neutral divergence is possible under stabilizing selection (Endler 1988; Barton 1989a; Mani and Clarke 1990)

Note that if the fitness function w is (approximately) quadratic, the resulting fitness landscape is similar to that of the Sherrington-Kirkpatrick spin glass model (Sherrington and Kirkpatrick 1975). The correlation function $\rho(d)$ for the latter and, thus, for the model considered here is

$$\rho(d) \approx \left(1 - 2\frac{d}{\mathcal{L}}\right)^2 \tag{4.9}$$

(Fontana et al. 1993). This correlation function decreases as a quadratic function of distance d. The correlation length is equal to a quarter of the overall number of loci, $l_c = \mathcal{L}/4$ and thus is large.

4.1.6   Models based on the Nk-model

The generalized Russian roulette model considered above provides an example of an extremely rugged fitness landscape with an enormous number of local peaks (see Box 2.2) and a total lack of correlation between fitnesses of similar genotypes. The multiplicative fitness model and the model of stabilizing selection on an additive trait lie at the other end of the spectrum. In both of these models, fitnesses of organisms with similar genotypes typically are similar. Kauffman (1993) came up with a class of models of random fitness landscapes in which the degree of correlation can be changed by varying a parameter.

In Kauffman's model,[2] there are $\mathcal{L}$ diallelic loci. Each locus interacts epistatically with a number k of other loci (for example, chosen at random). The contribution of locus i to fitness, w_i, is assumed to depend on the allele at this locus and the k loci $i_1, \ldots, i_k$ it interacts with. This assumption can be mathematically described as $w_i = w_i(l_i, l_{i_1}, \ldots, l_{i_k})$, where indicator variables l_i (equal to 0 or 1) specify the alleles at the corresponding loci. Kauffman chooses the values w_i to be independent random real numbers drawn from a standard uniform distribution. The genotype fitness is given by the average of the contributions of $\mathcal{L}$ loci: $w = \sum_i w_i / \mathcal{L}$. Parameter k controls the degree of ruggedness of the landscape, with small k corresponding to smooth landscapes and large k corresponding to very rugged landscapes. If $k = 0$, the loci do not interact and the model simplifies to the additive fitness model (see page 37) with random allelic contributions to fitness. If $k = \mathcal{L} - 1$, the model becomes similar to the generalized Russian roulette model discussed earlier in this section. The correlation function of the fitness landscapes arising in this model is

$$\rho(d) = \frac{(\mathcal{L} - k - 1)!(\mathcal{L} - d)!}{\mathcal{L}!(\mathcal{L} - k - 1 - d)!} \tag{4.10}$$

(Fontana et al. 1993; Campos et al. 2002). Figure 4.4 illustrates the dependence of the correlation function on k and d for $\mathcal{L} = 100$ loci. The distribution of fitness values is approximately normal with the mean equal to 0.5 and a very small variance equal to $1/(12\mathcal{L})$. The latter follows from the fact that an individual's fitness value is given by the average of $\mathcal{L}$ independent random variables, each of which has the standard uniform distribution (with the variance equal to $1/12$).

---

[2]This class of models is known as the "Nk-model" after the names of the two most important parameters: the overall number of the loci, which Kauffman denoted as N, and the average number of other loci with which each locus interacts epistatically, which Kauffman denoted as k. In describing the model, I will keep using $\mathcal{L}$ for the number of loci.

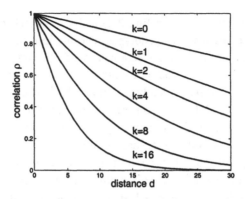

FIGURE 4.4. Correlation function $\rho(d)$ for the Nk-model for different k and $\mathcal{L} = 100$.

The complexity of fitness landscapes in Nk-models has so far prevented analytical studies of the properties of percolating nearly neutral networks. However, numerical studies have shown that these networks are both abundant and important. For example, Kauffman (1993) observed that populations diffused through the genotype space in a nearly neutral fashion. Ohta (1997, 1998) studied a modified version of the Nk-model numerically. Her results indicate that most mutations fixed in the process of divergence were nearly neutral, strongly suggesting that nearly neutral networks are widespread in this model. This conclusion is supported by analytical and numerical studies of two other modifications of the Nk-model. Newman and Engelhardt (1998) chose the values w_i to be integers in the range $0 \leq w_i < A$ (rather than real numbers in the interval from 0 to 1). This results in $\mathcal{L}A - \mathcal{L} + 1$ possible values for genotype fitnesses, which immediately implies a possibility for a high degree of neutrality with small A. In contrast, Barnett (1997) assumed that a proportion p of allelic combinations make no contribution to fitness. In his model, the degree of neutrality is controlled by the parameter p. The case $p = 0$ corresponds to the original Nk-model, while $p = 1$ corresponds to a flat landscape. Numerical simulations of Barnett (1997) and Newman and Engelhardt (1998) have demonstrated both the existence of neutral networks percolating through the genotype space and the importance of these networks in the evolutionary divergence of finite populations.

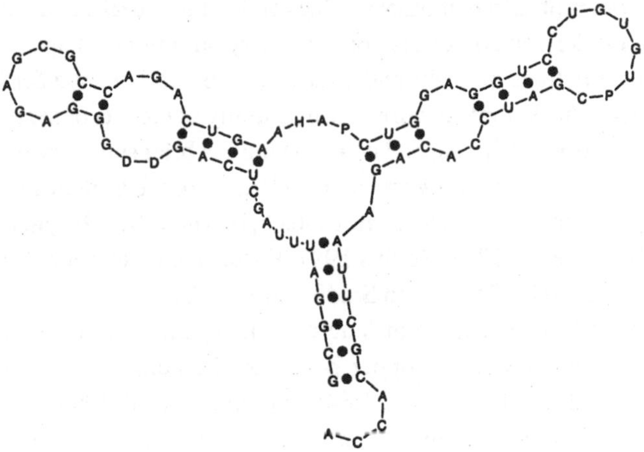

FIGURE 4.5. One possible secondary structure of phenylalanyl-transfer-RNA (t-RNA^{phe}).

4.2   NEUTRAL NETWORKS IN RNA LANDSCAPES

The relationship between the primary and secondary structures of RNA molecules has become the most studied example of realistic multidimensional landscapes. The primary structure of an RNA molecule is the sequence of nucleotides which can be of four different types: C, G, A, and U. The secondary structure is its pattern of complementary base pairings (Watson-Crick base pairs G-C and A-U, and the weaker G-U pairs). This pattern can be drawn as a planar graph that is free of knots and pseudoknots (see Figure 4.5). In the literature, specific secondary structures are often referred to as *shapes*. A secondary structure of an RNA molecule can be viewed as a coarse-grain approximation to the full three-dimensional structure of the molecule. The latter determines all relevant fitness components (e.g., free energy, melting temperature, kinetic constants, reproduction rate). The secondary structure can be predicted under a minimum free energy criterion using efficient numerical folding algorithms.

In the context of fitness landscapes, the primary structure is interpreted as "genotype" and the secondary structure is analogous to "fitness." In the fitness landscapes that we have considered above, the fitnesses were specified by real positive numbers that implied a natural relative order of different fitness values and defined peaks and valleys. No such order relation exists for secondary structures (shapes). Although the notions of peaks and valleys are not well defined for RNA landscapes, the neutral networks arise naturally as the sets of sequences that fold into a certain shape. The following is a brief summary of what is known about neutral networks in RNA landscapes (e.g., Schuster et al. 1994; Schuster 1995; Grüner et al. 1996a,b; Huynen 1996; Huynen et al. 1996; Reidys et al. 1997; Reidys 1997; Fontana and Schuster 1998a,b; Reidys et al. 2001; Reidys and Stadler 2001, 2002).

• For an RNA sequence of length $\mathcal{L}$, there can be $4^{\mathcal{L}}$ different sequences. An approximate upper bound on the number of secondary structures is $1.4848 \times \mathcal{L}^{-3/2} \times 1.8488^{\mathcal{L}}$ (Schuster et al. 1994; Schuster 1995). This implies that there are many more sequences than secondary structures. Therefore, many different sequences must fold into exactly the same structure.

• RNA landscapes are very rugged; randomly changing only 15% of the nucleotides results in a change in secondary structure comparable with the difference between random sequences.

• Typically there are relatively few common structures and many more different rare structures. For example, in a study of sequences formed by $\mathcal{L} = 30$ nucleotides that could be only of types G or C, it was found that more than 93% of all possible sequences folded into about 10% of all possible shapes (Grüner et al. 1996a,b). Overall, this study found 22,718 common shapes (corresponding to 999,508,805 sequences that fold into them) out of 218,820 possible shapes. Ranking different shapes according to the number of sequences folding into them usually yields a distribution resembling the Zipf law (Zipf 1949).

• The sequences folding into a common structure are distributed to a large extent uniformly in sequence space. No clustering is observed.

• The sequences folding into a common structure form neutral networks extending throughout the sequence space. For example, Schuster et al. (1994) and Schuster (1995) analyzed sequences formed by $\mathcal{L} = 100$ nucleotides. Two classes of nearest neighbors were allowed: those obtained by a single base exchange in unpaired stretches of the structure and those resulting from base pair exchanges. It was found that almost a quarter of common neutral networks reached a sequence that had no single base in common with a starting sequence, meaning that the corresponding cluster diameter is equal to the sequence length.

- The average distance between a sequence folding into a common structure and a sequence folding into a desired structure is very short relative to the average distance between pairs of sequences. This property is known as the *shape coverage principle*.

The existence of extended neutral networks in RNA sequence space was recently proved experimentally by Schultes and Bartel (2000), who described a close apposition of the neutral networks for the hepatitis delta virus (HDV) self-cleaving ribozyme and the class III self-ligating ribozyme. The sequences for these ribozymes (of length $L = 86$) have no more than the 25% sequence identity expected by chance. Schultes and Bartel (2000) identified a sequence that is within two or three point mutations from both neutral networks and re-created neutral paths (with the length of about 40 mutations) leading from the class III and HDV ribozymes into the neighborhood of this sequence.

4.3 NEUTRAL NETWORKS IN PROTEIN LANDSCAPES

Both the original claim about the importance of neutral networks in evolution and the first quantitative demonstration of their existence were made within the context of analyzing the relationships between amino acid sequences and functions of the corresponding proteins.

In a very short but groundbreaking paper that first[3] appeared in an edited volume entitled *The Scientist Speculates: An Anthology of Partly-Baked Ideas* in 1962, Maynard Smith argued that if evolution by natural selection is to occur, functional proteins must form a continuous network that can be traversed by unit mutational steps without passing through nonfunctional intermediates. He explained such networks by analogy with a word game where the goal is to transform one word to another by changing one letter at a time, with the requirement that all intermediate words are meaningful (as in the sequence of words WORD→WORE→GORE→GONE→GENE). Maynard Smith concluded that for such networks to exist, the average number of other meaningful proteins that are within one mutational step from a meaningful protein has to be larger than one. He expressed this condition as $fN > 1$, where N was defined as the overall number of proteins that

---

[3] A variant of this paper was published eight years later in *Nature* (Maynard Smith 1970) and has become a standard reference rather than the original 1962 paper.

are within one mutational step from a meaningful protein and f was the fraction of these that are meaningful.[4]

Thirty years later, Lipman and Wilbur (1991) performed a pioneering numerical study of neutral networks using a simple lattice model for proteins proposed by Lau and Dill (1989, 1990). Lipman and Wilbur worked with sequences with up to $\mathcal{L} = 19$ elements (amino acids), each of which could be of two kinds (hydrophobic and hydrophilic). They studied the distribution of the number of sequences among different phenotypes (which were defined by certain topological characteristics), the distribution of the number of neutral neighbors, the number of sequences in the largest network, and the lengths of certain paths in the sequence space. Lipman and Wilbur concluded that Maynard Smith's condition $fN > 1$ was satisfied and neutral networks were formed. They also noticed that although not all possible sequences with the same phenotype belonged to the same network, a larger fraction of them became linked into a single network with increasing sequence length.

More recent work provided both new empirical (Martinez et al. 1996; Rost 1997) and new theoretical (Bornberd-Bauer 1997; Bornberd-Bauer and Chan 1999; Bornberd-Bauer 2002; Rost 1997; Babajide et al. 1997; Govindarajan and Goldstein 1997; Babajide et al. 2001; Bastolla et al. 1999) evidence for the existence and importance of neutral networks in protein space. Overall the properties of protein landscapes appear to be similar to those of RNA landscapes discussed in the previous section, with a possible exception being the shape coverage principle. Bornberd-Bauer (1997, 2002) and Bornberd-Bauer and Chan (1999) argued that different neutral networks are denser and more separated in protein spaces than in RNA landscapes. This, however, might be an artifact of the model used by Bornberg-Bauer and Chan that allowed only for $\mathcal{A} = 2$ types of amino-acids (rather than $\mathcal{A} = 20$) and, thus, decreased the dimensionality of sequence space and made percolation more difficult (Bastolla et al. 1999).

Rost (1997) used a database of protein structures (which at that time had about 5,000 entries) to analyze all pairs of sequences with similar three-dimensional structures. His results showed that most pairs of

---

[4]This condition is equivalent to the condition separating the critical and subcritical regimes in the two Russian roulette models considered above. Maynard Smith's N is analogous to the dimensionality of the genotype space $\mathcal{D}$ and his f is analogous to the probability of being viable $\mathcal{P}$.

similar structures have sequence similarity as low as expected from randomly related sequences (8% to 9%). This strongly suggests that the corresponding neutral networks spread throughout the whole sequence space and that the time since the origin of life was sufficient for the proteins to spread across these networks.

4.4 OTHER EVIDENCE FOR NEARLY NEUTRAL NETWORKS

RNA and protein landscapes as well as landscapes in the simple models discussed above provide the most detailed information about (nearly) neutral networks extending through genotype space. Additional indirect evidence for such networks comes from a number of other studies.

• Biologists have documented more than 20 so-called ring species (e.g., Mayr 1942, 1963; Wake 1997; Irwin et al. 2001; Bowen et al. 2001; Demuth and Rieseberg 2001). A *ring species* is a chain of races (or subspecies) arranged in the shape of a ring with gradual morphological transitions and no RI between adjacent geographic races but abrupt morphological changes and RI where the terminal races come into contact. A classical, although still controversial, example of a ring species consists of herring gulls and lesser black-backed gulls, which are sympatric and reproductively isolated in Europe, but are connected by a chain of gradating forms that encircle the Arctic Ocean (Mayr 1942, 1963). The two most studied ring species are the salamanders of the genus *Ensantina* in northern and central California (Wake 1997), and the greenish warbler in Asia (Irwin et al. 2001). In ring species, networks of genotypes connecting two substantially different (judging from the presence of RI or significant morphological divergence) forms are re-created, at least to a significant degree, in a natural way. Therefore, the ring species can be considered as one of the best manifestations of extended (nearly) neutral networks in genotype space. Note that the lack of complete RI between the terminal forms and/or the lack of true continuity in the distribution of the intergrading forms observed for some ring species have no bearing on this interpretation.

• Strong artificial selection in a specific direction usually results in a desired response, but as a consequence of the genetic changes brought about by artificial selection, different components of fitness (such as viability or fertility) significantly decrease (e.g., Hill and Caballero 1992). Moreover, after relaxing artificial selection, natural selection usually tends to return the population to its original state. These observations stimulated Wright's view of species as occupying isolated peaks in a fitness landscape (Provine 1986). However, the size of the experimental populations under selection is usually very small – on the order of a few dozen individuals (Hill and Caballero 1992; Gavrilets and Hastings 1995a). Small populations will be characterized by low levels of genetic variation and may not find ridges in the fitness landscape even if they are present. Weber (1996) performed selection experiments using very large populations of *Drosophila melanogaster* with thousands of individuals, selecting for the ability to fly in a wind tunnel. He was able to change the selected trait (which is obviously nonneutral) from about 2 to 170 cm/sec but did not observe any significant reduction in fitness components nor any tendencies to return to the original state after selection was relaxed. A straightforward interpretation of Weber's results is that the large populations in his experiments were able to find a ridge of high-fitness genotypes in the fitness landscape.

• If two fitness peaks are separated by a valley, then intuitively one would expect that offspring from matings between individuals residing at different peaks will have an intermediate phenotype and will reside at the valley. Therefore, extensive and successful natural hybridization in animals and plants as documented by Bullini (1994); Rieseberg (1995), and Arnold (1997) is very difficult to reconcile with the idea of peaks separated by valleys. However, if there is a ridge (or ridges) of high-fitness values connecting two very different forms, then hybridization may occasionally result in offspring genotypes residing on the ridge(s). Such offspring is expected to have high fitness.

• The fossil record provides examples of fit intermediates between radically different morphologies (e.g., Carroll 1988). One can claim that there are no examples of unfit intermediates in the fossil record.

• Extended neutral networks have been observed in computational landscapes arising in experiments on evolving tone recognition circuits (Harvey and Thompson 1997; Thompson and Layzell 2000).

4.5 THE METAPHOR OF HOLEY FITNESS LANDSCAPES

The concepts of genotype, RNA, and protein spaces can be unified within a general notion of the *sequence space*. The sequence space is defined as the set of all possible sequences given certain restrictions on the length, $\mathcal{L}$, of sequences and the number, A, of different states

TABLE 4.1. Genome sizes as measured by the number of base pairs per chromosome (as of March 27, 2003) according to a database of the Center for Biological Sequence Analysis (*www.cbs.dtu.dk*).

Superkingdom	# of species	# of sequences	Range (in base pairs)
Archae	16	16	1,564,906–5,751,492
Bacteria	101	130	412,348–9,105,828
Eukaryotes	11	11	174,133–282,193,664

each element of the sequence could be at. The dimensionality $\mathcal{D}$ of sequence space is defined as the number of new sequences that one can get from a sequence by changing its single elements. Therefore, $\mathcal{D} = \mathcal{L}(\mathcal{A} - 1)$. Even the simplest organisms known have on the order of a thousand genes and on the order of a million DNA base pairs. Each of the genes can be at least at several different states (alleles). Thus, the dimensionality of biologically relevant sequence spaces is at least on the order of thousands. It is on the order of millions if one considers DNA base pairs instead of genes (see Table 4.1). As recognized already by Wright himself (Wright 1932), this huge dimensionality results in an astronomically large number of possible genotypes (or DNA sequences, or amino acid sequences), which is much higher than the number of organisms present at any given time or even cumulatively since the origin of life.

At the same time, the number of different fitness values is limited. For example, assume that all possible fitnesses are normalized to be between 0 and 1. Then if the smallest fitness difference that one can measure (or that is important biologically) is, say, 0.001, only 1,000 meaningfully different fitness classes are possible. One may argue, however, that from a theoretical point of view much smaller fitness differences than those measured experimentally could be important in evolution (e..g, Gillespie 1991). Assume that one wants to create an artificial fitness landscape for a set of binary sequences in a computer's memory by assigning a unique fitness value between zero and one to each sequence. Under double precision, each number in the computer's memory will be described by no more than 64 bits. This implies that the assignment of different numerical fitness values to different sequences is only possible for sequences with the length $\mathcal{L} < 64$. For longer sequences, one will just not have enough different numbers.

An important consequence of these observations is that many different genotypes are bound to have very similar (identical from any practical point of view) fitnesses. In other words, the many-to-one redundancy in the genotype-fitness relationship is unavoidable. Under rather general conditions discussed and illustrated above, genotypes with fitnesses within a certain fitness band form nearly neutral networks that extend throughout the sequence space. Generically, these networks tend to form if the dimensionality of the sequence space is sufficiently high and the overall frequency of sequences within the fitness band is not too small. The formation of these networks is largely independent of the degree of correlation characterizing the landscape. In fact, Reidys and Stadler (2001, 2002) have recently shown that ruggedness (as measured by the correlation length of the landscapes) and neutrality (as measured by the number of neutral neighbors) are to a large degree independent properties of fitness landscapes.

Among different percolating fitness bands, those at relatively high fitness levels are of particular importance. If they exist, then biological populations might evolve along the corresponding nearly neutral network by single substitutions and diverge genetically without ever going through any fitness valleys. To describe these neutral networks, Janko Gravner and I (Gavrilets and Gravner 1997) introduced the notion of holey landscapes. A *holey landscape* is defined as a fitness landscape where relatively infrequent high-fitness (or as Wright put it, "harmonious") genotypes form a contiguous set that expands throughout the genotype space. An appropriate three-dimensional image of such a landscape that focuses exclusively on the percolating network is a nearly flat surface with many holes representing genotypes that do not belong to the network (see Figure 4.6). The smoothness of the surface in this figure reflects close similarity between the fitnesses of the genotypes forming the corresponding nearly neutral network. The holes include both lower fitness genotypes (valleys and slopes) and very high fitness genotypes (the tips of the fitness peaks). Overstating just to make the point, tips are largely irrelevant because the population is almost never able to climb there (due to the absence of the right mutations and/or a high frequency of deleterious mutations pushing the population downhill). Valleys are largely irrelevant because natural selection will quickly move the population from there uphill, or the population goes extinct. What is most relevant for continuous genetic diversification and speciation are the

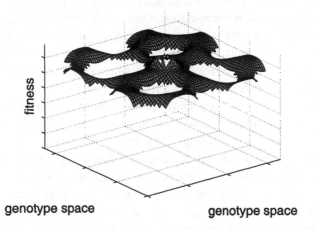

FIGURE 4.6. A holey fitness landscape formed by genotypes with fitnesses within a narrow fitness band. Evolution along a holey landscape is nearly neutral. (From Gavrilets (2003b), fig.5b.)

ridges of high-fitness genotypes that expand throughout the genotype space.

As was discussed above, each of the three classical metaphors of fitness landscapes emphasizes certain features of the landscapes and evolutionary dynamics while neglecting or de-emphasizing others. Wright's metaphor of rugged fitness landscapes emphasizes the existence of multiple high-fitness combinations of genes and the need for stochastic factors to overcome selection for continuous evolution. The metaphor of a single-peak landscape reflects Fisher's belief that there is a single "best" combination of genes and that selection alone is sufficient for evolutionary change and adaptation. The metaphor of flat fitness landscapes emphasizes Kimura's views on the importance of extensive genetic divergence by mutation and random genetic drift alone. In contrast, the metaphor of holey landscapes illustrated in Figure 4.6 emphasizes the percolating ridges of high-fitness genotypes at the expense of other features of multidimensional fitness landscapes (Gavrilets 1997b, 2003a).

Within the metaphor of holey landscapes, local adaptation and microevolution can be viewed as climbing from a hole towards a nearly neutral network of genotypes with fitnesses at a level determined by mutation-selection-random drift balance. The process of climbing occurs on a shorter time scale than that necessary for speciation, clade diversification, and macroevolution. Once a ridge is reached, the population will be prevented by selection from slipping off of this ridge to lower fitnesses and by mutation, recombination, and gene flow from climbing to higher fitnesses. Speciation occurs when a population evolves to a genetic state separated from its initial state by a hole.

The metaphor of holey landscapes puts the idea of alternative high-fitness combinations of genes, which is explicit in Wright's view, and the idea about the prevailing nearly neutral genetic changes, which is explicit in the view of flat landscapes, within a single unified picture. Within this metaphor, evolution mostly proceeds by fixation of weakly selected alleles which can be advantageous, deleterious, over- or underdominant, or apparently neutral, depending on the specific area of the genotype space the population passes through. Smaller populations will pass faster through the areas of genotype space corresponding to fixation of slightly deleterious mutations, whereas larger populations will pass faster through the areas corresponding to fixation of (compensatory) slightly advantageous mutations. That is, within the metaphor of holey fitness landscapes, most important substitutions are compensatory and nearly neutral. I note that, earlier, Hartl et al. (1985; see also Hartl and Taubes 1996) invoked the metabolic control theory of Kacser and Burnes (1973, 1981) to argue that natural selection can create a condition in which most mutations with small effects on protein structure are selectively nearly neutral. In a sense, by showing that nearly neutral mutations are sufficient for explaining continuous evolutionary changes, the metaphor of holey landscapes provides an additional justification for Ohta's theory of nearly neutral evolution (Ohta 1992, 1997, 1998). It also shows that significant genetic divergence and speciation do not require overcoming strong natural selection. The metaphor of holey landscapes identifies a simple reason for the decoupling of micro- and macroevolution, which is the many-to-one nature of the relationship between genes (microevolution) and fitness and other phenotypic characters (macroevolution). It shows that macromutations are not necessary for explaining the origin of even the most complex traits. The complexity of the system of ridges in a holey landscape and the existence of multiple possible paths between most points in the landscape explain both the improbability of reversed evolution and the importance of chance and contingency in evolutionary dynamics. A general conclusion is that macroevolution and speciation on biologically relevant fitness landscapes proceed according to the properties of the corresponding nearly neutral networks and holey fitness landscapes. That is, evolution along holey landscapes is the stuff of speciation and diversification.

In the next two sections I will consider some results on the evolutionary dynamics of asexual haploid populations on holey fitness landscapes. But first I would like to finish this section with the following comment. In this book, I build, analyze, and explore concrete mathematical models that rely not on metaphorical pictures and plausible verbal constructions, but on what is really known and supported by data. Therefore, most results and conclusions concerning the major topic of this book – speciation – are valid independently of how general the picture outlined in the last few paragraph will prove to be, although I, of course, am quite convinced it is general.

4.6 DETERMINISTIC EVOLUTION ON A HOLEY LANDSCAPE

Holey landscapes represent a new class of population genetic models for which not many mathematical results are known. Here I consider the implications of evolutionary dynamics on holey landscapes for the error threshold, the average fitness of the population, and genetic canalization. The analytical results to be presented assume that the population size is effectively infinite. This assumption allows one to neglect genetic drift and concentrate exclusively on the effects of selection and mutation. Some stochastic effects will be considered in the next section.

4.6.1 Error threshold

Let us consider an asexual population of haploid individuals. I will assume that the fitness landscape has a percolating neutral network of genotypes with fitness 1 and that all other genotypes have a reduced fitness $1 - s$, where $s > 0$. Let ν be the overall probability of mutation per genotype per generation. Asymptotically, the population is expected to reach a mutation-selection balance. Let variable ϕ stand for the overall frequency of genotypes with fitness 1 and let $\hat{N}$ be the average number of high-fitness one-step neighbors per high-fitness individual in the population. I will refer to $\hat{N}$ as the *population neutrality*. Then the results of van Nimwegen et al. (1999) can be used to show that if

$$\nu < \nu_c = \frac{s}{1 - \hat{\mathcal{N}}/\mathcal{D}}, \tag{4.11}$$

that is, if mutation rate is sufficiently small, then the equilibrium frequency of high-fitness genotypes is positive and can be written as

$$\phi^* = 1 - \frac{\nu(1 - \hat{\mathcal{N}}/\mathcal{D})}{s}. \tag{4.12}$$

If condition (4.11) is not satisfied, then the only equilibrium state is the one with no high-fitness genotypes present: $\phi^* = 0$. This shows that the optimum sequence can be maintained in the population only if ν is not too big (i.e., if there is minimal replication accuracy). If the mutation rate per locus μ or the sequence length $\mathcal{L}$ is too large, the optimum sequence will be lost. This transition between two different regimes is known as the *error threshold* (Schuster and Swetina 1982; Eigen et al. 1989; Higgs 1994). Both ν_c and ϕ^* increase with $\hat{\mathcal{N}}$. Thus, with neutrality, genotypes with the highest fitness can be maintained at larger mutation rates and at higher frequencies than in the case of no neutral mutations. In other words, the existence of neutral mutations increases the error threshold.

Using equation (4.12), one also finds the average fitness of the population at equilibrium:

$$\overline{w} = 1 - \nu(1 - \frac{\hat{\mathcal{N}}}{\mathcal{D}}). \tag{4.13}$$

The last equation shows that the average fitness of the population increases with increasing population neutrality $\hat{\mathcal{N}}$. Note that we reached these conclusions even without the knowledge of the exact value of the population neutrality $\hat{\mathcal{N}}$. The value of $\hat{\mathcal{N}}$ will be found in the next section, where we consider genetic canalization.

4.6.2 Genetic canalization

The term *canalization* reflects a general belief that biological systems ought to evolve to a state of greater stability (Schmalhausen 1949; Waddington 1957; Wagner et al. 1997). One usually distinguishes between *environmental canalization*, that is, insensitivity to environmental variation, and *genetic canalization*, which is insensitivity to genetic variation (e.g., due to mutations).

The metaphor of holey landscapes provides a general theoretical framework for understanding the evolution of genetic canalization (see Gavrilets and Hastings 1994 and Wagner et al. 1997 for models of environmental canalization). From general considerations, one should not expect complete homogeneity of fitness landscapes, which are supposed to have areas varying with respect to both the width and concentration of ridges of high-fitness genotypes. For example, in the Russian roulette model, the number of viable neighbors has a Poisson distribution which implies the existence of areas of high density of such genotypes (see equation (4.4)). It is obvious that a population inside such an area will be characterized by a smaller probability of deleterious mutations and by higher average fitness than a population in a randomly chosen part of a fitness landscape. If one thinks of the genotype space as a "universe," then the areas of high concentration of high-fitness genotypes can be viewed as "galaxies." Will evolution result in populations moving into the centers of "galaxies"? If yes, one should observe the evolution of genetic canalization (Wagner 1996) and a reduction in *evolvability*, that is, in the ability of mutation to produce phenotypic changes (Wagner and Altenberg 1996). Earlier numerical simulations did show that populations tend to spend more time in areas of high concentration of high-fitness genotypes (Huynen and Hogeweg 1994; Peliti and Bastolla 1994; Finjord 1996). Recent analytical results further advanced our understanding of these effects (van Nimwegen et al. 1999; Wilke 2001a,b; Kolář and Slanina 2003), which turned out to be controlled by the topology of the neutral network.

To illustrate this, I will use the same model as in the previous section, but now I will concentrate on the frequencies of different high-fitness genotypes, which will be denoted as ϕ_i, rather than on just the overall frequency $\phi = \sum \phi_i$ of such genotypes. Let $\phi^* = (\phi_1^*, \phi_2^*, \dots)^T$ be a column vector of equilibrium genotype frequencies. Let us also define a matrix M with the element M_{ij} equal to one if the genotypes i and j are one-step neighbors and equal to zero otherwise. In the terminology of graph theory, M is the *adjacency matrix* for the network of viable genotypes. An important result of van Nimwegen et al. (1999) is that vector ϕ^* solves the matrix equation

$$\hat{\mathcal{N}} \phi^* = M \phi^*. \tag{4.14}$$

Because the elements of matrix M are nonnegative and the network of viable genotypes is connected, matrix M is irreducible (Gantmacher 1959). Therefore, M has a single positive eigenvector corresponding to the largest positive eigenvalue, which gives the equilibrium values of ϕ^* and $\hat{\mathcal{N}}$, respectively. Thus, both the equilibrium population genetic structure (characterized by the vector of genotype frequencies ϕ^*) and the average number of viable neighbors (characterized by the population neutrality $\hat{\mathcal{N}}$) are functions only of the topology of the network of viable genotypes as determined by the adjacency matrix M. Soshnikov and Sudakov (2003) show that if the dimensionality of genotype space is very large, $\hat{\mathcal{N}} \approx \sqrt{\Delta(G)}$, where $\Delta(G)$ is the maximum degree of the graph G representing the network of viable genotypes. In biological terms, $\Delta(G)$ is the maximum number of neutral neighbors per genotype in the network.

As a simple illustration of these results, let us consider a network of viable genotypes depicted in Figure 4.7. In this neutral network, the average number of viable neighbors (network neutrality) is equal to 1.9. The largest eigenvalue of the corresponding adjacency matrix is $\hat{\mathcal{N}} = 2.83$. This shows that the population neutrality $\hat{\mathcal{N}}$ at mutation-selection balance is larger than the network neutrality. The eigenvector corresponding to the largest eigenvalue shows that at equilibrium the frequencies of genotypes with 5, 4, and 1 neutral neighbors are in proportions 1:0.57:0.20, that is, genotypes with more neutral neighbors are more common than genotypes with fewer viable neighbors. Thus, this model predicts that the population starting at a periphery of the "universe," say at genotype 14, will evolve into the center of the "galaxy" at genotype 0. One of the consequences of this evolution will be a reduced probability of deleterious mutation, that is, genetic canalization.

Wilke (2001a,b) has obtained some additional results on the evolution of genetic canalization in similar models. In particular, he allowed for multiple neutral networks (rather than a single giant component) and arbitrary mutation rates, and studied competition of populations residing on two different neutral networks.

4.7   STOCHASTIC EVOLUTION ON A HOLEY LANDSCAPE

4.7.1   Random walks

In Box 2.3, we considered some properties of random walks on a $\mathcal{L}$-dimensional hypercube. Here, I will consider the walks on binary and generalized hypercubes, assuming that some vertices are inaccessible (inviable), but that accessible (viable) vertices form an extended neutral

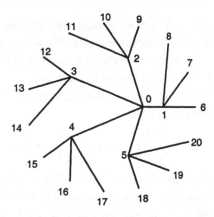

FIGURE 4.7. An example of a neutral network of viable genotypes. Each genotype is represented by a numbered vertex. The differences in the length of branches do not imply anything.

network. I will refer to such hypercubes as holey hypercubes. I will look at three features of random walks: equilibrium distribution, transient dynamics, and the dispersion ratio.

Equilibrium distribution Assume that at each time step the walk randomly choses a neighboring vertex. If this vertex is viable, the walk moves there. If it is not, the walk stays at the current vertex. In both cases, time increases by one. In the theory of random walks, this type of walk is referred to as the "blind ant in the labyrinth" (e.g., Huges 1996, Chapter 7). Another approach is to assume that the ant randomly chooses a vertex to move to only among viable vertices. This kind of walk is referred to as the "myopic ant in the labyrinth" walk. If steps are interpreted as fixations of neutral mutations (as we will do here), only the blind ant walk can be used as a model of mutation accumulation.

Asymptotically, the system will reach a state of stochastic equilibrium where viable vertices are visited with certain frequencies. In the case of the blind ant walk, each viable vertex is visited with equal probability. In the case of the myopic ant walk, each viable vertex is visited with probability directly proportional to the number of viable neighbors it has (e.g., Huges 1996; Norris 1997). The time to reach a stochastic equilibrium state is long (see Box 2.3). Therefore, it is important to analyze the transient dynamics.

Transient dynamics of the distance to an initial state In the case of a random walk on a binary hypercube with no holes, the distance d_t from the initial position asymptotically approaches $\mathcal{L}/2$ as predicted by equation (1) in Box 2.3. Let us consider the Russian roulette model, where a proportion $\mathcal{P}$ of vertices are viable. We will assume that $\mathcal{P} > \mathcal{P}_c$ so that there is a giant cluster of viable genotypes. How rapidly will a random walk on this cluster move away from its initial state? Campbell (1985) conjectured that on a holey hypercube, d_t will change according to a stretched exponential law:

$$d_t = \frac{\mathcal{L}}{2} \left\{ 1 - \exp\left[-\left(\frac{t}{\tau}\right)^{\beta} \right] \right\}, \qquad (4.15)$$

with the exponent β approaching 1/3 when $\mathcal{P}$ approaches the percolation threshold $\mathcal{P}_c$. Note that with no holes (i.e., if $\mathcal{P} = 1$), the exponent is $\beta = 1$ and the characteristic time is $\tau = \mathcal{L}/2$. If instead of the Hamming distance d one uses the *overlap* q related to d by the equality $q = (\mathcal{L}/2)(1 - d)$, then the stretched exponential law implies that

$$\ln(-\ln q_t) = \beta \ln t - \beta \ln \tau;$$

that is, there is a linear relationship between the double logarithm of the overlap $\ln(-\ln q_t)$ and the logarithm of time $\ln t$. This linearity property can be used for identifying the stretched exponential relationships.

In a series of later papers, Campbell with coauthors (Campbell et al. 1987; Lemke and Campbell 1996; de Almeida et al. 2000, 2001) studied blind ant walks on holey hypercubes numerically. Their simulations using hypercubes with $\mathcal{L}$ ranging from 10 to 24 confirmed that after some transient time (which was on the order of 10 to 100 time steps), equation (4.15) provides a good approximation. The decay of the overlap is closely approximated by the stretched exponential law. Reducing $\mathcal{P}$ towards the percolation threshold $\mathcal{P}_c$ resulted in β approaching 1/3 and in divergence (that is, indefinite increase) in τ.

Numerical results also showed that on a short time scale, the dynamics of d are closely approximated by the simple exponential law (see equation (1) in Box 2.3). During the exponential phase,

$$\tau = \frac{\mathcal{L}}{2} \times \frac{1}{\mathcal{P}}, \qquad (4.16)$$

which is the same as if there were no holes but the probability of a step reduced from 1 to $\mathcal{P}$. If a step is attempted with probability ν, then the effect of holes can be incorporated into the effective mutation rate

$$\nu_e = \nu \mathcal{P}. \qquad (4.17)$$

Thus, one might expect that many results valid for the evolution on flat landscapes will apply to the case of initial evolution on a holey hypercube if $\mathcal{P} > \mathcal{P}_c$. However, on a larger time scale, the results of Lemke and Campbell (1996) imply that formula (4.16) strongly underestimates τ. Note that equation (4.17) is the same as the one suggested by Ohta's (1992) theory of nearly neutral mutations.

Dispersion index   In the case of a random walk on a $\mathcal{L}$-dimensional binary hypercube with no holes, where a step is taken with probability ν, the dispersion index $\mathcal{I}$, that is, the ratio of the variance and the average of the number of steps taken per unit of time, is equal to 1 (see Box 2.2). In the case of a holey hypercube arising in the Russian roulette model, different vertices will have a different number of viable neighbors which will result in more irregularities in the dynamics of the random walk.

In the model under consideration, the number of steps taken within a certain time interval is a stochastic process that belongs to a class of the so-called renewal processes (e.g., Karlin and Taylor 1975, Chapter 5). If a viable genotype has k viable neighbors and the probability of mutation per locus is μ, the overall probability of a neutral mutation for this genotype is $k\mu$. The waiting time τ until the next step (that is, mutation) has an exponential distribution with parameter $1/(k\mu)$. In the Russian roulette model, the number of viable neighbors has a Poisson distribution with parameter $\overline{N}$(see equation (4.4)). Thus, the unconditional distribution of τ is given by the formula

$$f(\tau) = \sum_{k=1}^{\mathcal{L}} \frac{1}{k\mu} \exp\left(-\frac{\tau}{k\mu}\right) \exp(-\overline{N}) \frac{\overline{N}^k}{k!},$$

where the sum starts at $k = 1$ because only the vertices with at least 1 viable neighbor can belong to the giant cluster. Using this distribution, one can show that the mean and variance of the time till the next step are

$$\overline{\tau} \approx \frac{1}{\mu} \overline{N} \exp(-\overline{N}) \, H_{[1,1],[2,2]}(\overline{N}), \tag{4.18a}$$

$$\text{var}(\tau) \approx \frac{1}{\mu^2} \overline{N} e^{-\overline{N}} \left[\overline{N} e^{-\overline{N}} \, H_{[1,1],[2,2]}(\overline{N})^2 - 2H_{[1,1,1],[2,2,2]}(\overline{N}) \right], \tag{4.18b}$$

where $H_{[\cdot],[\cdot]}$ is the generalized hypergeometric function of the corresponding order (Abramowitz and Stegan 1965). Using the standard results on the renewal processes, the average number of steps per time interval t is equal to $\overline{\chi}_t = t/\overline{\tau}$, whereas the variance of the number of step is $\text{var}(\chi_t) = t \, \text{var}(\tau)/\overline{\tau}^3$. This implies that the dispersion index $\mathcal{I}$ can be written as $\mathcal{I} = \frac{\text{var}(\tau)}{\overline{\tau}^2}$, where $\overline{\tau}$ and $\text{var}(\tau)$ are given by

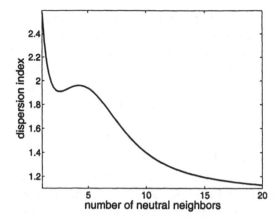

FIGURE 4.8. Dispersion index of a random walk in the Russian roulette model.

equations (4.18). Figure 4.8 illustrates the dependence of the dispersion index $\mathcal{I}$ on the average number of viable neighbors $\overline{N}$. One can see that $\mathcal{I}$ is always larger than 1. For small $\overline{N}$, the dispersion index $\mathcal{I}$ can be as large as 2 or higher. The observed values of $\mathcal{I}$ for protein evolution typically range from 2 to 7, and higher values are not unrealistic (Gillespie 1991; Ohta 1995). The Russian roulette model is an example of an uncorrelated landscape. I expect that introducing correlations between fitness values of similar genotypes will make $\mathcal{I}$ (much) larger.

4.7.2   Dynamics of haploid populations

The evolutionary dynamics of populations of asexual haploid individuals on fitness landscapes with neutrality were numerically studied in a number of papers dealing with RNA, protein, and Nk-type landscapes, as well as with landscapes arising from multiplicative selection (Amitrano et al. 1989; Huynen and Hogeweg 1994; Peliti and Bastolla 1994; Finjord 1996; Huynen et al. 1996; Woodcock and Higgs 1996; Barnett 1997; Ohta 1997, 1998; Newman and Engelhardt 1998; Bastolla et al. 1999; Fontana and Schuster 1998a,b; van Nimwegen et al. 1997a,b; van Nimwegen and Crutchfield 2000, 2001; Reidys et al. 2001; Wilke 2001a,b). In addition, Barton (1989a) and Mani and Clarke (1990) considered diploid populations with selection on a quantitative character. In all these models, the populations readily diffuse along the corresponding nearly neutral networks. Several other properties of evolutionary dynamics have also been observed:

(i) In the case of multiple neutral networks, the dynamics of adaptation proceeds in a stepwise fashion in which relatively short periods of transitions to a higher fitness level are interrupted by extended periods during which populations diffuse along nearly neutral networks maintaining the same fitness level (e.g., Huynen et al. 1996; Newman and Engelhardt 1998; Wilke 2001a).

(ii) The population typically maintains a cluster structure (Amitrano et al. 1989; Huynen et al. 1996), but with less fragmentation on average than in the neutral cases (see Box 3.3).

(iii) The populations tend to locate and spend more time in the areas with higher concentration of high-fitness genotypes ("galaxies"). This results in genetic canalization (Huynen and Hogeweg 1994; Peliti and Bastolla 1994; Finjord 1996; van Nimwegen et al. 1999; Wilke 2001a,b).

(iv) The dynamics of the accumulation of mutations is overdispersed, that is, the dispersion index $\mathcal{I} > 1$ (Bastolla et al. 1999).

Much more modeling work is still necessary to assess how general these patterns are.

4.8   SUMMARY

• Fitness landscapes describing real biological systems have enormous dimensionality with thousands and millions of dimensions. Simple mathematical arguments show that, with that many dimensions, genotypes with similar fitnesses must form nearly neutral contiguous networks expanding throughout the genotype space. In other words, extensive ridges at approximately the same height are a general feature of multidimensional fitness landscapes.

• There are many examples where the existence of nearly neutral networks is well established or where these networks are strongly implicated.

• The metaphor of holey fitness landscapes emphasizes ridges (i.e., nearly neutral networks) of high-fitness genotypes expanding throughout the genotype space. This view treats all other genotypes that do not belong to the nearly neutral network of high-fitness genotypes as holes. The holes include both largely inaccessible genotypes with very high fitness and low-fitness valleys and slopes. This metaphor combines the crucial features of rugged landscapes (namely, the existence of multiple high-fitness combinations of genes) and flat landscapes (namely, the importance of mutational order and random genetic drift in controlling evolutionary dynamics) in a single unified concept.

• The following are some features of evolution on holey landscapes:

(i) The majority of mutations underlying most pronounced genetic changes are nearly neutral (slightly advantageous, slightly deleterious and compensatory). That is, evolution on holey landscapes is fully compatible with the nearly neutral theory of molecular evolution (Ohta 1992, 1998). In a sense, the results on the widespread existence of holey fitness landscapes provide an additional justification for Ohta's theory.

(ii) Holey landscapes are expected to have multiple alternative paths connecting members of the network of high-fitness genotypes. This results in both the irreversibility of evolution and in the crucial importance of chance and contingency in evolutionary dynamics.

(iii) The huge discrepancy between the number of possible sequences and possible phenotypic states (or fitness values) results in a decoupling between evolution at the sequence level and morphological change.

(iv) Populations are expected to evolve towards areas of high density of high-fitness genotypes ("galaxies"). This leads to an apparent decrease in the ability of mutations to cause deleterious effects (i.e., genetic canalization).

(v) Evolutionary dynamics are expected to be episodic with periods of stasis and punctuation both at the sequence level and at the phenotypic level as new paths through genotype space and new neutral networks are discovered by the population via mutation.

4.9 CONCLUSIONS

Within Wright's metaphor of rugged fitness landscapes, the problem of crossing adaptive valleys is of fundamental importance. However, the multidimensionality of fitness landscapes results in the existence of nearly neutral networks that can be traversed without crossing any adaptive valley. This implies that the problem of crossing adaptive valleys is nonexistent. No fitness valleys have to be crossed for substantial genetic divergence of populations and speciation. This means that understanding speciation requires developing new approaches and models.

Part II

The

Bateson-Dobzhansky-Muller

Model

CHAPTER FIVE

Speciation in the
Bateson-Dobzhansky-Muller model

The existence of ridges of high-fitness genotypes extending throughout fitness landscapes has an extremely important implication for speciation. As Chapter 3 has illustrated, traditional models of evolution on rugged fitness landscapes have not been very successful in describing speciation. Speciation in these models required crossing deep fitness valleys, which is a very improbable process because of selection opposing any movement away from a fitness peak. However, if there are ridges of high-fitness genotypes, then biological populations can evolve along them without having to descend to low fitness levels, and the problem of crossing fitness valleys becomes nonexistent. As it turns out, a simplified version of this idea is already well established in speciation research but was not seriously studied until relatively recently. A number of authors, starting with Bateson (1909), Dobzhansky (1936, 1937), and Muller (1940, 1942), have argued that complete RI and speciation can result after a series of genetic substitutions, each of which is unopposed by selection and does not require crossing fitness valleys. In this chapter, I first define what in recent years has become a standard model of RI which utilizes this idea. Then, I consider the deterministic and stochastic features of this model, concentrating on the dynamics of allopatric and parapatric speciation.

5.1 THE BATESON-DOBZHANSKY-MULLER MODEL OF
REPRODUCTIVE ISOLATION

Following Dobzhansky (1937), let us consider a two-locus, two-allele diploid population initially monomorphic for a genotype, say $AAbb$.

Assume that this population is broken up into two geographically isolated parts. In one part, evolutionary factors cause substitution of B for
b and a subpopulation AABB is formed. In the other part, there is a
substitution of a for A, giving rise to a subpopulation aabb. Assume
that there is no RI among genotypes AAbb, Aabb, and aabb and
among genotypes AAbb, AABb, and AABB, that is, individuals
within these two groups are able to interbreed freely and produce viable
and fertile offspring. In contrast, let the cross of AABB and aabb be
difficult or impossible because alleles a and B are incompatible in the
sense that their interaction "produces one of the physiological isolating
mechanisms" (Dobzhansky 1937, p. 282). In this model, substitutions
of B for b in one population and a for A in the other population will
result in the emergence of two reproductively isolated populations.

There are several important observations regarding this scenario:

• RI arises as a *side effect* of accumulating genetic differences between the populations rather than as a response to factors directly promoting it.

• Alleles a and B are deleterious only in the presence of each other.
They can be perfectly fit in other genetic backgrounds.

• Substitutions at the locus under consideration can be driven by any
factor (or a combination of factors), including mutation, random genetic
drift, and various types of selection (e.g., natural or sexual, constant or
frequency-dependent).

• Another observation is about the biological meaning of "reproductive isolation" between genotypes aabb and AABB and "incompatibility" of alleles a and B. Bateson (1909) was concerned with sterility of hybrid progeny. Muller (1942) mostly concentrated on lethality
and sterility of hybrid progeny. However, as realized by Dobzhansky
(1937), this scenario is applicable to many other forms of RI. His "physiological isolating mechanisms" were defined rather broadly (p. 257)
and included (i) mechanisms that prevent parental forms from meeting (such as ecological isolation and seasonal or temporal isolation),
(ii) mechanisms that prevent hybridization if parental forms do meet
(such as sexual or psychological isolation, physical incompatibility of
reproductive organs, inability of sperm to reach the eggs or penetrate
the eggs, and inviability of offspring), and (iii) hybrid sterility.

• As realized by Muller (1942), RI can also evolve if one of the
populations experiences two substitutions in two different loci rather
than two populations experiencing one substitution each.

• Although the above scenario was initially described for diploid populations and for allopatric speciation, it is equally applicable for haploid populations and for parapatric and sympatric speciation.

The most crucial part of this scenario is a very specific assumption about the genetic architecture of RI, namely that there are two alleles at different loci that are incompatible. In what follows, I will refer to this assumption as the Bateson-Dobzhansky-Muller (BDM) model of RI. Although proposed a long time ago, this model was neglected for a number of years until it received renewed interest in the 1970s and early 1980s. For example, similar schemes were mentioned by Nei (1976), Maynard Smith (1983), Barton and Charlesworth (1984), and Kondrashov and Mina (1986). These earlier discussions were restricted to statements of the kind that if a specific type of genetic architecture exists, then the problem of crossing fitness valleys is solved. Verbal discussions were also complemented by some modeling attempts (e.g., Wills 1977; Christiansen and Frydenberg 1977; Bengtsson and Christiansen 1983; Nei et al. 1983). These papers laid the ground for extensive theoretical work performed in the 1990s. More than 65 years ago Dobzhansky observed that the scheme he proposed "may appear fanciful, but it is worth considering further since it is supported by some-well established facts and contradicted by none" (1937, p. 282). By now the BDM model is widely accepted and its basic components are supported by a growing amount of data (see below). This model is commonly referred to as a "standard model" for RI and speciation. Although there is no evidence that interactions between a single pair of genes underlie RI between species (because real species differ in many rather than just two genes; Wu 2001), the BDM model is a crucial element of any theory of speciation. As we will see in a moment, the BDM model provides the simplest example of a holey fitness landscape.[1]

5.1.1   Fitness landscapes in the BDM model

In this subsection, I consider a family of two-locus, two-allele haploid and diploid models illustrating the main idea of the BDM model in terms of fitness landscapes. I will concentrate exclusively on fitness

[1] It is kind of ironic that for many years Dobzhansky was promoting Wright's rugged landscapes without realizing that the mechanism of speciation that he favored implied a very different structure of fitness landscapes.

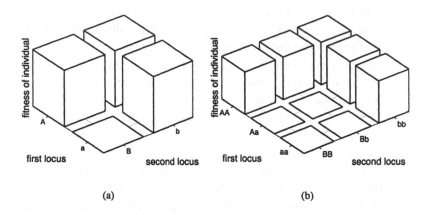

FIGURE 5.1. Fitness landscapes in the BDM model: selection against individual genotypes. (a) Haploid model. (b) Diploid model.

components underlying RI. Additional fitness components coming, for example, from selection for adaptation to biotic or abiotic environments will affect the evolutionary dynamics and can drive the divergence, but they are not considered here. I start with models describing selection acting on individuals and then turn to models incorporating selection on pairs of individuals.

Fitness components for individuals   In the haploid case, there are four different genotypes, **AB, Ab, aB**, and ab. Assume that genes a and B are incompatible in the sense that organisms with genotype aB have zero viability or are sterile. The other three genotypes have high fitness. The fitness landscape corresponding to this system is illustrated in Figure 5.1(a). In this model, one half of the recombinant genotypes produced as a result of matings between organisms **AB** and ab will have genotype aB and, thus, will be inviable or sterile. In the diploid case, there are ten different genotypes (note that there are two different types of double heterozygotes: those formed by gametes **AB** and ab, and those formed by gametes **Ab** and aB). Let genes a and B be incompatible in the sense that organisms that have these two genes simultaneously have zero viability or are sterile. The other combinations of genes are compatible, and the corresponding genotypes have high fitness. The fitness landscape corresponding to this system is illustrated in Figure 5.1(b). In this model, all offspring of matings between organisms **AABB** and **aabb** will have heterozygous genotype **AaBb** and, thus, will be inviable or sterile.

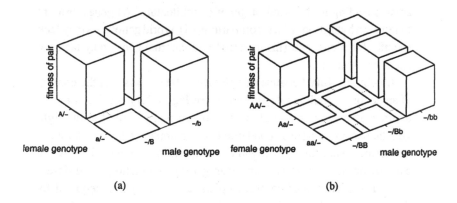

(a)   (b)

FIGURE 5.2. Fitness landscapes in the BDM model: selection against pairs of genotypes. "Fitness of pair" means the probability and/or fertility of specific mating. (a) Haploid model. (b) Diploid model.

Fitness components for mating pairs   The examples above concerned RI in the form of viability or fertility selection against hybrid organisms. However, exactly the same scheme is applicable to fertility selection against specific mating pairs and to sexual selection. For example, assume that the females bearing an allele a do not mate at all with the males bearing an allele B or that these matings are infertile. Assume that all other types of mating are not prevented and are fertile. The resulting fitness landscapes for mating pairs are illustrated in Figure 5.2. Thus, both fertility selection and sexual selection can be interpreted in terms of the BDM model.

The most important feature of fitness landscapes in Figures 5.1 and 5.2 is the existence of ridges of high fitness values. These landscapes also imply complete dominance of incompatibility in the sense that the joint presence of a single allele a and a single allele B is sufficient for a significant reduction in fitness. In general, this does not need to be the case, and the fitness effects may depend on the exact number of incompatible alleles and may vary both along the ridge and in the hole.

5.1.2   The mechanisms of reproductive isolation in the BDM model

The BDM model postulates a specific genetic architecture of reproductive isolation resulting in the existence of a ridge of high fitness values in the genotype space. There are at least three general classes of mechanisms leading to this kind of genetic architecture: (i) genes brought together in hybrids may be incompatible, (ii) hybrid genotypes may lack specific functions, and (iii) maternal and paternal genes may lack the necessary matching.

Incompatibility because of epistatic interactions   In the context of selection on individual genotypes (see Figure 5.1), the BDM model assumes that genes a and B, which are perfectly normal in the right genetic background (in a "coadapted gene complex"), become strongly deleterious when brought together in hybrid genotypes. This is the case of epistasis, that is, nonlinear genetic interactions at different loci. The existence of complex epistatic interactions as required by the BDM model is well documented experimentally (e.g., Dobzhansky 1937; Muller 1940, 1942; Macnair and Christie 1983; Thompson 1986; Cabot et al. 1994; Wu and Palopoli 1994; Whitlock et al. 1995; Davis and Wu 1996; Coyne and Orr 1998; Levin 2000; Wolf et al. 2000; Fishman and Willis 2001; Wu 2001). For example, inviability of hybrids between the platyfish species *Xiphophorus maculatus* and the swordtails *X. helleri* behaves as if controlled by a simple BDM model with incomplete dominance (Orr and Presgraves 2000). Spotted *X. maculatus* fish carry a sex-linked *Tu* locus that specifies macromelanophores and that is regulated by a major autosomal suppressor locus *R*. *Xiphophorus maculatus* fish lack both the *Tu* locus and suppressor alleles at the *R* locus. F_1 hybrids develop an increased number of exaggerated spots. When F_1 hybrids are backcrossed to *X. helleri*, half of the offspring inherit *Tu* and half of these lack *R* suppressors. As a result, they develop severe melanomas that are often lethal. The fitness matrix describing this system can be represented as

	BB	Bb	bb
AA	1	1	0
Aa	1	1	0
aa	1	1	1

where A (a) signifies the presence (lack) of the *Tu* locus, and B (b) signifies the presence (lack) of the suppressor alleles. Here, *X. maculatus* has genotype **AABB**, whereas *X. helleri* is represented by genotype **aabb**. More detailed work has shown that the *Tu* locus is actually composed of a set of tightly linked but separable loci.

Other examples of genes involved in incompatibilities are the *Odysseus* gene that contributes to the sterility of hybrid males produced by crosses of *Drosophila simulus* and *D. mauritiana* (Ting et al. 1998), the *Hybrid male rescue* gene involved in lethality and female sterility in hybrids among *D. melanogaster* and its sibling species (Barbash et al. 2003), and the *Nup98-Nup96* gene found on chromosome 3 of the fruit fly *D. simulans* that interacts with one or more unknown genes on the X chromosome of *D. melanogaster* to cause the death of male hybrid offspring (Presgraves et al. 2003).

Incompatibility of genes brought together in hybrids is also expected from general considerations. Indeed, the loss in fitness to species hybrids is no more surprising than the fact that a part from one car manufacturer does not function in a car from another manufacturer (Charlesworth 1990). This expectation is supported by recent mathematical models that show that negative epistatic interactions are a generic feature of gene regulatory networks (e.g., Johnson and Porter 2000; Kaneko and Yomo 2000; Omholt et al. 2000).

Lack of specific functions in hybrids Werth and Windham (1991) and Lynch and Force (2000) suggested a simple mechanism for the origin of genetic incompatibilities via gene duplication and subsequent loss of function in duplicate genes. Consider a pair of duplicate genes in an ancestral species. Assume that initially both genes are fixed for alleles **A** and **B** which are fully functional (and fully redundant in the sense that a single allele **A** or allele **B** is sufficient for a certain function). Suppose that two populations of this species become spatially isolated. The most common fate of the pair of gene duplicates is silencing of one member of a pair by degenerative mutations within a few million years (e.g., Lynch 2002). Let **a** and **b** be the corresponding nonfunctional (null) alleles that get fixed in the populations by degenerative mutations. Then, with 50% probability, one sister population will become fixed for genotype **AAbb** and the other for genotype **aaBB**. Hybridization between two such taxa would then lead to the F_1 hybrid progeny with genotype **AaBb**, which is normal. Assume that the loci are unlinked. Then one quarter of all gametes produced by F_1 hybrids will be **ab**, that is, will have null alleles. One-sixteenth of the F_2 progeny would be of genotype **aabb** and completely lacking in function, and another one-fourth would contain three null alleles and possibly have reduced function. In this model, the emergence of incompatibility between two species is totally driven by degenerative mutations. The model is fully compatible with the BDM model except that it does not invoke a change in gene action in a hybrid genetic background.

Lack of matching between maternal and paternal genes   The
BDM model implies both incompatibility of alleles a and B and no
incompatibility (or matching) of other gene combinations that do not
include these two alleles simultaneously. In fact, incompatibility and
matching are two sides of the same phenomenon: no matching means
incompatibility and vice versa. In the context of selection on pairs of
genotypes (see Figure 5.2), the BDM model assumes that gene a in
females and gene B in males do not match. It is well recognized that
matching of some genes (or traits) is required at all levels of the reproduction process (e.g., Tregenza and Wedell 2000), including sperm-egg
interactions (e.g., Vacquier 1998; Palumbi 1998; Howard 1999; Alipaz
et al. 2001), mate recognition and mating pair formation (e.g., Patterson 1985; Grant and Grant 1997; Grant 1999), the copulation process
(e.g., Arnqvist 1998), and postfertilization development. For example,
in dioecious isogamus Chlamydomonads, sexual isolation between and
within species is caused by nonoccurence of the initial contact between
noncompatible gamete types (Wiese and Wiese 1977). Initial contact
depends on a molecular complementarity between special glycopro-
teinaceous surface components at the gametic flagella tips. Gamete
contact is mediated by a carbohydrate ligand on one gamete type interacting with a trypsin-sensitive sugar-binding component on the other.
This and similar situations can be treated within the BDM model.

5.2   POPULATION GENETICS IN THE BDM MODEL

In this subsection, I consider the BDM model within the context of an
isolated population of a very large size. The assumption that the population size is large will allow us to neglect the stochastic effects of random
genetic drift and concentrate on effects of selection and mutation.

The BDM model belongs to a class of two-locus, two-allele models
with constant fitnesses. Typically, such models exhibit evolution towards an equilibrium (Crow and Kimura 1970; Ewens 1979; Hastings
1989; Nagylaki 1992; Christiansen 2000; Bürger 2000; see Hadeler and
Liberman 1975; Hastings 1981; Gavrilets 1998 for examples of cycling).
The questions of interest usually are, for example: how many different
equilibria are there? Does the eventual outcome of evolution depend on
initial conditions? Can genetic variation be maintained without mutation? What is the level of variation maintained by mutation? What are

the characteristics of transient dynamics? Selection schemes similar or identical to the ones implied in the BDM models have been studied by a number of authors (e.g., Nei and Roychoudhury 1973; Nei 1976; Wills 1977; Christiansen and Frydenberg 1977; Bengtsson and Christiansen 1983; Phillips and Johnson 1998). The following summarizes and expands some of these earlier results.

5.2.1 Haploid population

Let us consider a model of a sexual haploid population with fitnesses (viabilities) given by fitness matrix

	B	b
A	1	1
a	$1 - s$	1

where $0 \leq s \leq 1$. The fitness landscape in this model is similar to that in Figure 5.1(a). Here, I will refer to alleles a and B as *conditionally deleterious*. Let ϕ_1, ϕ_2, ϕ_3, and ϕ_4 be the frequencies of genotypes AB, Ab, aB, and ab. Assume that the order of events in the life-cycle is selection, recombination, and mutation. Let $p_1 = \phi_1 + \phi_2$ and $q_2 = \phi_2 + \phi_4$ be the frequencies of normal alleles A and b, and let r be the recombination rate. Assume first that there is no mutation. If linkage is not absolute (i.e., $r > 0$), then any population state in which one locus has lost a conditionally deleterious allele is an equilibrium. (If there is no recombination ($r = 0$), then any population state with no genotype aB present is an equilibrium.) In mathematical terms, the corresponding dynamics have two *lines of equilibria* at the boundaries of the set of possible allele frequencies: (i) $p_1 = 1$, q_2 arbitrary, and (ii) $q_2 = 1$, p_1 arbitrary. If the mutation rate is very small rather than zero, we expect the population to evolve to a neighborhood of these two lines. To confirm this expectation in precise mathematical terms, we need to analyze the model in more detail.

Assume that the initial population has fixed genotype AB. Let the mutation rates from A to a and from B to b be equal to μ. For now, we do not allow for backward mutations. The genotype frequencies after selection are described by equations

$$\phi_i' = \frac{w_i}{\overline{w}}\, \phi_i, \qquad\qquad (5.1a)$$

where $w_1 = w_2 = w_4 = 1, w_3 = 1 - s$, and the mean fitness of the population is $\overline{w} = 1 - s\phi_3$. The genotype frequencies after recombination can be written as

$$\phi_i'' = \phi_i + r\delta_i D', \tag{5.1b}$$

where $D' = \phi_1'\phi_4' - \phi_2'\phi_3'$ is the linkage disequilibrium after selection but before recombination, and $\delta_1 = \delta_4 = -1$ and $\delta_2 = \delta_3 = 1$ (see the above cited texts on multilocus models). The genotype frequencies after mutation (that is, in offspring) are described by equations analogous to equations (3) in Box 2.1.

The system of dynamic equations (5.1) can be analyzed using regular perturbation techniques (e.g., Holmes 1995). The main idea of these techniques is to approximate solutions as power series in a small parameter, assuming that these solutions are close to those arising when the parameter is exactly zero. In our case, a natural choice for a small parameter is the mutation rate μ. For the model under consideration, one can show that there is a line of mutation-selection balance equilibria at which deleterious genotypes aB are maintained at a low frequency

$$\phi_3^* = \frac{\mu}{s}. \tag{5.2}$$

The equation defining the line of equilibria is rather cumbersome and will not be presented here. I just note that this line of equilibria is indeed very close to the boundary lines $p_1 = 1$ and $q_2 = 1$. Numerical iteration of equations (5.1) shows that the evolutionary dynamics are characterized by fast initial movement towards the line of equilibria with a slowdown as the line is approached (see Figure 5.3(a)).

Dynamical systems possessing lines of equilibria are typically *structurally unstable* in the sense that a small change in model parameters can significantly change some features of the dynamics. Our model fits this general pattern. For example, if we allow for backward mutations at the same rate μ, then the line of equilibria collapses to a polymorphic equilibrium with $\phi_3^* = \mu/s$ and

$$\phi_1^* = \phi_4^* = \tilde{\phi} \approx \sqrt{\mu\frac{s+r-rs}{rs}}, \tag{5.3}$$

which apparently is *globally stable*; that is, the population evolves to this equilibrium for all initial conditions (see Figure 5.3(b)). If selection against the deleterious genotypes is very strong ($s \approx 1$), then $\tilde{\phi} \approx \sqrt{\mu/r}$. For example, with $\mu = 10^{-5}$ and $r = 0.1$, $\tilde{\phi} \approx 0.01$. If

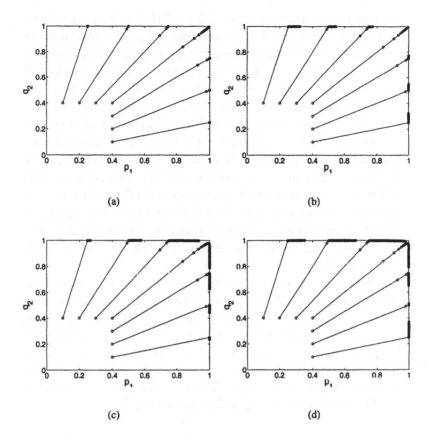

FIGURE 5.3. The dynamics of the haploid system on the (p_1, q_2)-plane. $s = 0.9, r = 0.5$, and, unless specified otherwise, $\mu = 10^{-5}$. Circles are plotted every 10th generation. Trajectories start in the lower left corner. (a) Evolution towards a line of equilibria. Equal mutation rates; no backward mutations. (b) Evolution towards a symmetric equilibrium. Equal forward and backward mutation rates. (c) Evolution towards an equilibrium with one locus polymorphic and another locus fixed. Mutation rates are 10^{-5} and 10^{-4}; no backward mutations. (d) Evolution towards an asymmetric equilibrium. Mutation rates are 10^{-5} and 10^{-4} and equal for forward and backward mutations.

mutations are one-directional but occur at different rates, then the system evolves to an equilibrium where the locus with the highest mutation rate is fixed for the conditionally deleterious allele, while at the other locus the conditionally deleterious allele is maintained at a very low

frequency by mutation (see Figure 5.3(c)). This is so if the recombination rate is much higher than the mutation rates. If r is comparable in
magnitude with the mutation rates, then numerical simulations suggest
that the population is able to maintain genetic variation in both loci. If
we allow for backwards mutations at a rate equal to that of forward mutations but assume that the mutation rates are different between the loci,
there is a polymorphic equilibrium that apparently is globally stable.
The location of this equilibrium is shifted towards fixation of the conditionally deleterious allele at the higher mutation locus relative to the
case of symmetric mutation (see Figure 5.3(d)). If we allow for small
differences in fitness among the genotypes on the ridge, there will be a
locally stable mutation-selection balance equilibrium with the highest
fitness genotype at a large frequency and other genotypes present at
much lower frequencies. In all cases increasing recombination speeds
up the dynamics.

Although the structure of equilibria differs between different cases,
in all these cases the dynamics are characterized by a relatively short
phase during which the deleterious genotypes aB are almost eliminated
and an extended phase of very slow evolution along the ridge of high-
fitness genotypes. During the first phase, the dynamics are driven by
strong selection. During the second phase, the dynamics are driven by
very weak forces of mutation. One can expect that the latter forces
would be easily overcome by random genetic drift.

5.2.2   Diploid population

Let us consider a model of a diploid population with fitnesses (viabilities) given by fitness matrix

	BB	Bb	bb
AA	1	1	1
Aa	$1 - s$	$1 - s$	1
aa	$1 - s$	$1 - s$	1

In this model, the population will be at Hardy-Weinberg proportions
in one generation, so that its genetic state can be characterized by the
frequencies of four gametes ϕ_1, ϕ_2, ϕ_3, and ϕ_4. If there is no mutation,
then, as before, the corresponding dynamical system has two lines of
equilibria corresponding to fixation of one of the two normal alleles (A

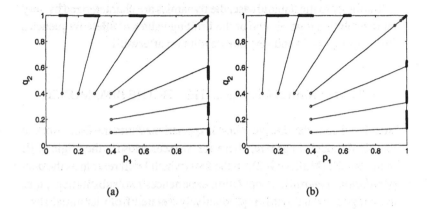

(a)   (b)

FIGURE 5.4. The projection of the dynamics of the two-locus, two-allele diploid system on the (p_1, q_2)-plane. (a) Evolution towards one of the two boundary equilibria. Equal mutation rates; no backward mutations. (b) Evolution towards a single symmetric equilibrium (located in the right upper corner). Equal forward and backward mutation rates. Parameters the same as in Figure 5.3.

and b). Assume now that the initial population has fixed haplotype **AB**. Let mutation rates from **A** to **a** and from **B** to **b** be equal to μ and no backward mutations be allowed. In this case, the population evolves to one of the two locally stable equilibria at which a conditionally deleterious allele is fixed at one locus, whereas at the other locus such an allele is maintained at a very low frequency at mutation-selection balance (Bengtsson and Christiansen 1983; see Figure 5.4(a)). The system also has a polymorphic equilibrium at which the frequency of the deleterious haplotype **aB** is very low,

$$\phi_3^* \approx \frac{\mu}{2s} \frac{r(1-s)}{s + 2r(1-s)}, \tag{5.4a}$$

and the frequencies of gametes **AB** and **ab** are

$$\phi_1^* = \phi_4^* \approx \sqrt{\frac{\mu}{s} \frac{s + r(1-s)}{s + 2r(1-s)}}. \tag{5.4b}$$

This equilibrium is unstable (which can be verified using equation 16 in Bengtsson and Christiansen 1983). However, if one allows for backward mutations at rate μ, then numerical simulations show that the equilibrium defined by equations (5.4) becomes stable (see Figure 5.4(b)).

Similarly to the haploid case, the dynamics are characterized by very rapid evolution towards one of the limiting values of allele frequencies ($p_1 = 1$ or $q_2 = 1$) and very slow evolution afterward.

5.3 DYNAMICS OF SPECIATION IN THE BDM MODEL

Next I will consider the dynamics of speciation within the framework of the BDM model. In general, there are two scenarios for the origin of RI in this model (Muller 1942). In the first (which I will refer to as the *one-population scenario*), a population experiences two substitutions, thus evolving to a state that is reproductively isolated from its initial state. In the second (which I will refer to as the *two-population scenario*), two populations experience substitutions at two different loci resulting in RI. Below I will treat these two scenarios separately.

In modeling allopatric and parapatric speciation, I will use a *weak mutation-weak migration approximation* neglecting within-population variation (e.g., Slatkin 1976, 1981; Lande 1979, 1985b; Tachida and Iizuka 1991; Barton 1993; Gavrilets 2000b). Under this approximation, the only role of mutation and migration is to introduce new alleles that quickly become fixed or lost. This is a reasonable assumption if the rate of mutation μ and the rate of migration m are small and the population size N is not too large, so that the time required for a mutant (or migrant) allele to become fixed in a population or to be lost is much smaller than the waiting time for a new mutant (or migrant) allele to enter the population. Mathematically this means that both $N\mu \ll 1$ and $Nm \ll 1$. While the first condition may be biologically realistic, the second condition appears to be too strict, as it requires very low migration rates. The effects of higher rates of migration will be considered in Chapters 6 and 7. An advantageous feature of the weak mutation-weak migration approximation is that its results are valid both for haploid and diploid populations as well as for selection acting both on the level of individuals and on the level of mating pairs.

The dynamics of speciation will be modeled as a biased random walk performed by the most abundant haplotype in the population along a ridge of high-fitness combinations of genes leading to a state of RI (see Figures 5.1 and 5.2). This approach is justified by the results of the previous section, which showed that after a short transient time, most of the population stays on the ridge, and by the fact that the deterministic forces that tend to move the population along the ridge are rather weak.

For simplicity, I will assume that mutations have equal rate μ in both loci and that no backward mutations are allowed. Allowing for backward mutations will change the results to some extent, but not dramatically. I will start by assuming that mutations are neutral, and then generalize the approach for the case of selection.

I will be interested in three quantitative characteristics: (i) the probability of speciation, u, (ii) the average waiting time to speciation, T, and (iii) the average duration of speciation, τ. The meaning of the average duration of speciation requires some clarification. Here, I will define τ as the average time that it takes to get from the ancestral state to the state of RI *without returning to the ancestral state*. Note that time τ is analogous to the average time during which an allele destined to be fixed segregates in the population (e.g., Kimura and Ohta 1969). Time τ can also be thought of as the average duration of the intermediate stages in the actual transition to a state of RI (speciation). Box 5.1 describes simple ways to determine u, T, and τ using the mathematical formalism of Markov chain models.

5.3.1 Allopatric speciation

Here, I will assume that two parts of a population become geographically isolated and that no migration of individuals between these two subpopulations is allowed ($m = 0$).

No selection for local adaptation In general, speciation by mutation and genetic drift represents a null hypothesis (model), against which different scenarios invoking selection have to be compared. Let us start with the one-population scenario, which is simpler to analyze mathematically. Initially, the population has haplotype ab fixed. Allele a mutates to state **A** and allele b mutates to state **B**. Both mutations have equal rates μ. Alleles **a** and **B** are assumed to be incompatible. We will model the evolutionary dynamics in this system as a Markov chain with three states we will denote as 0, 1, and Ω:

$$0 \longrightarrow 1 \longrightarrow \Omega$$

State 0 corresponds to the initial state of a population with haplotype ab fixed. State 1 is an intermediate state with haplotype **Ab** fixed. State Ω (speciation) corresponds to a population with haplotype **AB** fixed. When the population reaches this state, it becomes reproductively

isolated from a population at state 0. There is no state corresponding
to fixation of haplotype a**B** because alleles **a** and **B** are incompatible.
The arrows in the diagram above specify transitions that are possible in
the system. To complete this model, we need to define the probabilities
P_{ij} of transitions from state i to state j. In the absence of selection, the
probability of fixation of an allele is equal to its initial frequency, and the
rate of fixation is equal to the rate of neutral mutations (see Box 2.2).
Thus, the rates of transitions from state 0 to state 1 and from state 1
to state Ω are equal to the mutation rate: $P_{01} = P_{1\Omega} = \mu$. Because
mutations are irreversible, the system cannot return to state 0, once in
state 1, that is; $P_{10} = 0$. In the diagram above, this fact is reflected in the
absence of an arrow pointing from state 1 to state 0. The irreversibility
of mutations also implies that the population cannot leave state Ω, that
is, in the terminology of stochastic processes, this state is absorbing.

Because there is only one absorbing state, the system will definitely
reach it. Thus, in this system, speciation is certain: $u = 1$. The average
waiting time to speciation is

$$T = \frac{2}{\mu}. \tag{5.5a}$$

The last expression is very easy to understand. The average waiting time
until fixation of a neutral allele is approximately the reciprocal of the
mutation rate (e.g., Nei 1976). To evolve to a state with haplotype **AB**
fixed, the population has to fix two neutral alleles, which is reflected in
the expression for T above. Because in this model the system can never
return to state 0 once in state 1, the average duration of speciation is
equal to the average waiting time to speciation:

$$\tau = \frac{2}{\mu}. \tag{5.5b}$$

Note that because in the neutral case the rate of substitutions does not
depend on the population size, N does not enter equations (5.5).

In the two-populations scenario, assume that initially both allopatric
populations have haplotype **Ab** fixed. Allele **A** can mutate to state **a**
and allele **b** can mutate to state **B**. Both mutations have equal rates μ.
As before, alleles **a** and **B** are assumed to be incompatible. Speciation
occurs if one of the populations evolves to a state with haplotype **AB**
fixed, whereas the other population evolves to a state with haplotype
ab fixed. It is also possible that both populations evolve to a state with
the same haplotype fixed (**ab** or **AB**). In this case, speciation does

not occur. I will model the evolutionary dynamics in this system as a Markov chain with four states, which we will denote as 0, 1, Ω, and Ω':

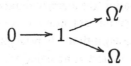

State 0 corresponds to the initial system state with haplotype Ab fixed in both populations. State 1 corresponds to system states in which one of the populations has fixed a mutant gene. That is, one population has ancestral haplotype Ab, whereas the other population has haplotype ab or AB. State Ω (speciation) corresponds to system states in which the populations have fixed the incompatible haplotypes. That is, one population has haplotype ab, whereas the other population has haplotype AB. State Ω' (no speciation) corresponds to system states in which both populations have fixed the same haplotype (AB or ab).

To complete this model we need to define the probabilities of transition between the states. The rate of fixation of neutral alleles is equal to the rate of neutral mutations. Thus, the rate of transition from state 0 to state 1 is $P_{01} = 4\mu$, where the factor 4 accounts for the fact that with two two-locus populations there are four mutable genes. Once in state 1, the system can move to state Ω or Ω', depending on which locus is changed by the next substitutions. Because the substitutions are neutral, the corresponding transition rates are $P_{1\Omega} = \mu$, $P_{1\Omega'} = \mu$. By the assumption that mutations are irreversible, the system cannot return to state 0 once in state 1, that is; $P_{10} = 0$. The irreversibility of mutations also implies that the system cannot leave the states Ω and Ω', that is; these states are absorbing.

One finds that speciation occurs with 50% probability: $u = 1/2$. This result has a simple explanation: once the system is in state 1, it moves to states Ω and Ω' with equal probabilities. In this model, the average waiting time to speciation is characterized by the average time to hit state Ω starting from state 0 conditioned on hitting state Ω rather than state Ω'. One finds that

$$T = \frac{3}{4\mu}. \tag{5.6a}$$

Because transitions are irreversible, the average duration of speciation is the same as the average waiting time to speciation:

$$\tau = \frac{3}{4\mu}. \tag{5.6b}$$

Note that although speciation is not certain in the two-population scenario, if it occurs, its duration is shorter than in the one-population scenario. In both scenarios, the average duration of speciation has the order of the reciprocal of the mutation rate per locus.

Selection for local adaptation   In principle, it is possible that new alleles affecting RI have selective advantage (or disadvantage) over ancestral alleles in a local environment (for example, due to pleiotropy). For example, Macnair and Christie (1983) identified a single gene present in a population of the yellow monkey flower *Mimulus guttatus* that interacted with a small number of genes from a different population to give seedling death at the four-leaf stage or later. This gene is the same as, or tightly linked to, the gene providing tolerance to excess copper in the soil. In a similar way, the three genes involved in hybrid incompatibility as discussed on page 122 (i.e., the *Odysseus* gene, the *Hybrid male rescue* gene, and the *Nup98-Nup96* gene) are all targets of direct natural selection, judging from their rapid evolution.

Below, I will assume that the new alleles contribute to fitness (viability) on an additive scale. Let us consider the one-population scenario, with the ancestral haplotype being ab. Assume that mutant alleles A and B are advantageous over alleles a and b. Let s be the corresponding selection coefficient and N be the population size. The Markov chain describing this model is the same as the one illustrated in the diagram on page 131. The expected number of mutations per generation in a diploid population is $2N\mu$. The probability that an advantageous mutation is fixed is $\approx 2s/[1 - \exp(-4Ns)]$ (see Box 3.1). The rate of fixation of new alleles can be approximated by the product of the average number of new alleles per generation and the probability of fixation. Thus, the average rate of transitions from state 0 to state 1 and from state 1 to state Ω can be written as $P_{01} = P_{1\Omega} = \mu \, \kappa_a$, where

$$\kappa_a = \frac{S}{1 - \exp(-S)} \tag{5.7}$$

is the rate of fixation of advantageous alleles relative to that of neutral alleles, and $S = 4Ns$. (Note that S has a different meaning from that in Chapter 3, where $S = Ns$.) Because transitions are irreversible, speciation is still certain: $u = 1$. But the average waiting time to speciation and the average duration of speciation change to

$$T = \tau = \frac{2}{\mu} \frac{1 - \exp(-S)}{S}. \tag{5.8}$$

This expression shows that selection for local adaptation can accelerate speciation. For example, increasing S from 0 to 10 decreases the waiting time to speciation to approximately 1/10 of that in the case of no selection for local adaptation.

To model the two-populations scenario, let us first assume that each of the incompatible alleles a and B is advantageous in both environments. This is the case of spatially uniform selection for local adaptation. As before, haplotype Ab is initially fixed in both populations. The Markov chain describing this model is the same as the one illustrated in the diagram on page 133. But now, because all mutations are advantageous, all nonzero transition rates are equal to $\mu\kappa_a$ rather than to μ. In this case, one can show that

$$u = \frac{1}{2}, \quad T = \tau = \frac{3}{4\mu}\frac{1 - \exp(-S)}{S}. \tag{5.9}$$

Comparing these equations with equations (5.6), one can see that spatially uniform selection does not change the probability of speciation but does make it faster.

To model spatially heterogeneous selection, let us assume that each of the incompatible alleles a and B is advantageous in one environment and neutral in the other environment. Let allele a be advantageous in the environment experienced by population I and allele B be advantageous in the environment experienced by population II. As before, we assume that A mutates to a and b mutates to B with rates equal to μ and that no backward mutations are allowed. Modeling speciation in this model using the Markov chain framework requires specifying five different states, which we will denote as states 0, 1, 2, Ω', and Ω:

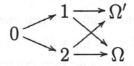

State 0 corresponds to the initial state in which both populations are fixed for haplotype Ab. State 1 corresponds to situations in which one of the populations has fixed the allele that is neutral in the local environment, whereas the other population still has the ancestral haplotype Ab. There are two possibilities here: the new haplotype ab is fixed in the first population and the ancestral haplotype Ab is still fixed in the second population, or ancestral haplotype Ab is still fixed in the first population and new haplotype AB is fixed in the second population.

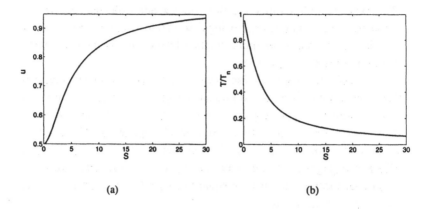

(a)   (b)

FIGURE 5.5. Speciation in the five-state model. (a) Probability of speciation u. (b) Average waiting time to speciation T relative to the waiting time with no selection for local adaptation T_n (as given by equation (5.6a)).

State 2 corresponds to situations where one of the populations has fixed the allele that is advantageous locally, whereas the other population still has the ancestral haplotype **Ab**. Again, there are two possibilities: new haplotype **AB** is fixed in the first population and ancestral haplotype **Ab** is still fixed in the second population, or ancestral haplotype **Ab** is still fixed in the first population and new haplotype **ab** is fixed in the second population. State Ω' (no speciation) corresponds to situations when both populations have fixed the same new haplotype (**AB** or **ab**). State Ω (speciation) corresponds to situations when the populations have fixed different haplotypes (**ab** in one population and **AB** in the other population).

In this model, the probabilities of transitions are as follows:

$$P_{01} = 2\mu, \ P_{02} = 2\mu\kappa_a, \ P_{1\Omega'} = P_{2\Omega'} = P_{1\Omega} = \mu, \ P_{2\Omega} = \mu\kappa_a.$$

These expressions account for the facts that transitions from state 0 to state 2 and from state 2 to state Ω are due to fixation of advantageous alleles, and that transitions from state 0 are due to fixations in either of the two populations. In this model, the probability of speciation is

$$u = 1 - \frac{2\kappa_a}{(\kappa_a + 1)^2} \approx 1 - \frac{2}{S}, \tag{5.10a}$$

and the average waiting time to and duration of speciation are

$$T = \tau = \frac{3}{2\mu(\kappa_a + 1)} \approx \frac{3}{2\mu S}, \tag{5.10b}$$

where the approximations assume that $S \gg 1$. Figure 5.5 illustrates the dependencies of the probability of speciation u and the waiting time to speciation T on S. Probability u quickly approaches 1 with increasing S, whereas time T decreases as $1/S$ for large S. These properties imply that spatially heterogeneous selection for local adaptation can both increase the overall probability of speciation and accelerate it.

5.3.2 Parapatric speciation

In the parapatric case, mutation is not the only source of new genes anymore. If the two populations have fixed alternative alleles in a locus, then immigrants will bring genes that are new for a current population state. This may trigger fixation of these alleles, which will increase the likelihood of the system becoming genetically uniform and will make speciation much less probable. In the allopatric case with no backward mutations, the waiting time to speciation T and the average duration of speciation τ are equal because the process of divergence is irreversible. With migration, populations can return to their original genetic state even after fixing new genes. In the process of evolution towards a reproductively isolated state, typically there will be many unsuccessful speciation attempts in which a population will genetically diverge to some extent from its original state only to lose new genes and return to the original state. As a result, in the parapatric case, T will typically be (much) larger than τ.

No selection for local adaptation In the one-population scenario, let us assume that the focal population receives immigrants from a source population at a constant rate m. All immigrants have ancestral haplotype ab. (Note that the one-population scenario can be viewed as describing peripatric speciation.) Now the diagram representing the corresponding Markov chain has an arrow pointing from state 1 back to state 0:

$$0 \rightleftarrows 1 \longrightarrow \Omega$$

Because the probability of fixation of a neutral allele is equal to its initial frequency, the transition rate from state 1 to state 0 is now equal to the rate of migration: $P_{10} = m$.

In this model, speciation is certain: $u = 1$. The waiting time to speciation is

$$T = \frac{m + 2\mu}{\mu^2} \approx \frac{m}{\mu^2}, \qquad (5.11a)$$

where the approximation is valid if $m \gg \mu$ (which should be true in most biologically realistic situations). The average duration of speciation is

$$\tau = \frac{m + 2\mu}{m + \mu} \frac{1}{\mu} \approx \frac{1}{\mu}. \qquad (5.11b)$$

Note that although speciation is certain, the average waiting time to speciation is large. However, when speciation occurs, the duration of intermediate stages (represented by state 1) is very short relative to the waiting time to speciation and has the order of the reciprocal of the mutation rate per locus. For example, if $m = 0.01$ and $\mu = 10^{-5}$, then $T \approx 10^8$ generations and $\tau \approx 10^5$ generations. The ratio of τ and T is very small:

$$\frac{\tau}{T} = \frac{\mu}{m + \mu} \approx \frac{\mu}{m}. \qquad (5.12)$$

This implies that observing the actual transition from state 0 to state Ω is very unlikely.

In the two-populations scenario, the diagram representing the corresponding Markov chain has an additional arrow pointing from state 1 back to state 0:

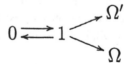

Migration will affect transition probabilities both from state 1 to state 0 and from state 1 to state Ω', which now are equal to $P_{10} = m$, $P_{1\Omega'} = \mu + m$. The last equality accounts for the fact that both migration and selection tend to push the population towards state Ω'.

In this model, the probability of speciation, the waiting time to speciation, and the average duration of speciation are

$$u = \frac{\mu}{m + 2\mu} \approx \frac{\mu}{m}, \qquad (5.13a)$$

$$T = \frac{m + 3\mu}{m + 2\mu} \frac{1}{2\mu} \approx \frac{1}{2\mu}, \qquad (5.13b)$$

$$\tau = \frac{m + 3\mu}{m + \mu} \frac{1}{4\mu} \approx \frac{1}{4\mu}. \qquad (5.13c)$$

For example, if $m = 0.01$ and $\mu = 10^{-5}$, then $u = 10^{-3}, T = 50,000$ generations, and $\tau = 25,000$ generations. Now, the probability of speciation is very small. The system will much more likely end up in a state with the same mutant allele being fixed in both populations. If speciation does occur, the waiting time to speciation has the order of the inverse of the mutation rate. The average duration of speciation is half of the average waiting time to speciation.

Selection for local adaptation What will be the effects of selection on the plausibility of parapatric speciation? In the one-population scenario, we assume as before that a diploid population of a constant size N, which is initially fixed for haplotype ab, is subject to immigration from an outside source at rate m. All immigrants have ancestral haplotype ab. Alleles a and b mutate to alleles A and B, respectively, with probability μ per generation. Alleles a and B are incompatible. We will consider two models of selection for local adaptation.

Model I (spatially uniform selection). Assume that both mutant alleles A and B are advantageous with selection coefficient s. With immigration, the population can return to state 0 after fixing a mutant allele. This requires fixing an ancestral allele brought by immigrants, which is deleterious in the local environment. The probability of fixing a deleterious allele is approximately $2s/[\exp(S) - 1]$, where $S = 4Ns$ (see Box 3.2), and there are approximately $2Nm$ deleterious alleles brought by immigrants each generation. Therefore, the probability of transition from state 1 to state 0 is $P_{10} = m\kappa_d$, where the rate of fixation of deleterious alleles relative to that of neutral alleles is

$$\kappa_d = \frac{S}{\exp(S) - 1}. \tag{5.14}$$

In this model, speciation is certain: $u = 1$, and the average waiting time to speciation and the average duration of speciation are

$$T = \frac{1}{\mu} \left(2 + \frac{m}{\mu e^S}\right) \frac{1 - e^{-S}}{S} \tag{5.15a}$$

$$\tau = \frac{1}{\mu} \frac{2 + m/(\mu e^S)}{1 + m/(\mu e^S)} \frac{1 - e^{-S}}{S} \tag{5.15b}$$

Table 5.1 shows that even relatively weak selection for local adaptation can decrease T by orders of magnitude. The strength of selection also affects τ, but to a lesser degree. The ratio of the average duration of speciation and the average waiting time to speciation is

$$\frac{\tau}{T} = \frac{1}{1 + m/(\mu e^S)}. \tag{5.16}$$

TABLE 5.1. Average waiting time T and average duration τ of speciation in
Model I with $\mu = 10^{-5}$ and $m = 0.01$.

S	T	τ
0	10^8	10^5
1	2.34×10^7	6.34×10^4
3	1.64×10^6	3.23×10^4
5	1.73×10^5	2.24×10^4

With the same values of μ and m and $S = 1, 3$, and 5, the ratio of τ and T is 1/369, 1/51, and 1/8. That is, the ratio τ/T is very small.

Model II (genetic barrier). Let us assume that new alleles **A** and **B** are neutral but that the population has already diverged from the source population in some other loci (or chromosomes). These can be genes controlling adaptation to local environmental conditions or some neutral genes that are incompatible with the ancestral genes (or chromosomes). Now immigrants or F_1 hybrids will have a reduced fitness, which will affect the fate of the neutral ancestral alleles a and b that they carry. In a sense, this will create a *genetic barrier* to the flow of genes into the population. The effects of selection induced on neutral alleles via their association with some locally deleterious alleles can be characterized in terms of the *gene flow factor* (see Box 5.2); small values of the gene flow factor imply that genetic barrier is strong. Let the gene flow factor be η. Then, using equation (1) from Box 5.2 for the effective migration rate, one finds that speciation is certain: $u = 1$, and that

$$T \approx \eta \, \frac{m}{\mu^2}, \quad \tau \approx \frac{1}{\mu}. \tag{5.17}$$

Thus, a strong genetic barrier to the neutral gene flow (i.e., low η) will significantly decrease the waiting time to speciation. For example, if $\eta = 10$, T decreases to 1/10 of that in the case of no selection.

In the two-population scenario, we will consider three different models of selection for local adaptation.

Model III (spatially uniform selection). Let us assume that the new, mutually incompatible alleles **A** and b have selective advantage s over the ancestral alleles a and **B** in both environments. In this case, the corresponding Markov chain can be represented as

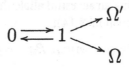

where states 0, 1, Ω, and Ω' describe the initial state, the state with one population fixed for a new allele and another population still fixed for an ancestral allele, the speciation state, and the state of no speciation, respectively. The transition probabilities are

$$P_{01} = 4\mu\kappa_a, P_{10} = m\kappa_d, P_{1\Omega} = \mu\kappa_a, P_{1\Omega'} = m\kappa_a,$$

where the relative rates of fixation of advantageous (κ_a) and deleterious (κ_d) alleles are defined by equations (5.7) and (5.14). In this model the probability of, the average waiting time to, and the average duration of speciation are approximately

$$u \approx \frac{\mu}{m}, \quad T \approx \frac{1}{4\mu}\frac{1}{\kappa_a}, \quad \tau \approx \frac{1}{4\mu}\frac{1}{\kappa_a}. \tag{5.18}$$

Comparing these equations with those for the case of no selection for local adaptation (see equation (5.13)), one can see that spatially uniform selection does not increase the overall probability of speciation, but does make it much faster if speciation occurs (assuming that κ_a is large).

Model IV (direct spatially heterogeneous selection). Let us assume that each of the incompatible alleles **A** and **b** is advantageous in one environment and neutral in the other environment. This is the same selection scheme as analyzed on page 135 for the allopatric case. Modeling speciation in this model using the Markov chain framework requires specifying five different states 0, 1, 2, Ω' and Ω, which were already defined above:

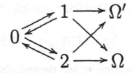

In this model, the nonzero transition probabilities are

$$P_{01} = 2\mu, \quad P_{02} = 2\mu\kappa_a,$$

where factor 2 is present because there are two subpopulations, each of which can undergo the substitution, and κ_a is defined by equation (5.7);

$$P_{10} = m, \quad P_{20} = m\kappa_d,$$

where the last expression accounts for the fact that moving from state 2 to state 0 implies fixing an ancestral allele that is locally deleterious, and κ_d is defined by equation (5.14);

$$P_{1\Omega'} = (\mu + m)\kappa_a, \quad P_{2\Omega'} = \mu + m,$$

which account for the fact that once the system is in state 1 or 2, both mutation and migration are pushing it towards a genetically uniform state; and

$$P_{1\Omega} = \mu, P_{2\Omega} = \mu\kappa_a,$$

which account for the fact that reaching speciation state Ω from state 1 requires fixing a neutral mutation, whereas reaching this state from state 2 requires fixing a locally advantageous mutation.

The exact expressions for u, T, and τ are rather cumbersome. However, if the mutation rate μ is very small and the coefficient S is at least moderately large (say, $S > 3$), then they can be approximated as

$$u \approx \frac{\mu}{m} S, \quad T \approx \tau \approx \frac{1}{2\mu} \frac{1}{2S}. \tag{5.19}$$

Thus, direct, spatially heterogeneous selection for local adaptation both increases the probability of speciation and decreases the waiting time to and the average duration of speciation. For example, if $S = 10$, u undergoes a tenfold increase.

Model V (genetic barrier). Finally, let us consider the effects of a genetic barrier to the neutral gene flow in the two-population scenario. That is, we assume that new incompatible alleles a and B are neutral, but that the population has already diverged from the source population in some other loci (or chromosomes), which has resulted in fitness reduction of immigrants or F_1 hybrids. Let η be the corresponding gene flow factor (see Box 5.2). One finds that approximately

$$u \approx \frac{\mu}{\eta m}, \quad T \approx \frac{1}{2\mu}, \quad \tau \approx \frac{1}{4\mu}. \tag{5.20}$$

Thus, strong genetic barriers to the neutral gene flow (i.e., low η) will significantly increase the probability of speciation, but will not affect T and τ relative to the neutral case (see equations (5.13)).

Table 5.2 summarizes the results of this section. The approximations given in this table assume that μ is very small and that S is large. The Roman numerals in the parentheses correspond to the different models of selection for the local adaptation introduced above. Note that

TABLE 5.2. Characteristics of the dynamics of speciation in the BDM model. Roman numerals refer to selection models defined in the text.

Scenario	Selection	Allopatric		Parapatric		
		u	$T = \tau$	u	T	τ
one-	neutral	1	$2/\mu$	1	m/μ^2	$1/\mu$
population	I	1	$2/\mu S$	1	$2/\mu S$	$2/\mu S$
	II	n/a	n/a	1	$\eta m/\mu^2$	$1/\mu$
two-	neutral	$1/2$	$3/4\mu$	μ/m	$1/2\mu$	$1/4\mu$
population	III	$1/2$	$3/4\mu S$	μ/m	$1/4\mu S$	$1/4\mu S$
	IV	$1 - 2/S$	$3/2\mu S$	$S\mu/m$	$1/2\mu S$	$1/2\mu S$
	V	n/a	n/a	$\mu/m\eta$	$1/2\mu$	$1/4\mu$

haploid populations are described by the same formula as the diploid populations, but with the coefficient S equal to $2Ns$ rather than $4Ns$.

5.4 SUMMARY

• The BDM model is a two-locus, two-allele model in which two alleles at two different loci are incompatible in the sense that their simultaneous presence in an offspring's genotype or in the genotypes of a (potential) mating pair reduces a corresponding fitness component (viability, fertility, or sexual attractiveness). In this model, strong RI can evolve without any need to cross fitness valleys. In the one-population scenario, (a part of) a population experiences two substitutions evolving to a state that is reproductively isolated from its initial state. In the two-population scenario, two populations experience substitutions at two different loci, resulting in RI.

• In terms of fitness landscapes, the BDM model is characterized by the presence of a ridge of high fitness values that goes around a fitness hole. The BDM model is one of the simplest models that possess a (nearly) neutral network of high-fitness genotypes and results in a holey landscape. There are at least three general classes of mechanisms leading to this kind of genetic architecture: genes brought together in hybrids may be incompatible, hybrid genotypes may lack specific functions, and maternal and paternal genes may lack the necessary matching.

• The evolutionary dynamics of very large populations in the BDM model are characterized by very rapid movement towards a ridge of high-fitness genotypes with very slow changes by weak factors afterward. After a relatively short period of time, most of the population resides on the ridge. Mutation will maintain low-fitness combinations of genes at a very low frequency (on the order of the mutation rate).

• If the population size is small and both mutation and migration are rather weak, then random genetic drift will remove genetic variation, and the population will be mostly monomorphic. Then the dynamics of speciation in the BDM model can be modeled as a biased random walk performed by the population along the ridge of high-fitness genotypes.

• The three important characteristics of the speciation process are the probability of speciation u, the average waiting time to speciation T, and the average duration of speciation τ. The following are conclusions emerging from studying a number of the BDM models:

- In the BDM model, speciation (both allopatric and parapatric) is possible by mutation and random drift only (that is, without selection for local adaptation). In general, speciation by mutation and drift represents a null model (hypothesis) against which different scenarios invoking selection have to be compared.

- Speciation is always certain in the one-population scenario ($u = 1$). In the two-population scenario, allopatric speciation has a very high probability, whereas the probability of parapatric speciation is approximately given by the ratio of the mutation rate per locus μ and the migration rate m. Spatially uniform selection does not affect the probability of speciation. Extrinsic, spatially heterogeneous selection (acting directly on the loci underlying reproductive isolation or acting indirectly by creating a genetic barrier to neutral gene flow) can significantly increase the probability of speciation. The same effect is expected if a genetic barrier between residents and immigrants is created by intrinsic genetic incompatibilities between other genes (or chromosomes).

- In the one-population scenario (which can be viewed as describing peripatric speciation) with no selection for local adaptation, the average waiting time to speciation T has order $1/\mu$ in the allopatric case and order m/μ^2 in the parapatric case. In the two-population scenario, speciation, although generally unlikely if it ever happens, is very fast and takes order $1/\mu$ generations. Direct selection for local adaptation can significantly decrease the waiting time to speciation. The presence of a genetic barrier does not affect T in the two-population case, but decreases T in the one-population case.

- With no selection for local adaptation, the duration of speciation τ is on the order of $1/\mu$ generations. In the allopatric case, τ is the same as the average waiting time T to speciation. In the parapatric case, τ is much smaller than T in the one-population scenario and is comparable in magnitude with T in the two population scenario. Direct selection decreases τ, whereas a genetic barrier does not affect τ.

- The average waiting time to speciation T and the average duration of speciation τ are comparable in the allopatric case. In the parapatric case, T is typically much larger than τ.

• The speciation models analyzed in this chapter are characterized by a high degree of symmetry and by only two fitness levels. More theoretical work is needed to evaluate if the above conclusions hold in more general situations.

5.5 CONCLUSIONS

The BDM model provides a simple tool for training our intuition about the dynamics of speciation. This model also represents one of the simplest examples of holey fitness landscapes. The BDM model shows that allopatric speciation is a very probable eventual outcome of genetic divergence. Parapatric speciation can be relatively rapid if selection for local adaptation is spatially heterogeneous and strong. The duration of parapatric speciation is expected to be very short.

BOX 5.1
HITTING PROBABILITIES AND HITTING TIMES IN DISCRETE-TIME
MARKOV CHAINS

Let us consider a discrete-time Markov chain with a finite number of states. Let P_{ij} be the probability of transition from state i to state j per unit of time. Assume that there are two absorption states: Ω and Ω'. In the main text, I interpret these states as speciation and no speciation, respectively. Then the probabilities π_i of hitting state Ω rather than state Ω', if starting at state i, are given by the solutions to a system of linear algebraic equations

$$\pi_i = \sum_j P_{ij}\pi_j, \pi_\Omega = 1, \pi_{\Omega'} = 0, \tag{1}$$

where the index j runs over all states.

Let T_i^* be the average waiting time to hit state Ω starting from state i conditioned on hitting state Ω rather than state Ω'. The average waiting times T_i^* are solutions to a system of linear algebraic equations

$$T_i^* = 1 + \sum_j p_{ij}^* T_j^*, \tag{2}$$

where the index j runs over all nonabsorbing states and $p_{ij}^* = P_{ij}\pi_j/\pi_i$ is the conditional transition probability from state i to state j per unit of time.

If there is only one absorption state Ω, then of course $\pi_i = 1$ for all i, and the average waiting times T_i to hit this state starting from state i solve a system of linear algebraic equations

$$T_i = 1 + \sum_j P_{ij}T_j. \tag{3}$$

These results can be found in a number of sources (e.g., Karlin and Taylor 1975; Ewens 1979; Norris 1997). In the main text, we are interested in the dynamics of speciation starting at state 0. This case corresponds to $u = \pi_0, T = T_0$, and $T^* = T_0^*$.

BOX 5.2
Genetic Barrier to Gene Flow

Consider a population subject to continuous immigration at a low rate. Assume that individuals carrying (some) immigrant genes have reduced fitness relative to that of residents. This can be because the immigrants do not have genes adapting individuals to the local environment or because resident and immigrant genes do not match. Because of selection against them, genes underlying reduced fitness will not increase in frequency in the population but will be maintained at a low level determined by the balance of selection and migration. If immigrants also differ from the residents in some neutral genes, these neutral genes will increase in frequency as they become incorporated into the local genetic background. The spread of neutral genes will, however, be delayed to some extent by selection against immigrants and hybrids. In a sense, selection against incoming genes will work as a genetic barrier to the spread of neutral genes. To characterize this effect, Bengtsson (1985) introduced the notion of the "gene flow factor," η, defined as the probability that a neutral gene brought by immigrants makes it to the local genetic background. Let m be the migration rate, that is, the proportion of the population replaced by immigrants per generation. With a genetic barrier, the proportion of resident neutral alleles replaced by immigrant neutral alleles per generation is

$$m_e = \eta m. \tag{1}$$

Equation (1) defines the effective migration rate of neutral alleles. The inverse of η is known as the "strength of genetic barrier" (e.g., Barton and Bengtsson 1986; Piálek and Barton 1997).

For example, assume that immigrating adults differ from the residents in two genes: a gene reducing the viability of F_1 hybrids to $1 - s$ (relative to viability 1 of the residents) and a neutral gene unlinked to the selected gene. Assume that mating is random. The probability that a neutral gene makes it to the next generation and, there, gets incorporated into the local genetic background is $(1 - s)/2$. The probability that this gene makes it to the next generation but remains associated with the deleterious gene is also $(1 - s)/2$. After one more generation, the probability of inclusion of the neutral allele into the local genetic background is $(1 - s)/2 + (1 - s)^2/2^2$, and the ultimate probability of inclusion (that is, after very many generations) is $\sum_{i=1}^{\infty} (1 - s)^i/2^i$, which simplifies to

$$\eta = \frac{1 - s}{1 + s}. \tag{2}$$

If the neutral gene is linked to the selected gene, then

(Box 5.2 continued)

$$\eta = \frac{r(1-s)}{1-(1-r)(1-s)},\tag{3}$$

where r is the recombination rate (Bengtsson 1985). For example, let $s = 0.5$. Then if $r = 0.5$, then $\eta = 0.33$; if $r = 0.05$, then $\eta = 0.09$. That is, close linkage with a negatively selected allele can significantly decrease the effective rate of immigration of neutral alleles. The above expressions for η are also applicable if the immigrants differ from residents in their karyotype, so that F_1 hybrids have a reduced fitness $1 - s$.

Assume next that not only F_1 hybrids have reduced viability, but also that the average number of offspring produced by matings between the immigrant adults and residents, α, and by matings between F_1 hybrids and residents, β, are reduced relative to that of matings among residents (which we normalize to be 1). This can happen if there is fertility and/or sexual selection against immigrants and/or hybrids. Then

$$\eta = \frac{r(1-s)\alpha\beta}{1-(1-r)(1-s)\beta}\tag{4}$$

(Gavrilets and Cruzan 1998). For example, let $s = \alpha = \beta = 0.5$. Then if $r = 0.5$, then $\eta = 0.07$; if $r = 0.05$, then $\eta = 0.016$. That is, joint action of viability selection, fertility selection, and assortative mating can significantly decrease the effective rate of immigration of neutral alleles even if they are unlinked to the locus under selection.

If there is a number of unlinked loci with equal effects contributing to viability in a multiplicative way, then the gene flow factor for a neutral locus unlinked to the loci under selection is

$$\eta = (1-s)^2,\tag{5}$$

where $1 - s$ is the fitness of the F_1 hybrids (Bengtsson 1985).

For the BDM model with fitnesses as in the fitness matrix given on page 128, and assuming that all three loci (two selected and one neutral) are unlinked, Bengtsson (1985) showed that the gene flow factor is

$$\eta = \frac{1-s}{1+s}\tag{6}$$

For other examples of multilocus models see Gavrilets (1997c), Piálek and Barton (1997), and Navaro and Barton (2003a).

CHAPTER SIX

Multidimensional generalizations of the BDM model

In this chapter, I consider a number of speciation models that can be viewed as multidimensional generalizations of the BDM model. I start by describing a series of multiallele models proposed by Nei et al. (1983) and then move on to multilocus models. First, I describe a model introduced by Walsh (1982) in which RI results as a byproduct of a chain of underdominant substitutions at different chromosomes (or loci). Then, I consider a model of divergent degeneration of duplicated genes advanced by Werth and Windham (1991) and Lynch and Force (2000). After that I analyze two specific models with three and four loci that require one to specify the genetic architecture of RI explicitly. Finally, I study a series of models in which RI results from incompatibilities between different groups of loci or from cumulative divergence over a specific set of loci. In developing these models, I will apply both the Markov chain approach used in the previous chapter, and more elaborated approaches that combine Markov chains with methods from dynamical systems theory. As in the previous chapter, the main focus will be on the probability of speciation, on the average waiting time to speciation, and on its average duration. I will consider both the allopatric and parapatric cases. In this chapter, my analysis of the parapatric case will be limited by modeling a population subject to immigration from a source population (as implied in the peripatric speciation scenario).

6.1   ONE- AND TWO-LOCUS, MULTIALLELE MODELS

The original BDM model was formulated in terms of two diallelic loci. Nei et al. (1983) introduced a series of multiallele versions of the BDM model. Using these models, Nei et al. (1983) were able to deduce several dynamical patterns of the speciation process which later proved to be rather general. I start by specifying the models used by these authors and then describe their results.

One-locus diploid model of postzygotic RI   With multiple alleles, the main idea underlying the BDM model works even in the case of a single locus. Following Nei et al. (1983), let us consider the stepwise mutation model of Ohta and Kimura (1973). Assume that there is an ordered sequence of alleles $\ldots, A_{-2}, A_{-1}, A_0, A_1, A_2, \ldots$ such that allele A_i can mutate only to alleles A_{i-1} and A_{i+1}. Assume that alleles separated by more than one mutational step are incompatible in the sense that the corresponding heterozygous genotypes (e.g., genotypes $A_i A_{i-3}, A_i A_{i-2}, A_i A_{i+2}, A_i A_{i+3}$, etc.) are inviable or sterile. All homozygotes $A_i A_i$ as well as heterozygotes $A_i A_{i+1}$ and $A_{i-1} A_i$ are assumed to be completely fertile and viable. The fitness landscape corresponding to this model was illustrated in Figure 2.4(a) on page 29. Speciation occurs if two populations fix alleles separated by at least two mutational steps. Another model considered by Nei et al. (1983) uses the infinite site model of Kimura and Crow (1964), assuming that each allele A_i is compatible only with its ancestral allele A_{i-1}, its descendant alleles A_{i+1}, and with itself. In this version, the subscript gives the number of mutations that have occurred in a lineage under consideration.

Two-locus diploid model of postzygotic RI   Here, the stepwise mutation model is used for each of the two loci A and B. It is assumed that gamete $A_i B_j$ is compatible only with the nine gametes that can be formed by alleles A_{i-1}, A_i, A_{i+1} at the first locus and alleles B_{i-1}, B_i, B_{i+1} at the second locus. Diploid zygotes that consist of compatible gametes are assumed to be viable and fertile, whereas all others are inviable or sterile. In this model any two substitutions at a locus cause reproductive incompatibility.

Two-locus haploid model of prezygotic RI   Here, there are two unlinked haploid loci A and B that control male-limited and female-limited characters, respectively. The stepwise mutation model is used for each of the loci. It is assumed that the males having allele A_i mate only with the females having alleles B_{i-1}, B_i, or B_{i+1}, and the females having allele B_i mate only with the males having alleles A_{i-1}, A_i, or A_{i+1}. As in the previous model, any two substitutions at a locus result in complete RI. This model, studied by Wu (1985) within the context of sympatric speciation, will be considered in more detail in Chapter 9.

To study allopatric speciation, Nei et al. (1983) used stochastic numerical simulations. The following are some of their findings:

• RI evolves faster in small populations than in large populations. This happens because mutant alleles that are two or more steps apart from the most common alleles are selected against when rare. That is, selection at the gene level is frequency-dependent and acts against rare alleles. Therefore, random genetic drift is essential for overcoming selection and fixing mutant alleles.

• In spite of the continuous presence of incompatible alleles in the population, the average within-population fertility and viability are always very high.

• Evolution of RI is a very slow process and it takes thousands to millions of generations if the mutation rate is on the order of 10^{-5} per locus per generation. However, once the process starts, the substitutions leading to RI occur rather quickly.

• RI evolves faster with two loci than with one locus. This seems to be happening because more loci imply higher overall mutation rates and more possible directions for divergence.

• Dynamics of postzygotic RI and prezygotic RI are similar.

• The rate of accumulation of RI decreases in time.

• The characteristic time scale for speciation appears to be on the order of 10 to 100 times the inverse of the mutation rate per locus for populations with up to 50,000 individuals. For the populations with less than 500 individuals, the time scale for speciation is on the order of the reciprocal of the mutation rate.

In these models, allopatric populations evolved in different directions and/or at different rates along a ridge of high-fitness combinations of genes (illustrated in Figure 2.4(a)) and became reproductively isolated when the genetic distance at specific genes exceeded a threshold (which was equal to one or two mutational steps). Nei et al. (1983) also used a one-locus stepwise mutation model of postzygotic RI to study parapatric speciation in a system of two populations. They noticed that even a small amount of gene flow retards the evolution of RI considerably.

6.2 MULTILOCUS MODELS

Next, I consider a series of multilocus models concentrating on underlying fitness landscapes and on the dynamics of allopatric and parapatric speciation.

6.2.1   The Walsh model

I start with the model introduced by Walsh (1982), who studied the dynamics of accumulation of chromosome rearrangements in a finite diploid population. His results are also applicable to situations in which RI is due to multiplicative viability selection against hybrid genotypes. I will describe his results in terms of underdominant loci rather than chromosomes.

Let there be $\mathcal{L}$ unlinked diallelic loci, each affecting fitness in an underdominant way. Assume that the fitness contributions of homozygous loci are equal and normalized to be 1 and those of heterozygous loci are $1 - s$ (with $0 < s < 1$). Let the overall fitness be given by the product of the contributions of individual loci, so that the fitness of an organism heterozygous at d loci is

$$w(d) = (1 - s)^d \approx e^{-sd}, \tag{6.1}$$

where the approximate equality is valid if s is small. Fitness of hybrids $w(d)$ is a measure of the degree of RI between the homozygous parental forms. Figure 6.1(a) shows that $w(d)$ decreases exponentially with d. This means that RI increases the fastest initially, with a slowdown afterward. Function $w(d)$ can be thought of as the probability of no complete RI between two populations at genetic distance d. Below, I will refer to such a function as *function of reproductive compatibility*.

The stochastic dynamics of allele frequencies at a single underdominant locus in the presence of mutation and random genetic drift were studied in Chapter 3. There, I showed that the corresponding dynamical system has two metastable states, in which one allele is common and another allele is rare (and maintained by mutation), and that random genetic drift results in occasional stochastic transitions back and forth between these two states. Similar behavior is expected in the multilocus case. However, now the number of metastable states is $2^{\mathcal{L}}$ rather than just 2. As a result of random genetic drift and of mutational order (see page 39), the system will wander across these $2^{\mathcal{L}}$ states by occasionally fixing an initially rare allele introduced by mutation at a random locus.

Let us consider two allopatric populations of the same size N that initially have the same haplotype fixed. Assume that the number of loci is very large but the mutation rate per locus μ is very small, so that the average number of mutations $\nu = \mu\mathcal{L}$ per gamete per generation is small ($\nu \ll 1$). Then the two populations will diverge at a rate

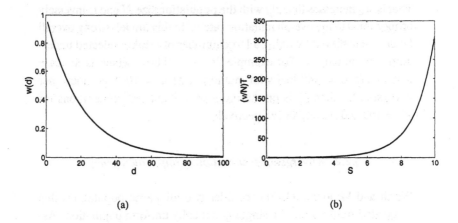

(a)   (b)

FIGURE 6.1. Evolution of RI in the Walsh model. (a) Fitness of heterozygote at d loci (with $s = 0.05$). (b) Characteristic time for the evolution of RI expressed as $\frac{\nu}{N} T_c$.

of $2\nu\kappa_u$ substitutions per generation, with κ_u being the probability of fixation of underdominant alleles relative to that of neutral alleles (see equation (3.2) on page 55). The number of substitutions separating the populations will grow as $d = 2\nu\kappa_u t$, that is, linearly in time. The fitness of an F_1 hybrid after t generations of divergence will be

$$w(t) \approx e^{-2s\nu\kappa_u t};\qquad (6.2a)$$

that is, it will decrease exponentially in time. Using expression (3.2), equation (6.2a) can be written as $w(t) \approx \exp(-t/T_c)$, where

$$T_c = \frac{N}{\nu}\,\frac{\sqrt{\pi}}{4}\,\frac{e^S\,\mathrm{erf}(S)}{S\sqrt{S}}\qquad (6.2b)$$

is the characteristic time for the evolution of RI. Here $S = Ns$ as in Chapter 2. The dependence of T_c on S is illustrated in Figure 6.1(b). T_c always increases with increasing population size N and decreasing overall mutation rate ν. Because ν increases with the number of loci, increasing $\mathcal{L}$ will decrease the characteristic time T_c. The dependence of T_c on the strength of selection at a single locus is more complicated: for each N there is an "optimum" value $s_{opt} \approx 1.028/N$ that minimizes T_c, and T_c rapidly increases as s deviates from s_{opt} in either direction. Assuming that $s = s_{opt}$, T_c can be approximated as

$$T_c \approx 1.015\,\frac{N}{\nu}.\qquad (6.2c)$$

That is, T_c increases linearly with the population size N and is inversely proportional to the overall mutation rate ν. In this model, strong overall RI can be achieved by fixing a large number of weakly selected underdominant mutations. For example, let $\nu = 0.001$ (which is the case if there are $\mathcal{L} = 100$ loci with mutation rate $\mu = 10^{-5}$ per locus per generation). Then T_c is on the order of 10^5 and 10^6 generations for $N = 100$ and $N = 1,000$, respectively.

6.2.2 Divergent degeneration of duplicated genes

Werth and Windham (1991) consider two allopatric populations that originated from a split of a single genetically uniform population. Assume that the initial population had c pairs $A_i B_i$ of duplicated and functional genes ($i = 1, \ldots, c$). Let μ be the rate of mutation to nonfunctional states a_i and b_i. After a period of time on the order of $1/\mu$ generations, each gene will be present in one functional copy and one degenerate copy. Because ancestral (A_i) and descendant (B_i) copies of genes have equal chance of being degenerated by mutation, on average $c/2$ pairs of genes will experience divergent degeneration, so that the corresponding haplotypes will be $A_i b_i$ in one population and $a_i B_i$ in the other population.

Next, consider the first generation of hybrids (F_1). The genotype of F_1 hybrids at each pair of genes that have experienced divergent degeneration will be $A_i a_i / B_i b_i$. Assuming that genes are unlinked, one quarter of gametes produced by these hybrids will have a pair of nonfunctional genes $a_i b_i$. Then the proportion of gametes that have no pairs of nonfunctional genes is

$$w = \left(\frac{3}{4} \right)^{c/2} \tag{6.3}$$

(Werth and Windham 1991). This proportion can be interpreted as the fitness (fertility) of hybrids. Correspondingly, $1 - w$ is a measure of reproductive isolation between the populations. If $c = 20$, then $w = 0.06$, and if $c = 40$, then $w = 0.003$; that is, resulting reproductive isolation can be rather strong even with moderate values of c.

This approach can be extended to situations where gene duplications occur *after* the populations become spatially separated (Lynch and Force 2000). Let the rate of gene duplication be ω per genome. Assuming that all but one copy of each duplicated gene eventually become nonfunctional, the rate at which genes that have experienced divergent degeneration accumulate in a pair of populations is $2\omega \times 1/2 = \omega$, where the coefficient 2 is because there are two populations and the coefficient $1/2$ is because only 50% of pairs of loci undergo divergent degeneration. If initially there were no duplicated genes, then after t generations of divergence the proportion of F_1 gametes that have no pairs of nonfunctional genes is

$$w = \left(\frac{3}{4}\right)^{\omega t}. \tag{6.4}$$

A conservative estimate of the rate of duplication is on the order of 0.1% per gene per 1 My (Lynch 2002). Assuming a moderate genome size of 20,000 loci, ω is on the order of 20 per 1 My. Then after 1 My the expected fitness of F_1 hybrids is $w = 0.003$, that is, very small.

Overall, the rate of duplication per gene appears to be comparable with (or higher than) the rate of mutation per gene (which is on the order of $10^{-5} - 10^{-6}$ per generation). For example, in *C. elegans* the rate of gene duplication is estimated to be ~ 250 duplication events per 1 My (Lynch and Conery 2000). With higher values of ω, reproductive isolation will accumulate even faster. This suggests that divergent degeneration of duplicated genes can be a very important mechanism for the evolution of reproductive isolation (Lynch and Force 2000; Lynch and Conery 2000; Lynch 2002). Although Lynch and Force (2000) expand the model considered here in several important directions, more theoretical work is still needed.

6.2.3 Three- and four-locus models

The Markov chain approach used in the previous chapter can be applied to multilocus models as well. I will illustrate this approach using a three-locus model and a four-locus model for both allopatric and parapatric speciation within the one-population scenario. Recall from the previous chapter that in this scenario speciation is certain.

Three-locus model I assume that there are three unlinked diallelic loci and that the population is monomorphic for haplotype abc initially. Let allele a be incompatible with the pair of alleles BC in the sense that genotypes that carry a gamete aBC have reduced fitness or that individuals of different sexes carrying allele a and alleles BC have reduced probability of mating. Note that a similar type of gene interactions is observed for genes underlying male sterility in hybrids between *Drosophila sechellia* and *D. simulans*, where two genetic factors on the X chromosome from *D. sechellia* interact with a genetic factor from *D. simulans*; each of the genetic factors alone is ineffective in causing sterility, or even a partial spermatogenic effect (Cabot et al. 1994). Assume that the mutation rates from **a** to **A**, from **b** to **B**, and from **c** to **C** are equal to μ. As before, I will neglect backward mutations. In this system, the population will evolve to a state of complete RI from its initial state if haplotype **ABC** is fixed. The diagram below illustrates possible paths between the initial and final states:

$$
\begin{array}{ccc}
 & \mathbf{Abc} \longrightarrow \mathbf{ABc} & \\
\mathbf{abc} \rightrightarrows \mathbf{aBc} \quad \quad \mathbf{AbC} \longrightarrow \mathbf{ABC} \\
 & \mathbf{abC} \quad \quad \mathbf{aBC} &
\end{array}
$$

Note that the population cannot pass through the state with haplotype **aBC** fixed because of the incompatibility of alleles **a** and **BC**.

Allopatric speciation. The average waiting time T to allopatric speciation can be found using the general approach described in Box 5.1. There is also a simple intuitive way to find T. Because there are three mutable alleles, the average waiting time until the first mutant allele is fixed is $1/(3\mu)$. The average waiting time between the first and second fixations is $1/(2\mu)$ if the population state after the first fixation is **Abc**. It is $1/\mu$ if the population state after the first fixation is **aBc** or **abC**. In either case, the waiting time between the second and third fixation is $1/\mu$. Thus, the average waiting time to allopatric speciation is

$$
T = \frac{1}{3\mu} + \frac{1}{3} \times \left(\frac{1}{2\mu} + \frac{1}{\mu} + \frac{1}{\mu} \right) + \frac{1}{\mu} = \frac{13}{6\mu}. \tag{6.5}
$$

This time is slightly longer than the average time to allopatric speciation in the original BDM model which is $2/\mu$ (see equations (5.5) on page 132).

Parapatric speciation. The parapatric speciation case can be treated in a similar way, but the treatment is somewhat more cumbersome because one has to analyze a seven-state Markov chain. In the parapatric case, the above diagram will have additional arrows pointing backward to the preceding states (excluding the absorption state **ABC**). The rates of all forward transitions are equal to the mutation rate μ, whereas the rates of all backward transitions are equal to the migration rate m. If there is no selection for local adaptation,

$$T = \frac{2m^3 + 10\mu m^2 + 19\mu^2 m + 13\mu^3}{2\mu^3(2m + 3\mu)} \approx \frac{m^2}{2\mu^3}, \qquad (6.6)$$

where the approximation is valid if $\mu \ll m$. For example, if $\mu = 10^{-5}$ and $m = 0.01$, then the average time to parapatric speciation is $T \approx 5 \times 10^{10}$ generations. This time is much longer than the waiting time to speciation in the corresponding BDM model (see equations (5.11) on page 138). This is expected because more mutations have to be fixed for RI.

Four-locus model Here, I assume that there are four diallelic loci and that the population is monomorphic for haplotype abcd initially. Let allele a be incompatible with allele B and allele c be incompatible with allele D; that is, there are two incompatible pairs of alleles. Assume that the mutation rates from the alleles represented by the lowercase letters to alleles represented by the uppercase letters are equal to μ. As before, I will neglect backward mutations. In this system, the population will evolve to a state of complete RI from its initial state if one of the following four genotypes is fixed: ABcd, abCD, ABCd, or AbCD. The diagram below illustrates possible paths between the initial and final states:

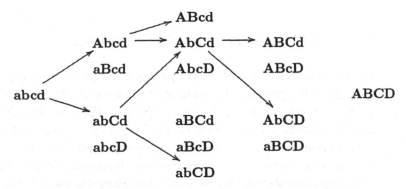

Note that the population cannot pass through a number of the states because of the assumptions of incompatibility, and that complete RI does not require fixing mutant alleles at all four loci (that is, reaching the state ABCD).

Allopatric speciation. Because there are two mutant alleles that can be fixed, the average waiting time until the first mutant allele is fixed is $1/(2\mu)$. The waiting time between the first and second fixation is

$1/(2\mu)$ as well. If the third fixation is necessary for RI (which is the case if the population state after the second fixation is AbCd), then the waiting time between the second and third fixation is $1/(2\mu)$. The probability that the population passes through the state AbCd is one half. Thus, the average waiting time to allopatric speciation is

$$T = \frac{1}{2\mu} + \frac{1}{2\mu} + \frac{1}{2} \times \left(\frac{1}{2\mu} + 0\right) = \frac{5}{4\mu}. \qquad (6.7)$$

This time is smaller by almost 40% than the time to allopatric speciation in the original BDM model (see equations (5.5) on page 132).

Parapatric speciation. To find the waiting time to parapatric speciation, one has to analyze an eight-state Markov chain. One can show that with no selection for local adaptation,

$$T = \frac{m^2 + 4\mu m + 5\mu^2}{2\mu^2(m + 2\mu)} \approx \frac{m}{2\mu^2}. \qquad (6.8)$$

This time is equal to half of that in the BDM model (see equations (5.11) on page 138).

Notice that in most models considered so far the waiting time to speciation decreases with increasing the number of loci involved. This is due to the simple fact that speciation in these models is driven by mutation and that more loci means higher overall mutation rates.

6.2.4 Accumulation of genetic incompatibilities

The examples considered above (as well as in the previous chapter) show that if the genetic architecture of RI is known, the dynamics of speciation can be modeled, at least in principle. However, first, the genetic architecture of RI is never known completely, and, second, the mathematical treatment becomes extremely complicated as the number of genes involved increases. This complexity is unavoidable because in real organisms there are at least hundreds of genetic factors affecting RI at the earliest steps of its evolution, and these factors can interact in a very complicated way (e.g., Wu and Palopoli 1994; Davis and Wu 1996; Naveira and Masida 1998; Wu and Hollocher 1998; Coyne and Orr 1998; Orr and Presgraves 2000; Orr and Turelli 2001; Wu 2001). Therefore, rather than studying dynamics of speciation *given a specific genetic architecture*, it becomes much more fruitful to look at the dynamics of speciation expected on average. One such approach was introduced by Orr (Orr 1995; Orr and Orr 1996; Orr and Turelli 2001), who used it to study allopatric speciation. The essence of this approach is to reduce the complexity of fitness landscapes underlying RI to simpler effects of genetic incompatibilities of certain types that arise with certain probabilities. In what follows, two (or more) alleles are called incompatible if their joint presence in an individual's genotype or in the genotypes of a (potential) mating pair results in reduction of a fitness component. I will build on Orr's earlier results to analyze how RI between populations is expected to depend on the genetic distance between them. I will also consider the properties of fitness landscapes implied by these models.

Let us consider two isolated diploid populations that have evolved from a common ancestor and now differ in d loci. Assume that divergence was occurring by single locus substitutions and involved neither low-fitness intermediate states nor multiple substitutions in the same loci. In this case, the genotypes present in the populations at different stages of their divergence can be arranged in an evolutionary path that can be represented without any loss of generality as:

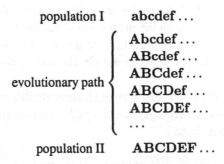

Here, the lowercase letters represent alleles present in population I and the uppercase letters represent alleles present in population II. The (unknown) ancestral genotype is one of the genotypes on the evolutionary path. All combinations of alleles present on the evolutionary path are compatible by definition. For example, allele A must be compatible with allele c and the pair of alleles AB must be compatible with the pair of alleles cd because they all are present in genotype $ABcdef \ldots$ which is one of the genotypes on the evolutionary path. However, other combinations of alleles can potentially be incompatible and can result in RI between populations I and II (e.g., alleles a and B or pairs ab and CD). Some of these combinations may have never been produced during evolutionary divergence while some others may have been produced, only to be rejected by natural selection.

Pairwise incompatibilities Following Orr (1995), let us assume that each pair of alleles not present on the evolutionary path have a fixed probability ε of being incompatible. Overall, there are $d(d-1)$ pairs comprising alleles from different populations. Observe that allele A must be compatible with $d-1$ alleles b, c, d, e, f, ..., allele B must be compatible with $d-2$ alleles c, d, e, f, ..., and so on. Summing up, one finds that $d(d-1)/2$ pairs of alleles must be compatible. The remaining $d(d-1)/2$ pairs of alleles can potentially be incompatible. The actual number of incompatibilities between the populations is a random variable that has a binomial distribution with parameters ε (the success probability) and $d(d-1)/2$ (the number of tries). Therefore, the expected number of pairwise incompatibilities is

$$J = \varepsilon \, \frac{d(d-1)}{2}. \qquad (6.9)$$

This number increases quadratically with the genetic distance d between the populations. Orr (1995) called this rapid increase in the number of incompatibilities the "snowball effect." Note that the above expression for J does not depend on the proportion of substitutions that occur in each population nor on the order of substitutions. How does the snowball effect translate into RI between the populations? I will consider two models.

Model I. Assume first that a single incompatibility causes complete RI. Then the probability $w(d)$ that two populations at distance d are not reproductively isolated is

$$w(d) = (1-\varepsilon)^{d(d-1)/2} \approx \exp\left(-J\right) \approx \exp\left(-\varepsilon \, \frac{d^2}{2}\right), \qquad (6.10\mathrm{a})$$

where the approximations assume that ε is small, whereas d is large. Equation (6.10a) shows that the dependence of the function of reproductive compatibility w on J is approximately exponential (compare to equation (6.1) for the Walsh model) and that on d is closely approximated by a Gaussian curve. The probability of RI is the probability that at least one incompatibility occurs and is given by $1 - w(d)$.

How many substitutions are necessary for RI? The probability $f(k)$ that RI is caused by the k-th substitution is equal to $w(k-1) - w(k)$, which can be approximated by the derivative of $1 - w(d)$ with respect to d evaluated at $d = k$. Then the average number of substitutions, K,

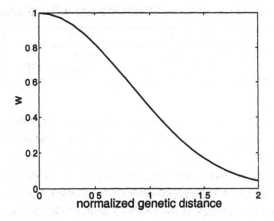

FIGURE 6.2. Probability of no RI w as a function of normalized genetic distance, d/K, for the model of speciation due to a single incompatibility between a pair of alleles

required for RI is approximately

$$K = \sum_{k=1}^{\infty} k \, f(k) \approx \int_0^{\infty} x f(x) dx \approx \sqrt{\frac{\pi}{2\varepsilon}}. \qquad (6.10b)$$

For example, with $\varepsilon = 10^{-3}$, $K \approx 40$ substitutions and with $\varepsilon = 10^{-5}$, $K \approx 400$. Figure 6.2 illustrates the shape of the function of reproductive compatibility $w(d)$ using a normalized genetic distance d/K. On this scale, $w(d)$ does not depend on ε.

Of course, the actual number of substitutions until reproductive isolation is a random variable. It is easy to show that the variance of the number of substitutions required for speciation is

$$\mathrm{var}(K) \approx \int_0^{\infty} (x - K)^2 f(x) dx = \frac{4 - \pi}{2\varepsilon}. \qquad (6.10c)$$

The corresponding coefficient of variation is

$$\mathrm{CV}(K) = \frac{\sqrt{\mathrm{var}(K)}}{K} \approx \sqrt{\frac{4 - \pi}{\pi}} \approx 0.52, \qquad (6.10d)$$

which is large. Notice that $\mathrm{CV}(K)$ does not depend on ε.

What are the properties of fitness landscapes implied by this model? If one randomly chooses a homozygous genotype, the probability that it is "fit" (i.e., does not have any incompatible pairs of gene) is

$$\mathcal{P} = (1 - \varepsilon)^{\frac{\mathcal{L}(\mathcal{L}-1)}{2}} \approx \exp\left(-\varepsilon \frac{\mathcal{L}(\mathcal{L}-1)}{2}\right). \qquad (6.11)$$

TABLE 6.1. The probability of being fit $\mathcal{P}$ predicted by equation (6.11).

$\mathcal{L}$	$\varepsilon = 10^{-2}$	10^{-3}	10^{-4}	10^{-5}	10^{-6}
100	2.48×10^{-20}	0.01	0.61	0.95	0.99
1,000	6.02×10^{-2180}	9.15×10^{-216}	2.02×10^{-20}	0.01	0.61

With probability $1 - \mathcal{P}$, a randomly chosen genotype has at least one pair of incompatible genes and, thus, has zero fitness. Table 6.1 illustrates the dependence of the probability of being fit $\mathcal{P}$ on the number of loci $\mathcal{L}$ and the probability of pairwise incompatibility ε. Notice that a change in $\mathcal{L}$ or ε can cause a disproportionately large change in $\mathcal{P}$.

The fitness landscape in this model resembles the one in the Russian roulette model (see Chapter 4) and is "holey". However, in contrast to the Russian roulette models, here the fitness landscape is correlated. To get ideas about the correlation structure, let us first consider the probability $\mathcal{P}_d$ that a randomly chosen genotype at distance d from a fit reference genotype is fit as well. Because there are $d(d-1)/2$ pairs comprising genes not present in the reference genotype and $d(\mathcal{L} - d)$ pairs comprised by one gene present and one gene not present in the reference genotype, probability $\mathcal{P}_d$ can be written as

$$\mathcal{P}_d = (1 - \varepsilon)^{\frac{d(d-1)}{2} + d(\mathcal{L}-d)} \approx \exp(-\varepsilon \mathcal{L} d), \qquad (6.12)$$

where the approximate equality assumes $d \ll \mathcal{L}$.

Two randomly chosen genotypes at distance d are both fit with probability $\mathcal{P}\mathcal{P}_d$, and the overall average and variance in fitness are $\mathcal{P}$ and $\mathcal{P}(1 - \mathcal{P})$, respectively. Therefore, in the model under consideration, the correlation function of the fitness landscape (defined by equation (4.1) on page 83) can be represented as

$$\rho(d) = \frac{\mathcal{P}_d - \mathcal{P}}{1 - \mathcal{P}} \approx \frac{\exp(-\varepsilon \mathcal{L} d) - \mathcal{P}}{1 - \mathcal{P}}. \qquad (6.13)$$

Figure 6.3 illustrates the decay of the correlation function with d. The decay is very rapid if $\varepsilon > 0.001$.

The correlation length of the fitness landscape, which was defined by equation (4.2) on page 83, is approximately

$$l_c = \frac{1}{\varepsilon \mathcal{L}}. \qquad (6.14)$$

The expression in the denominator is approximately the average number of incompatibilities between two genotypes that are one-step neighbors. If $\varepsilon \mathcal{L}$ is small, l_c is large and the landscape is highly correlated.

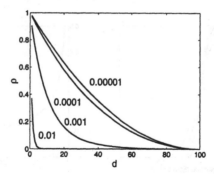

FIGURE 6.3. The correlation function of the fitness landscape for different values of ε implied by the model in which a single pairwise incompatibility causes complete RI.

One can also find the neutrality $\overline{\mathcal{N}}$ of the fitness landscape (which was defined as the average number of fit one-step neighbors per fit genotype; see page 88). Because each genotype has $\mathcal{L}$ one-step neighbors each of which is independently fit with probability $\mathcal{P}_1$ (see equation (6.12)),

$$\overline{\mathcal{N}} = \mathcal{L}\,(1 - \varepsilon)^{\mathcal{L}-1} \approx \mathcal{L}\exp(-\varepsilon\mathcal{L}). \qquad (6.15)$$

I conjecture that in the limit of large $\mathcal{L}$, a fit genotype will be connected to a giant component (similar to those in the Russian roulette models) if $\overline{\mathcal{N}} > 1$. The last inequality can be expressed in terms of the probability of incompatibility ε as

$$\varepsilon < \varepsilon_c = \frac{\ln \mathcal{L}}{\mathcal{L}}. \qquad (6.16a)$$

This shows that the higher the dimensionality of the genotype space, the smaller ε has to be for the giant component of fit genotypes to form. Notice that inequality (6.16) expresses the percolation threshold in terms of the probability that a randomly chosen pair of alleles is incompatible, whereas for the Russian roulette models considered in Chapter 4 it was expressed in terms of the probability that a randomly chosen genotype is viable. Using equation (6.11), inequality (6.16a) can be rewritten as

$$\mathcal{P} > \mathcal{P}_c = \mathcal{L}^{-\mathcal{L}/2}. \qquad (6.16b)$$

The percolation threshold $\mathcal{P}_c$ decreases exponentially with $\mathcal{L}$.

Model II. The derivations above assume that a single incompatibility results in complete RI. Orr (1995) and Orr and Turelli (2001) considered more complex models in which complete RI occurs from cumulative effects of multiple pairwise incompatibilities. Assume, as before, that any

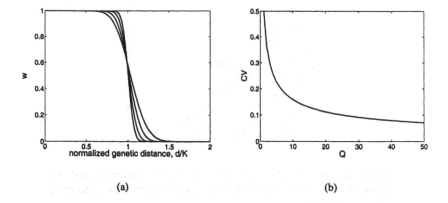

(a)   (b)

FIGURE 6.4. Multiple pairwise incompatibilities. (a) The probability of no
RI. Different lines correspond to $Q = 10$ (the shallowest), $20, 40, 80$ (the
steepest). (b) Coefficient of variation $CV(K)$ of the number of substitutions
required for speciation.

two alleles can be incompatible with probability ε. Let complete RI occur when Q incompatibilities separate the populations. Consider two
populations that differ at d loci. If d is large and ε is small, the distribution of the number of pairwise incompatibilities is approximately
Poisson with parameter J given by equation (6.9). Then the probability
of no complete RI between the populations is

$$w(d) = \sum_{i=0}^{Q-1} \exp(-J)\frac{J^i}{i!} = \frac{\Gamma(Q, J)}{\Gamma(Q)} \approx \frac{\Gamma\left(Q, \varepsilon\frac{d^2}{2}\right)}{\Gamma(Q)}, \qquad (6.17)$$

where $\Gamma(\cdot)$ is the gamma function and $\Gamma(\cdot, \cdot)$ is the incomplete gamma
function (e.g., Gradshteyn and Ryzhik 1994).

The average, variance, and the coefficient of variation of the number
of substitutions required for RI are

$$K = \sqrt{\frac{2}{\varepsilon}} \frac{\Gamma(Q + 1/2)}{\Gamma(Q)} \approx \sqrt{\frac{2Q}{\varepsilon}}, \qquad (6.18a)$$

$$\text{var}(K) = \frac{2}{\varepsilon}\left[Q - \frac{\Gamma(Q + 1/2)^2}{\Gamma(Q)^2}\right] \approx \frac{1}{2\varepsilon}, \qquad (6.18b)$$

$$CV(K) = \sqrt{\frac{\Gamma(Q + 1)\Gamma(Q)}{\Gamma(Q + 1/2)^2} - 1} \approx \frac{1}{2\sqrt{Q}}, \qquad (6.18c)$$

where the approximate equalities assume that Q is not small (> 5).
Notice that $CV(K)$ does not depend on ε. Figure 6.4 illustrates the

dependence of the function of reproductive compatibility $w(d)$ on normalized genetic distance d/K (on this scale, $w(d)$ does not depend on ε) and of the coefficient of variation $CV(K)$ on the number of incompatibilities required for complete RI. If Q is large, the coefficient of variation is small, and function $w(d)$ has a steplike form. I will call this property the *threshold effect*.

Analyzing the correlation structure of the fitness landscape implied by this model would require one to specify fitnesses of genotypes with different numbers of incompatibilities. I will not attempt this here.

Complex incompatibility Generalizing the above approach for more complex incompatibilities is straightforward. Assume that any combination of k alleles each taken from a different locus can be incompatible with probability ε_k if it is not present on the evolutionary path (see the diagram on page 159). As before, *incompatibility* means epistatic interactions between the alleles resulting in a loss of fitness.

The overall number of different combinations of k alleles from d diallelic loci is $2^k \binom{d}{k}$, where $\binom{d}{k}$ is the binomial coefficient.[1] Of these, $2\binom{d}{k}$ combinations comprise alleles from the same population and, thus, must be compatible. Another $(k-1)\binom{d}{k}$ combinations are present on the evolutionary path and, thus, are compatible as well. For example, let $k = 3$. Consider three loci A, B, and C. Out of the eight possible combinations of alleles, which are abc, Abc, aBc, abC, ABc, AbC, aBC, and ABC, two combinations, abc and ABC, comprise alleles from the same population, and two other combinations, Abc and ABc, are present on the evolutionary path. Thus, the overall number of potentially incompatible combinations is $\binom{d}{k}(2^k - k - 1)$ and the expected number of incompatibilities between two populations that differ at d loci is

$$J_k = \varepsilon_k \binom{d}{k}(2^k - k - 1), \qquad (6.19)$$

(Welch 2004). I note that the equation for J_k reported by Orr (1995, Eq. 10) and later used by Gavrilets (2003b, Eq. 1) missed the last factor of the product and, thus, is erroneous. However, because the missed term does not depend on d, the error did not affect the qualitative conclusions

---

[1] $\binom{d}{k} = \frac{d!}{k!(d-k)!}$ and gives the number of combinations of k objects chosen from a set of d objects.

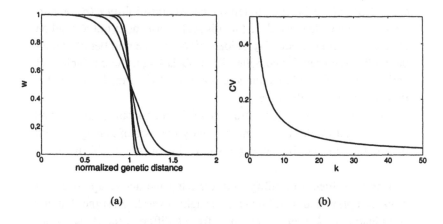

(a)   (b)

FIGURE 6.5. Single complex incompatibility model. (a) The probability of
no RI. Different lines correspond to $k = 5$ (the shallowest), $10, 20, 30$ (the
steepest). (b) Coefficient of variation $CV(K)$ of the number of substitutions
required for speciation as a function of k.

of these authors. The value of J_k increases as the k-th order of genetic
distance d. That is, the snowball effect is much more pronounced with
larger values of k. I consider two models of complex incompatibilities.

Model III. Assume that a single complex incompatibility results in
complete RI. Then the probability of no RI after the d-th substitution is

$$w(d) = (1 - \varepsilon_k)^{\binom{d}{k}(2^k - k - 1)} \approx \exp\left(-J_k\right) \approx \exp\left(-\nu_k \, d^k\right), \quad (6.20)$$

where $\nu_k = \varepsilon_k(2^k - k - 1)/k!$ and the approximate equalities assume
that ε_k is small and d is large.

The average, variance, and the coefficient of variation of number of
substitutions required for RI are

$$K = \nu_k^{-1/k} \frac{\Gamma(1/k)}{k}, \quad (6.21a)$$

$$\mathrm{var}(K) = \nu_k^{-2/k} \frac{2k\Gamma(2/k) - \Gamma(1/k)^2}{k^2}, \quad (6.21b)$$

$$CV(K) = \frac{\sqrt{2k\Gamma(2/k) - \Gamma(1/k)^2}}{\Gamma(1/k)} \approx \frac{1}{k}, \quad (6.21c)$$

where the approximate equality is very good for all k. Notice that
$CV(K)$ does not depend on ε_k. Figure 6.5(a) illustrates the shape of
$w(d)$ using a normalized genetic distance d/K. As before, on this scale

TABLE 6.2. The probability $\mathcal{P}$ that a randomly chosen homozygous genotype is fit.

$\mathcal{L}$	k	ε	$\mathcal{P}$
100	4	10^{-4}	0.51×10^{-170}
		10^{-5}	0.93×10^{-17}
		10^{-6}	0.0198
	8	10^{-9}	0.15×10^{-80}
		10^{-10}	0.83×10^{-8}
		10^{-11}	0.1555
1,000	4	10^{-8}	0.13×10^{-179}
		10^{-9}	0.10×10^{-17}
		10^{-10}	0.0159
	8	10^{-17}	0.19×10^{-104}
		10^{-18}	0.34×10^{-10}
		10^{-19}	0.0897

$w(d)$ does not depend on ε_k. With large k, the coefficient of variation $\mathrm{CV}(K)$ is small, and function $w(d)$ has a steplike form (threshold effect). Figure 6.5(b) illustrates the dependence of $\mathrm{CV}(K)$ on k.

Characteristics of the fitness landscapes implied by this model can also be found. The probability that a randomly chosen homozygous genotype is fit (that is, has no incompatible genes) is

$$\mathcal{P} = (1 - \varepsilon_k)^{\binom{\mathcal{L}}{k}} \approx \exp\left[-\varepsilon_k \binom{\mathcal{L}}{k}\right]. \qquad (6.22)$$

Table 6.2 shows that the value of $\mathcal{P}$ is very sensitive to parameters.

The probability of being fit at distance d from a fit reference genotype can be written as

$$\mathcal{P}_d = (1 - \varepsilon_k)^{\binom{\mathcal{L}}{k} - \binom{\mathcal{L}-d}{k}} \approx \exp\left\{-\varepsilon_k \left[\binom{\mathcal{L}}{k} - \binom{\mathcal{L}-d}{k}\right]\right\}. \qquad (6.23)$$

The correlation function ρ is still given by equation (6.13) and can be evaluated numerically. Figure 6.6 illustrates the dependence of ρ on genetic distance d. This figure shows that the correlation function drops quickly to zero if $\varepsilon_k \geq 10^{-5}$.

The correlation length of the landscape can be approximated as

$$l_c = \left\{\varepsilon_k \left[\Psi(\mathcal{L}+1) - \Psi(\mathcal{L}+1-k)\right]\binom{\mathcal{L}}{k}\right\}^{-1} \approx \left[\varepsilon_k \binom{\mathcal{L}}{k-1}\right]^{-1}, \qquad (6.24)$$

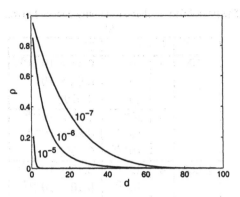

FIGURE 6.6. The correlation function of the fitness landscape for different ε in the model of RI caused by a single complex incompatibility.

where $\Psi(\cdot)$ is the digamma function (Gradshteyn and Ryzhik 1994). As in the case of pairwise incompatibilities, the correlation length is approximately the inverse of the expected number of incompatibilities between one-step neighbors. For example, if $\mathcal{L} = 100, k = 4$, and $\varepsilon_k = 10^{-3}$, then $l_c = 0.06$. In this case, the correlation length is very short. Notice that decreasing ε_k by one order of magnitude will increase l_c by one order of magnitude.

One can also find the neutrality $\overline{\mathcal{N}}$ of the fitness landscape. Using the same reasoning as in deriving equation (6.15), one finds that

$$\overline{\mathcal{N}} = \mathcal{L} \left(1 - \varepsilon_k\right)^{\binom{\mathcal{L}-1}{k-1}} \approx \mathcal{L} \, \exp\left[-q\binom{\mathcal{L}-1}{k-1}\right]. \tag{6.25}$$

I conjecture that in the limit of large $\mathcal{L}$, a fit genotype will belong to the giant component if $\overline{\mathcal{N}} > 1$. This condition can be rewritten as

$$\varepsilon_k < \varepsilon_c = \frac{\ln \mathcal{L}}{\binom{\mathcal{L}-1}{k-1}}. \tag{6.26}$$

The denominator of the ratio in the right-hand side of equation (6.26) can be thought of as the number of within-genotype incompatibilities that a given locus can be potentially involved in. The percolation threshold ε_c decreases dramatically with k. For example, if $\mathcal{L} = 100$, then $\varepsilon_c = 0.047, 0.94 \times 10^{-3}, 0.29 \times 10^{-4}$, and 0.12×10^{-5} for $k = 2, 3, 4$, and 5, respectively. This implies that complex incompatibilities may prevent formation of the giant cluster.

Model IV. The above derivations can be generalized for the case of multiple complex incompatibilities. Assume that any set of k alleles can be incompatible with probability ε_k. If the genetic distance betwen the two populations d is large and ε_k is small, the distribution of the number of incompatibilities is approximately Poisson with parameter J_k given by equation (6.19). Then the probability of no complete RI is approximately

$$w(d) = \sum_{i=0}^{Q-1} \exp(-J_k)\frac{J_k^i}{i!} = \frac{\Gamma(Q, J_k)}{\Gamma(Q)}, \tag{6.27}$$

which has the same form as equation (6.17) describing the case of pairwise incompatibilities. The average, variance, and the coefficient of variation of the number of substitutions required for RI are

$$K = \nu_k^{-1/k}\frac{\Gamma(Q+1+1/k)}{\Gamma(Q+1)} \approx \left(\frac{k!}{\varepsilon_k}\right)^{1/k} Q^{1/k}, \tag{6.28a}$$

$$\mathrm{var}\, K = \nu_k^{-2/k}\frac{\Gamma(Q+1+\frac{2}{k})\Gamma(Q+1) - \Gamma(Q+1+\frac{1}{k})^2}{\Gamma(Q+1)^2}, \tag{6.28b}$$

$$\mathrm{CV}(K) = \sqrt{\frac{\Gamma(Q+1+2/k)\Gamma(Q+1)}{\Gamma(Q+1+1/k)^2} - 1} \approx \frac{1}{k\sqrt{Q}}, \tag{6.28c}$$

where the approximations assume that Q is sufficiently large (≥ 4). Because increasing k decreases $\mathrm{CV}(K)$, the transition from no reproductive isolation to complete RI occurs faster with larger k (that is, the threshold effect is much more pronounced). Figure 6.7 illustrates the dependencies of fitness function w and of the coefficient of variation $\mathrm{CV}(K)$ on parameters. Notice that, as before, neither the coefficient of variation, nor the function $w(d)$ (if expressed using the normalized genetic distance d/K) depend on ε_k.

To find the correlation structure of the fitness landscapes in this model, one would need to specify fitnesses for genotypes with a different number of incompatibilities, which I will not attempt here.

Comments Model IV includes the other three models as special cases. Model I corresponds to $k = 2$ and $Q = 1$, model II corresponds to $k = 2$ and arbitrary Q, and model III corresponds to $Q = 1$ and arbitrary k.

The derivations above concentrated on fitness of homozygotes and did not specify fitnesses of the much more numerous heterozygotes. In principle, if heterozygotes have low fitness, then the evolution along a percolating (nearly) neutral network of high-fitness homozygotes can

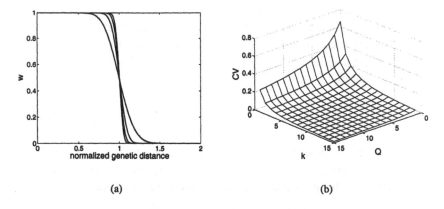

(a)   (b)

FIGURE 6.7. Multiple complex incompatibility model. (a) The probability
of no RI with $Q = 10$. Different lines correspond to $k = 2$ (the shallowest),
$4, 6, 8$ (the steepest). (b) Coefficient of variation, $CV(K)$ of the number of
substitutions required for speciation as function of Q and k.

be prevented by selection even if such a network exists. However, this
would require underdominance. Existing data on genetic incompatibilities show that alleles causing intrinsic postzygotic isolation tend to act
recessively in hybrids (Wu and Palopoli 1994; Orr 2001). For example,
according to Haldane's rule, if in F_1 hybrids one sex is absent, rare,
or sterile, that sex is nearly always the one having heterogametic sex
chromosomes (Haldane 1922). This fact is most easily explained by
the recessive effects of sex-linked genes (e.g., Muller 1942; Orr 1997).
In the absence of underdominance, heterozygotes between two fit homozygotes at distance $d = 1$ are expected to be fit as well. Therefore,
one does not have to worry about heterozygotes when computing, for
example, the percolation threshold and correlation function.

   Genetic distance d between two homozygous diploid populations
controls the number of heterozygous loci in hybrid offspring. Therefore,
the function of reproductive compatibility $w(d)$ is similar to fitness
functions implied by the diploid models of constant viability selection in
which fitness depends on the number of heterozygous loci (e.g., Hastings
1989; Bürger 2000; Christiansen 2000). This is very convenient because
one can use standard results on the latter models in analyzing the models
based on incompatibilities.

The models considered in this section share the following properties:

(i) The expected number of incompatibilities increases as a power function of the number of genetic differences (the snowball effect; see equations (6.9) and (6.19)).

(ii) The probability that two genotypes are not reproductively isolated decreases as an exponential function of the expected number of incompatibilities between them (see equations (6.10a) and (6.20a)).

(iii) As genetic divergence increases, the strength of reproductive isolation is expected to undergo a rapid transition from weak to strong especially if RI is due to multiple complex incompatibilities (the threshold effect; see Figures 6.4, 6.5, and 6.7).

(iv) The underlying fitness landscapes belong to a class of correlated holey landscapes.

The snowball effect implies that the probability that a genetic introgression from a different species results in an incompatibility increases with genetic distance between the two species. Kondrashov et al. (2002) used data on human pathogenic mutations to estimate this probability for single amino acid substitutions. Their data show that about 10% of amino acids that are "wild type" in nonhuman proteins would be seriously deleterious if introduced into the homologous site of the human protein. Contrary to the expectations based on the snowball effect, this probability was independent of the genetic distance between the human and nonhuman proteins. Both the snowball effect and the threshold effect are the consequences of the explicit assumption that alleles (or amino acids) forming an incompatible combination with a given allele (or amino acid) can come from any other locus (site). In contrast, if an allele from a specific locus can be incompatible only with an allele from another specific locus, the number of incompatibilities will grow linearly with genetic distance. In this case, the probability that an introgression results in incompatibility will not depend on genetic distance (Kondrashov et al. 2002). Alternatively, if all relevant loci interact in determining fitness (i.e., if $k = d$), fitness landscape will be uncorrelated and there will be no correlation between the number of incompatibilities (or fitness) and genetic distance.

Can one estimate the number of incompatibilities? If the probability of an incompatibility (pairwise or complex) is very low, a small genetic change will be expected to result in no more than one incompatibility. In this case, the number of genetic changes resulting in fitness reduction represents an estimate of the number of incompatibilities. For example, a powerful experimental approach for studying genetics of RI is to introduce segments of chromosomes from one species (or race) into the genetic background of another closely related species or race (e.g., Wu and Palopoli 1994; Wu 2001). The number of segments resulting in reduction of fitness components can be used as a lower boundary on the number of incompatibilities between the two species (Orr 1995). For example, more than 100 genetic factors affect the sterility of male hybrids between *D. simulans* and *D. mauritiana* (Coyne 1984; Wu et al. 1996). Thus, in this case $J \approx 10^2$.

How many incompatibilities are required for speciation? In trying to answer this question one should keep in mind that incompatibilities continue to accumulate even after the threshold genetic distance K has been exceeded. Therefore, the number of incompatibilities between two species can be used only as the upper boundary on K.

Estimating the probability of incompatibility ε requires additional assumptions. If all incompatibilities are pairwise, then ε can be found from genetic distance d separating two species:

$$\varepsilon \approx \frac{2J}{d^2}. \tag{6.29}$$

Using this estimate and assuming that $d = 1,200$ (which was thought to be typical for different species of *Drosophila*) and $J = 1$, Orr (1995) estimated that $\varepsilon \approx 10^{-6}$. Orr and Turelli (2001) attempted to estimate the probability of incompatibility between two nonsynonymous site substitutions. Using $J = 120$ genetic factors as estimated by Wu et al. (1996) and estimating that the two species of *Drosophila* have diverged by $d \approx 48,000$ replacement substitutions, they estimate that $\varepsilon \approx 10^{-7}$.

These estimates of ε should be treated with extreme caution. First, all numerical values involved are no more than very crude approximations. Second, it is generally unknown whether the incompatibilities are pairwise or complex. The knowledge of this is crucial for a meaningful interpretation of data. For example, using the approach of Orr and Turelli (2001) but assuming that incompatibilities are within sets of $k = 3$ loci, one finds that $\varepsilon_3 = 6.5 \times 10^{-12}$, a change of five orders of magnitude relative to the case of $k = 2$. If $k = 4$, then $\varepsilon_4 = 5.4 \times 10^{-16}$ which is a change of four more orders of magnitude. Third, there is no guarantee that the assumption of equal probability of incompatibility between all possible sets of loci is valid. To see the effects of the violation of this assumption, let us assume that the genome is divided into

C compartments of interacting loci, such that incompatibilities can occur only between substitutions within the same compartment (Orr and Turelli 2001, Appendix). Assuming that interactions are pairwise, the overall number of incompatibilities can be written as

$$J = \sum_i \varepsilon_{(i)} \frac{d_i^2}{2}, \tag{6.30}$$

where $\varepsilon_{(i)}$ is the probability of a pairwise incompatibility between the loci in the i-th compartment, and d_i is the genetic distance between the two populations computed over the i-th compartment. Let $f_i = d_i / \sum d_i$ be the proportion of substitutions that occur in compartment i. Then the following equality is true: $2J/d^2 = \varepsilon_e$, where

$$\varepsilon_e = \sum \varepsilon^{(i)} f_i^2 \tag{6.31}$$

is the *effective probability of incompatibility* (Orr and Turelli 2001). For example, if approximately the same number of substitutions occurs in each compartment (i.e., $f_i = 1/C$), then $\varepsilon_e = \bar{\varepsilon}/C$, where $\bar{\varepsilon}$ is the average of $\varepsilon_{(i)}$. That is, the average probability of incompatibility will be C times larger than suggested by the ratio $2J/d^2$. Next assume that $f_i = 1/C$, but incompatibilities are possible only within $c \, (< C)$ compartments with equal probability (that is, $\varepsilon_{(i)} = \varepsilon$ for c compartments and $\varepsilon_{(i)} = 0$ for the remaining $C - c$ compartments). Then $\varepsilon_e = (\varepsilon/C) \times (c/C)$, which is (much) smaller than the estimate of ε_e in the previous model. Changing c and/or C (which are generally unknown) can easily change ε_e by orders of magnitude. All these limitations make any meaningful estimation of the probability of incompatibility extremely difficult.

The general structure of the models considered in the previous subsection can be represented as

genetic differences $\rightarrow$ incompatibilities $\rightarrow$ fitness reduction.

This resembles the structure of models describing dynamics of quantitative characters under selection, which can be represented as

genetic differences $\rightarrow$ phenotypic differences $\rightarrow$ fitness differences.

A major difference between these two approaches concerns the intermediate stage: while in quantitative genetics, phenotypic differences can be observed (and measured) independently of fitness, incompatibilities are defined with respect to their effects on fitness and thus cannot be observed (or measured) independently. In quantitative genetic models, what really controls the evolutionary dynamics of populations is the relationship between genotype and fitness mediated via phenotypic differences. In a similar way, what really controls dynamics of speciation is the relationship betwen genetic differences and fitness reduction represented by the function on reproductive compatibility $w(d)$.

What are the general properties of this function? In the models considered above, w is a decreasing function of d. In fact, a general feature of *all* speciation models is that the probability of RI between two organisms increases with genetic divergence over a set of specific loci. This is exactly what our intuition tells us: combining genes from increasingly different organisms should result in increasing the probability of malfunctioning of the resulting mating pairs or hybrids.

Direct empirical data on $w(d)$ are scarce. An important example is provided by studies of relationships between genetic divergence and recombination rates in *Bacillus* (Roberts and Cohan 1993; Cohan 1995; Zawadzki et al. 1995; Majewski and Cohan 1998, 1999), yeast (Datta et al. 1997), and *E. coli* (Vulić et al. 1997). For these organisms, the rate of recombination can be interpreted as a measure of genetic compatibility between different organisms. The data show an exponential decrease in the rate of recombination with sequence divergence. Therefore, sexual isolation increases with genetic distance, and the probability of compatibility can be written as an exponential function $w(d) \approx \exp(-ad)$ with $a > 0$, which is similar to that in the Walsh model and in the model of divergent degeneration of duplicated genes.

Finally, I note that Coyne et al. (2000), Orr and Turelli (2001), and Turelli et al. (2001) criticized the approach for speciation modeling based on functions of reproductive incompatibility $w(d)$ and simultaneously praised the models built on the notion of incompatibility. These authors apparently did not realize that both classes of models are equivalent from both mathematical and biological points of view.

6.2.5 *Allopatric speciation*

The results on incompatibility models described above show that the degree of RI between an ancestral population state and its current state is basically a function of the genetic divergence between these two states over a specific set of genes, and that sufficient genetic divergence will result in strong reproductive isolation. Having specified how the degree of RI depends on genetic distance d, one can use standard population genetic approaches describing divergence by different evolutionary factors to study the dynamics of speciation. This is especially straightforward in the case of allopatric speciation.

Neglecting within-population genetic variation The simplest approach is to completely neglect genetic variation within the populations and to assume that the populations accumulate substitutions at a constant rate, say ω substitutions per generation (Orr 1995; Orr and Orr 1996; Gavrilets 2000b; Orr and Turelli 2001). In this case, genetic distance d can serve as a proxy for time and the qualitative features of the dependences of the degree of RI on time and on genetic distance should be similar. If, on average, K substitutions are required for complete RI between two populations, the average waiting time to speciation is approximately

$$T = \frac{K}{2\omega} \tag{6.32}$$

generations, where the coefficient 2 represents two populations accumulating substitutions.

Thus, the waiting time to allopatric speciation can be predicted on the basis of K alone without the need to specify fitness function $w(d)$. For example, if RI results from Q pairwise incompatibilities, then K is given by equation (6.18) and the average waiting time to speciation is

$$T = \frac{1}{\omega} \sqrt{\frac{Q}{2\varepsilon}}. \tag{6.33}$$

Orr (1995) and Orr and Turelli (2001) derived a similar expression using a different approach.

One can also find the variances of these waiting times. If substitutions accumulate at a constant rate, then the coefficient of variation CV_T of the time to speciation is the same as the coefficient of variation of the number of substitutions necessary for complete reproductive isolation (see equation (6.18c)). That is,

$$CV_T \approx \frac{1}{k\sqrt{Q}}. \tag{6.34}$$

Orr and Turelli (2001) derived a similar expression for $k = 2$ using a different approach. Interestingly, the above analysis shows that speciation times of the organisms characterized by large Q are expected to have both larger averages and narrower relative ranges (Orr and Turelli 2001).

How can one approximate the rate of substitutions ω? Assume that genes underlying RI have no other pleiotropic effects. In this case, genetic divergence will be driven exclusively by mutation and random drift, which represents a general null model. Let ν be the mutation rate per gamete per generation. If the within-population genetic variation is absent (or very low), each new mutation that is compatible with the current genetic state can be treated as selectively neutral. Then the rate of accumulation of substitutions can be written as $\omega = 2\nu_e$, where the effective mutation rate $\nu_e = \nu P$ accounts for the fact that not all one-step neighbors are fit and/or compatible (see equation (4.17) on page 110). For example, assume that there are $\mathcal{L} = 5,000$ loci that can affect RI on the earlier stages of divergence. Assuming $P \approx 1$ and $\mu = 10^{-6}$, the rate of fixation is $\omega = 0.01$. Therefore, if divergence in $K = 500$ loci results in complete RI, then, using equation (6.32), the time to speciation is only $T = 25,000$ generations.

An interesting question is how the population size affects the dynamics of speciation. Because the rate of accumulation of neutral substitutions does not depend on the population size, the approach used here predicts that the dynamics of speciation driven by mutation and random drift will not depend on the population size either. Therefore, one can argue that the average waiting time to speciation driven by mutation and random drift in a system of allopatric populations is the same whether there are two large populations or many small populations (Orr and Orr 1996; Coyne and Orr 1998).

We can also consider the case when mutant alleles underlying RI have some other pleiotropic effects resulting in them being locally advantageous. For example, in Darwin's finches of the Galápagos Islands, there is positive genetic correlation between beak and body size and characteristics of songs, such as rates of syllable repetition and frequency bandwidths (Podos 2001). Therefore, morphological adaptation may drive mating signal evolution and RI. Another example is provided by two highly specialized host races of the pea aphid, in which genes controlling adaptation to hosts also pleiotropically control (or are very tightly linked to genes controlling) assortative mating through habitat choice (Hawthorn and Via 2001). To model such situations, let us assume that each mutant allele has the same selective advantage s. In this case, one can write ω as $\omega = \mu_e \kappa_a$, where κ_a is the rate of fixation of advantageous mutations relative to that of neutral mutations (see equation (5.7) on page 134). Obviously, with selection for local adaptation, speciation is predicted to be (much) faster than in the neutral case. Large populations are more responsive to selection than small populations. Therefore, the average waiting time to speciation driven by selection in a system of allopatric populations will be shorter if there are two large populations than if there are many small populations (Orr and Orr 1996).

Comments How does the above approach fare with regard to existing data? The amount of time necessary for achieving RI has been estimated for *Drosophila* (Coyne and Orr 1989, 1997; Gleason and Ritchie 1998), frogs (Sasa et al. 1998), lepidoptera (Presgraves 2002), and birds (Price and Bouvier 2002). The most common method to evaluate time since divergence is to calculate Nei's (1972) genetic distance D_n using allozyme data. D_n is intended to measure the number of codon substitutions per locus; the value of $D_n = 1$ roughly corresponds to 5 My of divergence (Nei 1987). Postzygotic isolation is usually measured by the average of 4 binary (i.e., equal only to 0 or 1) indices indicating the presense or absence of (i) complete inviability in F_1 male offspring, (ii) complete inviability in F_1 female offspring, (iii) complete sterility in F_1 male offspring, and (iv) complete sterility in F_1 female offspring. Prezygotic isolation is characterized as one minus the ratio of the frequencies of heterospecific and homospecific matings. In *Drosophila*, allopatric speciation on average requires $D \approx 0.54$, which corresponds to $T \approx 2.7$ My (Coyne and Orr 1997). (An intriguing conclusion of Coyne and Orr (1997), the significance of which is still not clear, is that for sympatric pairs of species the required value of D is much smaller, ≈ 0.04, which corresponds to $T \approx 200,000$ years of divergence.) In frogs and lepidoptera, the required value of D is close to 1 (Sasa et al. 1998; Presgraves 2002). In birds, the loss of hybrid fertility occurs on the time scale of 5 to 15 My, whereas hybrid inviability occurring over longer time scales with some viable hybrids produced between taxa that appear to have been separated for more than 55 My (Price and Bouvier 2002).

These results suggest that equation (6.32) greatly overestimates the rate of speciation, predicting much smaller values of T than observed. One possible reason for this discrepancy is that our approximation neglected within-population genetic variation. As we will see below, a more elaborate treatment of the process of fixation in this model shows that mutant alleles are weakly selected against when rare, which significantly decreases the rate of divergence.

The dependence of the strength of RI on time revealed in the studies described above appears to be smooth and continuous rather than threshold type, which may be seen as a fact undermining the relevance of the snowball and threshold effects. However, the data represent the averages of four separate measures of RI. Moreover, the data were collected for different pairs of species, each of which most likely is characterized by its own dynamics of accumulation of RI. Therefore, the variation in times to RI revealed in these studies more likely reflects the variation among different pairs of species and among different components of RI rather than the shape of a single common function $w(d)$.

Estimating $w(d)$ requires information on the relationship between *within-species* genetic differentiation and RI. Unfortunately, existing data, which do show positive relationships between measures of RI and genetic distances, are not detailed enough to evaluate $w(d)$ quantitatively. For example, in a study of salamanders, the partial correlation between prezygotic RI and genetic divergence is statistically insignificant (Tilley et al. 1990). In studies of postzygotic RI in tidepool copepods (Edmands 1999) and subshrubs (Montalvo and Ellstrand 2001), it is not possible to distingush a threshold type dependence from a smooth one because pairs of populations at intermediate genetic distances are missing. Therefore, although a positive relationship between measures of RI and genetic divergence (or time) is well established (Edmands 2002), its exact form is yet unknown.

Effects of within-population genetic variation   Let us consider a finite population of sexual haploid individuals with discrete, nonoverlapping generations following Gavrilets (1999a). The restriction to haploids is for algebraic simplicity. I assume that individuals are different with respect to a large number $\mathcal{L}$ of possibly linked diallelic loci with alleles A_i and a_i for $i = 1 \ldots \mathcal{L}$. In line with the approach used above, I posit that two individuals at genetic distance d are not reproductively isolated with probability $w(d)$. Below I describe an approximate approach for describing evolutionary dynamics of the average genetic distance within populations, $\overline{d}_w$, and between populations, $\overline{d}_b$, under selection, recombination, mutation, migration, and genetic drift. For simplicity of notation, I will drop the overbar, using d_w and d_b instead. The approximations to be used will be based on three assumptions: (i) alleles at different loci are independent (linkage equilibrium approximation), (ii) genetic variation within each locus is small (rare allele approximation), and (iii) population size is not too small (diffusion approximation).

The distribution of d. Let p_i be the frequency of allele A_i at the i-th locus in a population, $q_i = 1 - p_i$. Then the average genetic distance within the population can be represented as

$$d_w = \sum_i 2 p_i q_i. \tag{6.35}$$

The distribution of d in the population is approximately Poisson with parameter d_w and, thus,

$$\Pr(d = i) = \exp(-d_w) \frac{d_w^i}{i!}. \tag{6.36}$$

Average fitness of the population. In the model under consideration, the mean fitness of the population, $\overline{w}$, can be defined as the proportion of pairs of individuals that can mate and produce fertile and viable offspring (Nei et al. 1983), that is,

$$\overline{w} = \sum_{i=0}^{\mathcal{L}} w(i) \Pr(d = i). \tag{6.37}$$

Dynamics of allele frequencies under mutation and selection. The dynamics of the general model of fertility selection and prezygotic isolation in a haploid population considered here are identical to those of a symmetric viability selection model for a diploid population with viabilities $w(d)$, depending on the number of heterozygous loci d. In particular, the expected change in allele frequency per generation brought by selection against incompatible genotypes can be rewritten as

$$\Delta p_i = s p_i q_i (p_i - q_i), \tag{6.38}$$

with the coefficient s being equal to the negative of the derivative of $\ln \overline{w}$ with respect to d_w. This coefficient characterizes the strength of selection experienced by each individual locus. If the average fitness $\overline{w}$ decreases with increasing genetic variation (which is characterized by d_w), the coefficient s is positive. This is always the case if $w(d)$ is nonincreasing, which is the situation we are interested in here. If $w(d)$ is nondecreasing, $s \leq 0$. Equation (6.38) shows that selection experienced by individual loci is similar to that in the underdominant selection model (compare with equation (3.4) on page 57). Note that in contrast to that model, here the coefficient s is not constant but depends on the population genetic structure.

Dynamics of d_w under selection, mutation, and genetic drift. Because d_w can be expressed in terms of allele frequencies (see equation (6.35)), the equations for allele frequencies dynamics can be used to describe the expected changes in d_w. Under the joint action of selection, mutation, and genetic drift, the change in d_w per generation in an isolated population of size N is approximately

$$\Delta d_w = -sd_w + 2v - \frac{d_w}{N}, \qquad (6.39)$$

where, as before, $v = \mu\mathcal{L}$ is the rate of mutation per gamete per generation. The first and the third terms in the right-hand side of equation (6.39) describe reductions in d_w caused by selection against incompatible genotypes and genetic drift, and the second term describes an increase in d_w by mutation. The value of d_w at the mutation-drift-selection equilibrium can be written as

$$d_w^* = \frac{2v}{s^* + \frac{1}{N}}, \qquad (6.40a)$$

which simplifies to

$$d_w^* = 2vN, \qquad (6.40b)$$

if selection induced by RI is very weak ($s^* \ll 1/N$), and to

$$d_w^* = \frac{2v}{s^*}, \qquad (6.40c)$$

if the population size is very large ($N \gg 1/s^*$). Here, s^* is the equilibrium value of s. Equation (6.40b) defines d_w at the mutation-drift balance, whereas equation (6.40c) defines d_w at the mutation-selection balance. Note that, because s^* depends on the population state, equations (6.40a) and (6.40c) define the equilibrium value of d_w implicitly rather than explicitly.

Genetic divergence between isolated populations. The population will keep evolving even after the average genetic distance d_w has reached an equilibrium level d_w^*, as different loci keep experiencing substitutions. As a consequence, if the number of loci $\mathcal{L}$ is large, allopatric populations will continuously diverge genetically. (If the number of loci is small, the populations will wander on a small set of metastable equilibria diverging and converging sporadically.) The rate of stochastic divergence between

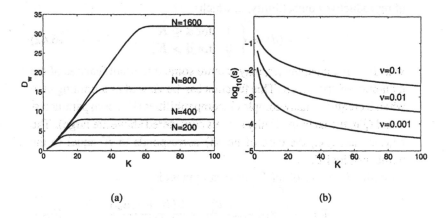

FIGURE 6.8. (a) Average genetic distance d_w maintained by mutation-selection-drift balance in an isolated population of size N as a function of K for $v = 0.01$. The bottom line corresponds to $N = 100$. (b) Effective selection coefficient s in an infinite population for three values of v.

two populations can be found using the results from Chapter 3. This rate is approximately equal to the expected number of new mutations in the system of two populations, which is $2vN$, multiplied by the probability that a given one will be fixed, u. It follows from results in Chapter 3 that the asymptotic rate of divergence (that is, the average number of substitutions per generation) of two isolated populations of size N each can be written as

$$\Delta d_b = 2v\kappa_u, \qquad (6.41)$$

where κ_u is given by equation (3.2) on page 55 with $S = Ns^*/2$. Note that in the definition of S we divide by 2 because the derivations here are for haploid populations rather than for diploid populations as in Chapter 3.

Example 1. To illustrate the above results we need to choose a specific form for the function of reproductive compatibility $w(d)$. In the absence of reliable data, a legitimate criterion for choosing a model is mathematical simplicity. A common feature of *many* models considered above is that the transition from no RI to complete RI is rather sharp (the threshold effect). Therefore, the simple threshold function of reproductive compatibility, in which

$$w(d) = \begin{cases} 1 & \text{for } d \leq K, \\ 0 & \text{for } d > K, \end{cases} \tag{6.42}$$

with K being a parameter, may capture some important features of the dynamics of speciation. This function can be viewed as a limiting case when RI requires many complex incompatibilities (i.e., both parameters k and Q in the incompatibility models considered above are large). The neutral case (i.e., the case of no RI) corresponds to K equal to the number of loci $\mathcal{L}$.

With this choice of $w(d)$, the mean fitness is

$$\bar{w} = \sum_{i=0}^{K} \exp(-d_w) \frac{d_w^i}{i!} = \frac{\Gamma(K+1, d_w)}{\Gamma(K+1)}. \tag{6.43}$$

The corresponding coefficient s characterizing the strength of selection induced on a locus can be written as

$$s = \frac{e^{-d_w} d_w^K}{\Gamma(K+1, d_w)}. \tag{6.44}$$

Figure 6.8(a) illustrates the dependence of the equilibrium value d_w^* on the parameters of the model. This figure indicates that d_w^* is close to the corresponding neutral predictions (6.40b) if $K > 5\nu N$. For small K, d_w increases linearly with K. Figure 6.8(b) gives the values of the effective selection coefficient s. With moderately large K (that is, with $K \geq 10$), s is very small. Thus, very strong selection on the whole genotype (implied by the existence of complete RI at finite values of K) results in very weak selection at the level of individual loci.

Figure 6.9 illustrates the dependence of the relative rate of divergence κ_u on model parameters. In the neutral case, the rate of genetic divergence $\Delta d_b = 2\nu$ and does not depend on the population size. In contrast, with RI, the rate of divergence decreases with increasing population size. This happens because genetic divergence is driven by random drift which is necessary to overcome selection against rare genotypes. For example with $K = 10$ and $\nu = 0.01$, a population of size $N=1,000$ will accumulate about 2 substitutions per 1,000 generations. Ten thousand generations will be sufficient for d_b to exceed K significantly. In contrast, a population of size $N=10,000$ will need on average 10^{12} generations to fix a substitution. Figure 6.9 also indicates that the rate of substitutions is close to the corresponding neutral predictions if $K > 5\nu N$. Note that an implicit assumption of equation (6.41)

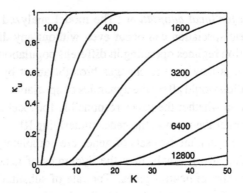

FIGURE 6.9. The rate of divergence κ_u relative to the neutral case in an isolated population as a function of K for $\nu = 0.01$ and different N.

is that genetic distance d_b is small relative to the number of loci $\mathcal{L}$. In the general case, d_b approaches $\mathcal{L}/2$ asymptotically (see Box 3.3). Numerical results reported in Gavrilets (1999a) confirm the validity of the approximations used here.

Example 2. Assuming the threshold function (6.42) makes it very easy to accumulate substitutions because new alleles start "feeling" selection against them only after genetic variation becomes sufficiently large. If $w(d)$ decreases (even slightly) for small values of d, the accumulation of substitutions will occur much slower. To illustrate this, let $\tilde{s}$ be the reduction in the probability of compatibility w as a result of a single substitution: $\tilde{s} = w(0) - w(1)$. Then if mutation rate is low ($\nu \ll \tilde{s}$), one expects a very low level of variation maintained in the populations, so that the distribution of d will be dominated by the frequencies corresponding to $d = 0$ and $d = 1$. In this case, using the formulas above, one can show that the strength of selection on an individual locus is $s \approx \tilde{s}$. With results from Chapter 3, the rate of divergence will be very low if $\tilde{s}$ is not vanishingly small. For example, if w decreases as a linear function of d, then $\tilde{s} = 1/K$. The values of K smaller than, say, 100 will result in values of $\tilde{s} > 0.01$ and very slow divergence.

These two examples show that large populations will diverge much more slowly than small populations if divergence is driven by mutation and drift. The reason for this is selection against rare alleles induced by selection against rare, reproductively incompatible organisms.

Selection for local adaptation. The model analyzed above shows that allopatric speciation can occur even without any differences between selection regimes operating in different populations. An important question is how genetic changes brought about by selection for local adaptation would affect the dynamics of speciation. These effects will depend on whether the genes responsible for local adaptation are different from or the same as the genes underlying RI.

Assume first that the two sets of genes are completely different. If the population size is small, then the dynamics of genes underlying RI resembles that of neutral genes. The rate of substitution of neutral genes is not affected by selection on other loci for any recombination rates (Birky and Walsh 1988). Therefore, selection for local adaptation will not affect the dynamics of speciation. If the population size is not too small, then the dynamics of genes underlying RI resembles those of underdominant genes. Close linkage to advantageous alleles increases the rate of fixation of detrimental mutations, but only slightly (Birky and Walsh 1988). This suggests that unless the two sets of loci are very tightly linked, effects of selection for local adaptation on the rate of speciation will not be significant.

Assume now that the loci under consideration pleiotropically affect both survival in a given environment and RI. Let s_{LA} be the average strength of selection per locus induced by selection for local adaptation. Then the expected change in p_i due to selection can be rewritten as

$$\Delta p_i = s p_i q_i (p_i - q_i) + s_{LA} p_i q_i, \qquad (6.45)$$

where the first term specifies the effect of selection against incompatible organims and the second term specifies the effects of selection for local adaptation. The right-hand side of this equation is similar to that of equation (3.8) on page 58, which describes the case of underdominant selection with nonequal peaks. This fact can be used to show that the relative rate κ_{LA} of fixation of new mutations in an isolated haploid population of size N can be approximated by equation (3.9) on page 59 with $S = N(s - s_{LA})/2$ and $\beta = 2s_{LA}/(s - s_{LA})$. Increasing s_{LA} always increases κ_{LA}. Thus, as expected, direct selection for local adaptation always increases the rate of divergence of two populations, $2\nu\kappa_{LA}$, and promotes speciation. Note that with sufficiently strong selection for local adaptation (i.e., if $s_{LA} > s$), the net effect of new alleles will be advantageous and their frequencies will tend to increase even when rare. In this case, the rate of accumulation of genetic differences can be significant and the time to speciation T can be short.

6.2.6 Parapatric speciation

Treating parapatric speciation is much more difficult because now one has to account for the effects of migration. In this subsection, I will use a multilocus generalization of the Markov chain approach that was used in the previous chapter for analyzing the BDM model and at the beginning of this chapter for analyzing a three- and a four-locus model. Recall that this approach neglects the effects of within-population genetic variation. Following Gavrilets (2000b), I will consider a one-population scenario in which a (peripheral) diploid population is subject to immigration from a source population. This scenario corresponds to peripatric speciation.

Markov chain approximation   Assume that the number of loci is very large. All immigrants have a constant ancestral haplotype. Let m be the rate of immigration (that is, the proportion of the peripheral population substituted by immigrants each generation) and ν be the rate of mutation per haplotype per generation. We assume that both m and ν are small and that the population size N is not too large. The latter assumption allows one to neglect genetic variation within the population. In this case, the system state is completely characterized by the (average) genetic distance d_b between the genotypes in the population and the ancestral genotype. Genetic distance d_b takes integer values $0, 1, 2, \ldots$. The dynamics of speciation can be modeled as a biased random walk performed by genetic distance d_b on a set of integers. The changes in d_b are controlled by two opposing forces. Fixation of a mutant allele in the population increases d_b by one. Fixation of an ancestral allele brought by immigration at the loci that had previously fixed mutant alleles decreases d_b by one. We will assume that the dependence of the degree of RI on genetic distance has a threshold form defined by equation (6.42). This is the worst-case scenario for parapatric speciation because immigrants have absolutely no problems mating with the residents unless the genetic distance d_b exceeds K.

No selection for local adaptation. With no selection for local adaptation and neglecting within-population genetic variation, the process of fixation is approximately neutral. The probability of fixation of an allele is equal to its initial frequency. The average frequency of new mutant alleles per generation is approximately equal to the mutation rate ν. If the immigrants differ from the residents at $d_b = i$ loci, there are i loci that can fix ancestral alleles brought by migration. The average frequency of such alleles per generation is im. Thus, the probabilities

TABLE 6.3. Exact expressions for the average waiting time to speciation, T, and the average duration of speciation, τ, for small K with no selection for local adaptation and a threshold function of reproductive compatibility. $R = m/\nu$.

	Allopatric case	Parapatric case	
K	$T = \tau$	T	τ
1	$\frac{2}{\nu}$	$(2+R)\frac{1}{\nu}$	$\frac{2+R}{1+R}\frac{1}{\nu}$
2	$\frac{3}{\nu}$	$(3+3R+2R^2)\frac{1}{\nu}$	$\frac{3+4R+2R^2}{1+R+2R^2}\frac{1}{\nu}$
3	$\frac{4}{\nu}$	$(4+6R+8R^2+6R^3)\frac{1}{\nu}$	$\frac{4+8R+13R^2+6R^3}{1+R+2R^2+6R^3}\frac{1}{\nu}$

of stochastic transitions increasing and decreasing d_b by one are approximately $P_{i,i+1} = \nu, P_{i,i-1} = im$. Speciation occurs when d_b hits the (absorbing) boundary $K + 1$.

Because this is a one-population scenario, speciation is certain ($u = 1$). Using general results on Markov Chains (e.g., Karlin and Taylor 1975), one can also find the average waiting time to speciation, T, and the average duration of speciation, τ. With small K, the exact expressions for T and τ are relatively compact (see Table 6.3). With larger K, the approximate equations are more illuminating. The average waiting time to speciation is approximately

$$T \approx \frac{1}{\nu}\left(\frac{m}{\nu}\right)^K K!$$ (6.46)

The average duration of speciation is approximately

$$\tau \approx \frac{1}{\nu}\left(1 + \frac{\Psi(K+1)+\gamma}{m/\nu}\right),$$ (6.47)

where γ is Euler's constant (≈ 0.577), and $\Psi(\cdot)$ is the psi (digamma) function (Gradshteyn and Ryzhik 1994). (Function $\Psi(K+1)+\gamma$ slowly increases with K and is equal to 1 at $K = 1$, to 2.93 at $K = 10$, and to 5.19 at $K = 100$.) For example, if $m = 0.01, \nu = 0.001$, and $K = 5$, then the waiting time to speciation is very long: $T \approx 1.35 \times 10^{10}$ generations; but if speciation does happen, its duration is relatively

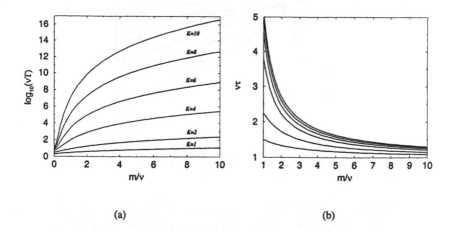

FIGURE 6.10. The average waiting time to speciation, T, and the average duration of speciation, τ, in the case of speciation driven by mutation and random drift. The horizontal axes give the ratio of migration and mutation rates. (a) The vertical axis gives the product of T and the mutation rate ν on the logarithmic scale. (b) The vertical axis gives the product of τ and the mutation rate ν on the linear scale. The lines correspond to $K = 1, 2, 4, 6, 8$, and 10 (from bottom to top). (Using composite variables for the vertical axes allows one to represent relevant dependencies in two dimensions. (From Gavrilets (2003b), fig.7.))

short: $\tau \approx 1{,}236$ generations. Figure 6.10 illustrates the dependence of T and τ on model parameters (using an exact rather than approximate formula). Notice that τ is order $1/\nu$ across a wide range of parameter values.

Direct selection for local adaptation. Assume that new alleles also improve adaptation to the local conditions. Let s be the average selective advantage of a new allele over the corresponding ancestral allele. Each generation there are $2N\nu$ such alleles supplied by mutation. The probability of fixation of an advantageous allele is given by equation (4) in Box 3.1. Immigration into a population that has already diverged from the source population in i loci brings approximately $2Nmi$ ancestral alleles at the loci that have previously fixed new alleles. These alleles are deleterious in the new environment. The probability of fixation of a deleterious allele is approximately given by equation (5) in Box 3.1. The probabilities of stochastic transitions increasing and decreasing d_b

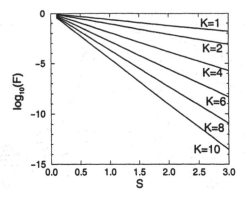

FIGURE 6.11. Effects of direct selection for local adaptation on the waiting time to speciation. The figure shows the proportion F (on the logarithm scale) by which selection for local adaptation decreases T ($F = \exp(-KS)\frac{1-\exp(-S)}{S}$; see equation (6.48)).

by one are approximately $P_{i,i+1} = \nu\kappa_a$ and $P_{i,i-1} = im\kappa_d$, where κ_a and κ_d are defined by equations (5.7) and (5.14) on pages 134 and 139, respectively. The waiting time to speciation is approximately

$$T_s \approx T \, \exp(-KS) \, \frac{1 - \exp(-S)}{S}, \qquad (6.48)$$

where T is given by equation (6.46) and $S = Ns$. The average duration of speciation is approximately

$$\tau_s \approx \frac{1}{\nu}\left(1 + \frac{\Psi(K+1)+\gamma}{(m/\nu)e^{-S}}\right)\frac{1-\exp(-S)}{S}. \qquad (6.49)$$

For example, if $m = 0.01, \nu = 0.001, K = 5$, and $S = 2$, then $T_s \approx$ 27,400 generations and $\tau_s =$ 2,170 generations. Thus, selection for local adaptation dramatically decreases T (in the numerical example, by the factor $\approx 50,000$). Figure 6.11 illustrates the effect of selection for local adaptation on T in more detail. Selection for local adaptation somewhat increases τ relative to the case of speciation driven by mutation and genetic drift. Intuitively, with selection for local adaptation counteracting effects of migration, the population can afford more backward steps on its route to a state of complete RI than when such selection is absent. These additional backward steps and the steps necessary to compensate for them increase the average duration of speciation.

Effects of genetic barrier. Assume that alleles underlying reproductive isolation are neutral but that the population has already diverged from the source population in some other loci underlying adaptation to local conditions. Let η be the corresponding gene flow factor resulting from selection for local adaptation (see Chapter 5 and Box 5.2). In this case, migration rate m has to be substituted by the effective migration rate $m_e = \eta m$. Now the average waiting time to speciation can be approximated as

$$T_\eta \approx T \, \eta^K, \tag{6.50}$$

where T is given by equation (6.46). Even a weak genetic barrier can significantly decrease the waiting time to speciation if there are many loci underlying reproductive isolation. For example, if $\eta = 0.1$ and $K = 5$, then T_η is reduced by five orders of magnitude relative to the neutral case. The average duration of speciation is approximately

$$\tau_\eta \approx \frac{1}{\nu} \left(1 + \frac{\Psi(K+1)\gamma}{\eta m / \nu} \right), \tag{6.51}$$

and is increased by the barrier.

Equation (6.50) shows that genetic barriers can have profound effects on the dynamics of parapatric speciation if there are multiple genes underlying RI. Smaller values of η dramatically increase the plausibility of parapatric speciation. The gene flow factor η decreases with increasing linkage to gene(s) that have already diverged (see Box 5.2) These observations provide a theoretical justification for two recent predictions about the dynamics of genetic divergence in the presence of gene flow. The genes underlying RI are more likely to diverge if they are physically close to the genes that have already diverged. Therefore, rather than being distributed randomly across the genome, genes underlying reproductive isolation between closely related species are more likely to be found in clusters (Wu 2001). Chromosomal rearrangements can reduce recombination along a large section of the genome (White 1978; King 1993). Therefore, such rearrangements create conditions favoring further accumulation of genes underlying reproductive isolation within the rearranged areas (Noor et al. 2001; Rieseberg 2001; Navaro and Barton 2003a,b).

Several other conclusions emerge from the analysis presented in this subsection. Most species consist of geographically structured populations, some of which experience little genetic contact for long periods of time (e.g., Ward et al. 1992; Hamrick and Godt 1996; Friesen et al. 1996; Brown et al. 1997; Avise 2000). Different mutations are expected to appear and increase in frequency in different populations, necessarily resulting in some geographic differentiation even without any variation in local selection regimes because of random genetic drift and of mutational order (see page 39). An interesting question is whether mutation and drift alone are sufficient to result in parapatric speciation. The results presented above provide an affirmative answer to this question. However, the waiting time to speciation is relatively short only if a very small number of genetic changes is sufficient for complete RI. For example, T is on the order of 10 to 1,000 times the inverse of the mutation rate if $K = 1$ or 2. With higher values of K, the waiting time to speciation increases dramatically. With large K, direct selection on the loci underlying RI or indirect selection through a genetic barrier to gene flow are necessary for parapatric speciation to occur in a biologically reasonable time.

The waiting time to speciation T is extremely sensitive to parameters: changing a parameter by a small factor, say two or three, can increase or decrease T by several orders of magnitude. Looking across a range of parameter values, T is either relatively short (if the parameters are right) or effectively infinite. Most of the parameters of the model (such as the migration rate, intensity of selection for local adaptation, the population size, and, probably, the mutation rate) directly depend on the state of the environment (biotic and abiotic) that the population experiences. This suggests that speciation can be triggered by changes in the environment, as argued by Eldredge (2003). If it is an environmental change that initiates speciation, the populations of many different species inhabiting the same geographic area should be affected. In this case, one should expect more or less synchronized bursts of speciation in a geographic area – that is, a *turnover pulse* in the terminology of Vrba (1985). For example, Vrba (1985) described antelop faunas from eastern and central Africa that displayed a long period of morphological stasis and ecological similarity prior to dramatic climate cooling occurring approximately 2.8 My ago. This cooling apparently caused a relatively abrupt reorganization of African ecosystems after about 300,000 years, with many wet-woodland-adapted species (e.g., antelope) disappearing and animal species adapted to open savannas soon appearing. (These included two new hominid species.) Brett and Baird (1995) have documented eight successive faunas of marine invertebrates in the Appalachian basin of the Middle Paleozoic. Each fauna survived an average of 5 to 7 My. Whereas most species persisted unchanged throughout the period, on average only 20% of the species managed to survive to the next faunal interval, and the species that made up the next succeeding marine regional system were either newly evolved or migrated in from adjacent regions. The results described above provide a theoretical justification for these patterns.

The results on the duration of speciation lead to two important generalizations. First, the average duration of parapatric speciation, τ, is much smaller than the average waiting time to speciation, T. This is compatible with patterns observed in the fossil record, which form the empirical basis of the theory of punctuated equilibrium (Eldredge 1971; Eldredge and Gould 1972; Gould 2002). The second generalization concerns the absolute value of τ. The waiting time to speciation changes dramatically with slight changes in parameter values. In contrast, the duration of speciation is on the order of one over the mutation rate over a set of the loci affecting reproductive isolation for a wide range of migration rates, population sizes, intensities of selection for local adaptation, and the number of genetic changes required for RI. Given a typical mutation rate on the order of $10^{-5} - 10^{-6}$ per locus per generation (e.g., Griffiths et al. 1996; Futuyma 1998) and assuming that there are at least on the order of 10 to 100 genes involved in the initial stages of the evolution of RI (e.g., Singh 1990; Wu and Palopoli 1994; Coyne and Orr 1998; Naveira and Masida 1998), the duration of speciation is predicted to range between 10^3 and 10^5 generations.

Effects of within-population genetic variation The effects of genetic variation within populations on the dynamics of parapatric speciation can be incorporated in a way similar to that used above in analyzing allopatric speciation. The general approach is to model the dynamics of speciation as a chain of stochastic transitions between different metastable equilibria that differ in the number of loci k that have diverged between the populations (Gavrilets 1999a). Unfortunately, the approximations that I have developed so far are not particularly satisfying. These approximations require that the migration-selection balance equilibria for allele frequencies are locally stable at each value of k. A necessary condition for this stability is that the (effective) migration rate m_e is sufficiently small relative to the strength of selection s induced on individual loci. For this condition to be satisfied, either the effective migration rate has to be very small, or the selection coefficient has to be large enough. In the former case (which takes place if migration rate is small or if there is a sufficiently strong barrier to the flow of neutral genes), the approximations predict reasonable values for the waiting time to speciation. However, in the latter case, transition probability $P_{k,k+1}$ becomes very small if the population size is at least moderately large. If selection is weak and migration is not too weak, the approximations break down. Alternative methods need to be developed.

6.3 SUMMARY

• In multidimensional versions of the BDM model, RI emerges as a byproduct of a chain of substitutions, each of which is neutral or nearly neutral. This process can be visualized as a biased random walk performed by the population along a ridge of high fitness values in a holey landscape.

• In small populations, strong RI can accumulate after a chain of weakly selected underdominant mutations or chromosomal rearrangements. The characteristic time scale for this process is on the order of 10^5 to 10^6 generations.

• Divergent degeneration of duplicated genes can result in RI between allopatric populations. The characteristic time scale for this process is on the order of the reciprocal of the rate of gene duplication. For the evolution of very strong RI, 10^5 to 10^6 generations can be sufficient .

• There are two general approaches for modeling speciation. The first approach requires one to specify the genetic architecture of RI explicitly. However, this kind of information is typically missing and the complexity of the corresponding mathematical treatment quickly increases with the number of genes. The second approach is to specify some general rules underlying RI and then deduce the properties of the speciation process expected under these rules on average.

• A powerful application of the second approach is to reduce RI to cumulative effects of genetic incompatibilities that are observed with certain probabilities between certain groups of loci. The following is a list of properties of models based on this approach:

- The number of incompatibilities increases as a power function of the number of genetic differences (the snowball effect).

- The probability that two genotypes are not reproductively isolated decreases as an exponential function of the expected number of incompatibilities between them.

- As genetic divergence increases, the strength of reproductive isolation undergoes a rapid transition from very weak to very strong. The transition is especially rapid if RI is due to multiple complex incompatibilities (the threshold effect).

- The underlying fitness landscapes belong to the class of correlated holey landscapes. If the probability of incompatibility is low, the adaptive landscape is both highly correlated and allows for substantial neutral divergence.

• Selection against incompatible genotypes induces selection on individual loci resembling frequency-dependent selection against rare alleles and underdominant selection. This implies that if substitutions are driven by genetic drift and mutation, then small populations will evolve and speciate faster than large populations.

• Selection for local adaptation (acting directly on the loci underlying RI or mediated via the genetic barrier to the flow of neutral alleles) can dramatically accelerate the rate of speciation. If selection for local adaptation acts directly on the loci underlying RI and is sufficiently strong, speciation occurs deterministically and very rapidly.

• In the case of allopatric speciation, the average waiting time until two populations become reproductively isolated is approximately the number of genetic changes required for reproductive isolation divided by twice the rate of substitutions per generation. In the case of speciation driven by mutation and drift, the rate of substitutions is approximately equal to the effective mutation rate. Selection for local adaptation can significantly increase the rate of substitutions and, thus, the rate of allopatric speciation.

• For parapatric speciation, the following conclusions apply:

- The average waiting time to speciation T is extremely sensitive to parameters (such as the migration rate, intensity of selection for local adaptation, the population size, the mutation rate, and the number of genetic changes required for strong RI). In looking across a range of parameter values, T is either relatively short (if the parameters are right) or effectively infinite. Extreme dependence on parameters also implies that one has to exercise caution when extrapolating from numerical studies utilizing only a small number of parameter values.

- Most of the parameters of the model affecting T directly depend on the state of the environment (biotic and abiotic) that the population experiences. This suggests that speciation can be triggered by changes in the environment. If it is an environmental change that initiates speciation, the populations of many different species inhabiting the same geographic area should be affected in a similar way. In this case, one should expect more or less synchronized bursts of speciation in a geographic area, that is, a turnover pulse.

- The average duration of speciation is much smaller than the average waiting time to speciation. This is compatible with the patterns observed in the fossil record, which form the empirical basis of the theory of punctuated equilibrium. Given typical mutation rates and assuming that there are at least on the order of 10 to 100 genes involved in the initial stages of the evolution of RI, the duration of speciation is predicted to range between 10^3 and 10^5 generations.

- Genetic barriers can have profound effects on the dynamics of parapatric speciation, especially if there are multiple genes underlying RI. The plausibility of speciation is dramatically increased by small values of the gene flow factor η, which decreases with increasing linkage to gene(s) that have already diverged. Therefore, the genes underlying RI are more likely to diverge if they are physically close to the genes that have already diverged. Consequently, rather than being distributed randomly across the genome, genes underlying reproductive isolation between closely related species are more likely to be found in clusters. Chromosomal rearrangements can reduce recombination along a large section of the genome, creating a barrier to gene flow. Therefore, such rearrangements create conditions favoring further accumulation of genes controlling RI within the rearranged areas.

• Empirical studies of the reproductive compatibility function $w(d)$ will be very valuable in determining whether particular models provide a realistic view of speciation. Equally important will be empirical studies aiming to map the loci that have diverged across genomes of sister species.

6.4   CONCLUSIONS

Multilocus versions of the BDM model lead to conclusions similar to those reached on the basis of the two-locus model. In particular, allopatric speciation is a very probable eventual outcome of genetic divergence. Parapatric speciation can occur if selection for local adaptation is spatially heterogeneous and strong. Genetic barriers to gene flow formed in hybrid zones greatly promote parapatric speciation. The duration of parapatric speciation is expected to be very short. Mathematical models provide strong theoretical support for the patterns of punctuated equilibrium and coordinated stasis.

CHAPTER SEVEN

Spatial patterns in the BDM model

Most species are composed of many local populations experiencing little genetic contact for long periods of time (Avise 2000). A simple tool for describing such systems of parapatric populations is provided by spatially structured multideme models. In this chapter I consider the dynamics of the evolution of RI and diversification in multideme models in the presence of low migration.

Several recent studies modeled the joint dynamics of speciation, extinction, and colonization in a spatially explicit framework using phenomenological descriptions of speciation. These studies did not consider any underlying genetics and simply postulated that a new species with a certain number of individuals (one or more) emerges with a certain probability from the ancestral species (e.g., Bramson et al. 1996; Durrett and Levin 1996; Allmon et al. 1998; Barraclough and Vogler 2000; Hubbell 2001). Although these earlier approaches are very useful for training our intuition about the process of diversification in metapopulations and for providing a basis for additional numerical and analytical work, the phenomenological treatment of speciation is not satisfactory. Excluding processes like polyploidy and major chromosomal changes, speciation does not occur instantaneously. Changes in at least several loci are required for a change in RI or morphology necessary for assigning an individual or a population to a new species. The probability that this happens instantaneously is extremely small. For example, if 10 genes have to be changed, then, using the standard estimates of mutation rates (10^{-5} to 10^{-6}), the probability of instantaneous speciation is on the order of 10^{-50} to 10^{-60}. Hubbell's (2001) numerical work shows that the probability of speciation, the initial number of individuals of the new species, and their initial spatial distribution have dramatic effects on various dynamic characteristics of the system. However, the phenomenological approaches cannot provide information on the appropriate values or ranges of these parameters. Hubbell (2001) and others argued convincingly that species-area curves (i.e., the relationships between the number of species found in a region and its area) must be derived from the underlying processes of extinction, speciation, and dispersal rather than postulated to follow a certain statistical distribution. However, in a similar way, the dynamics of speciation must be derived from the underlying microevolutionary processes rather than postulated to follow a certain statistical distribution. In this chapter, I expand the previous approaches by explicitly considering multiple genetic loci underlying RI and species differences within the framework of the multidimensional version of the BDM model.

The methods used in the previous two chapters for analyzing one- and two-deme systems become too difficult to apply in a multideme context, which prompts the need for alternative approaches. I start by describing results of *individual-based simulations* in which RI emerges as a consequence of a localized spread of mutually incompatible neutral genes. Because of computational limitations, these simulations are limited both by duration (which is up to 10,000 generations) and by the overall number of individuals considered (which is on the order of up to several thousands). Their main emphasis is on the waiting time until the population splits into two reproductively isolated groups for the first time and on the location (e.g., central or peripheral) of the first split. I will concentrate on the effects of mutation rate, migration rate, local subpopulation size, genetics of RI, and the spatial structure of the overall population. These simulations are intended to describe an initial stage of diversification either after colonizing a new environment or the emergence of a key innovation that allows the existing environment to be used in a new way.

Next, I consider a series of *deme-based models*, in which the basic units are demes rather than individuals. The shift from individuals to demes allows one to extend both spatial and temporal limits of the system, as well as to incorporate the effects of local extinction and recolonization. It has also allowed me to obtain some illuminating results analytically. The main emphasis of the deme-based models will be on the number and structure of different genetic clusters (species) emerging as a result of diversification, as well as on their turnover rate. I will concentrate on the effects of mutation and migration rates, local extinction and recolonization rates, genetics of RI, and the spatial structure of the overall population. In most results to be presented here, clustering of the initially uniform population results from the localized spread of mutually incompatible neutral genes. At the end I consider recent models in which incompatible genes are advantageous throughout the system. As expected from general considerations as well as from results discussed in the previous two chapters, the effects of selection on the dynamics of speciation and diversification can be rather profound.

7.1 INDIVIDUAL-BASED MODELS: SPREAD OF MUTUALLY INCOMPATIBLE NEUTRAL GENES

Here I describe the results of individual-based simulations performed by Gavrilets et al. (1998, 2000b).

7.1.1 Model

I consider a finite subdivided population of sexual haploid individuals that are different with respect to a large number of linked diallelic loci. The restriction to haploids is made for computational simplicity. Evolutionary factors included are mutation, recombination, migration, genetic drift, and RI. To decrease the number of parameters, I make a number of symmetry assumptions. Mutation rates are equal for forward and backward mutations and across loci. Recombination rates between adjacent loci are equal and recombination events take place independently of each other. The population is subdivided into subpopulations of equal size. I assume that there are no systematic differences in environmental conditions between different subpopulations and no asymmetries in migration regimes. Migration is restricted to neighboring populations only and the migration rate between any two neighboring subpopulations is the same. The only difference between peripheral and central subpopulations is that the former have a smaller number of neighboring subpopulations than the latter. As a consequence, in peripheral demes individuals encounter individuals from the same population more often than in central demes. To describe RI, I posit that an encounter of two individuals can result in mating and viable and fecund offspring only if the individuals are different in no more than K loci. Otherwise the individuals do not mate (prezygotic RI) or their offspring is inviable or sterile (postzygotic RI). That is, I use the threshold function of reproductive

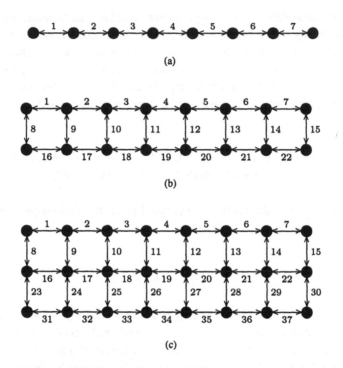

FIGURE 7.1. Examples of 1D and 2D stepping-stone systems. (a) 1×8 system. (b) 2×8 system. (c) 3×8 system. The numbers label connections between neighboring populations that potentially can be broken as a result of speciation.

compatibility (6.42) introduced on page 182. Recall that this function arises when complete RI requires many complex incompatibilities.

7.1.2   Parameters

The following is a list of parameter values used:

• Probability of mutation per locus per generation: $\mu = 0.00005, 0.0001$, and 0.0002.

• The threshold level of genetic compatibility: $K = 10, 20$, and 30.

• Subpopulation size: $N = 50, 100$, and 200 individuals.

• Migration rate: $m = 0.04, 0.06$, and 0.08.

• The geometric structure of the habitat: $1 \times 8, 2 \times 8$, and 3×8 stepping-stone systems (see Figure 7.1).

The migration rate m was defined as the probability that an individual from a central deme encounters an individual from a different subpopulation. Thus, the probability of encountering an individual from a specific neighboring subpopulation was $m/2$ in 1×8 stepping-stone models, $m/3$ in 2×8 stepping-stone models and $m/4$ in 3×8 stepping stone models. Two parameters did not change throughout the simulations: the number of loci, which was $\mathcal{L} = 386$, and the recombination rate between adjacent loci, which was $r = 0.005$. Theoretical arguments (Gavrilets 1999a) and additional simulations suggest that changing $\mathcal{L}$ and r will not strongly affect the conclusions. Note that the values of μ used here are higher than current estimates of the mutation rates per locus, which are typically on the order of 10^{-5} to 10^{-6} (Griffiths et al. 1996; Futuyma 1998), whereas the values of K are within the range of estimates of the minimum number of genes involved in RI (Singh 1990; Wu and Palopoli 1994; Coyne and Orr 1998). The number and size of the subpopulations had to be relatively small and the mutation rate per locus had to be relatively large because of the computational constraints. The values of m chosen reflect the interest in the case of relatively low migration.

7.1.3   Numerical procedure

Each individual was represented as a binary string of length $\mathcal{L}$. The following procedure was used for generating the next generation for each subpopulation. First, randomly pick up an individual (interpreted as female) from the subpopulation. Then, choose a subpopulation from which a second individual (interpreted as male) of the potential mating pair will be chosen. This can be the same or a different subpopulation. Then, randomly pick up an individual from that subpopulation. Calculate the number d of loci at which the two individuals differ. Disregard both individuals and return to the first step, if $d > K$ (the chosen individuals are reproductively isolated). Do crossover producing an offspring if $d \leq K$. Place each offspring in its mother's subpopulation. (Note that this assumption effectively means that only males are allowed to migrate.) Repeat until all N offspring for the current subpopulation are generated. Repeat for all other subpopulations.

The simulations started with all individuals identical. During the first 1,000 generations there were no restrictions on migration between subpopulations, and the whole population evolved as a single randomly mating unit. One thousand generations was sufficient for the population to reach a state of stochastic equilibrium. This was apparent from the behavior of the average genetic distance, the number of segregating loci, and the number of different genotypes in the system (data not presented). Starting with generation 1,000, migration was restricted to neighboring subpopulations.

Gavrilets et al. (2000b) started by performing 10 preliminary runs for each of the parameter configurations. Each run ended whenever the average genetic distance between any two *neighboring* subpopulations exceeded $d_c = K + 10$, or at generation 10,000. The former outcome was interpreted as breakage of the population into reproductively isolated groups (i.e., speciation). Experience with running the model strongly suggests that this value of d_c guarantees that breakage is irreversible (for the parameter values used). For each parameter configuration with at least one breakage event in 10 preliminary runs, 90 additional runs were made, and the times to the first breakage and the exact connection(s) broken (see Figure 7.1) were recorded. To reduce computation time, the average genetic distances were computed only in certain generations, typically every 32nd generation. As a consequence, some runs exhibited more than one breakage as new breaks occurred after the first one before the run was stopped.

7.1.4   Results

Time till speciation   The distribution of the time till speciation (i.e., the time to the first breakage) is nonnormal, strongly asymmetric, and truncated (at time 9,000). For these reasons one has to use order statistics of location and dispersion (e.g., Sokal and Rohlf 1995). Table 7.1 shows a sample of the results on the median waiting time T_m till the first split, together with the 10th and 90th percentiles, for $N = 50$. Table 7.2 shows results corresponding to $N = 100$. With $N = 200$, speciation was observed only for five combinations of parameters, each of which assumed the smallest migration rate ($m = 0.04$).

As expected, reducing mutation rate μ and/or increasing migration rate m always increases T_m. Effects of subpopulation size N, the amount of genetic change required for RI K, and the system size vary. In general, increasing N increases T_m. However, if mutation rate μ is

TABLE 7.1. The median waiting time to speciation T_m with the 10th and 90th percentiles (in brackets) for N=50. "–" means $T_m > 9,000$.

System	K	m	$\mu = 2 * 10^{-4}$	$\mu = 0.5 * 10^{-4}$
1×8	10	0.04	169 (121;,217)	601 (409;857)
		0.08	281 (185;409)	1305 (729;2,425)
	30	0.04	185 (153;249)	3,145 (1,785;7,225)
		0.08	297 (217;441)	– (–;–)
2×8	10	0.04	153 (121;185)	537 (345;729)
		0.08	281 (185;409)	3,113 (1,241;6,521)
	30	0.04	153 (121;185)	1129 (825;1,593)
		0.08	281 (217;377)	5,385 (2,265;–)
3×8	10	0.04	121 (121;121)	377 (281;505)
		0.08	217 (153;249)	1,465 (761;2,361)
	30	0.04	121 (121;153)	777 (569;985)
		0.08	185 (153;249)	2,041 (1,177;4,057)

small and the population lacks the genetic variation necessary to initiate divergence, then increasing N might help speciation. For example, in 1×8 systems with $K = 30, m = 0.04$, and $\mu = 0.00005$, the median time till speciation $T_m = 2,249$ if $N = 100$, but $T_m = 3,145$ if $N = 50$. This difference is statistically significant ($P < 0.01$, G-test). Increasing K increases the median waiting time to speciation T_m in systems with small subpopulation size ($N = 50$), which may lack genetic variation, but decreases T_m in systems with large subpopulation size ($N = 100$). Increasing K implies that more genetic divergence is necessary for speciation, but it also implies that selection acting against rare alleles becomes weaker (see below), which accelerates divergence. In general, increasing the system size (from 1×8 to 2×8 to 3×8) decreases T_m. The reason for this appears to be the increase in the overall level of genetic variation and the increase in the opportunity for speciation, that is, the number of connections between neighboring populations to be broken (which is 7, 22, and 37 in 1×8, 2×8 and 3×8 systems, respectively; see Figure 7.1). Increasing population subdivision while keeping the overall population size constant significantly decreases T_m.

For many species, there is a strong positive correlation between local population density and species range: species with larger ranges usually have higher local densities as well (e.g. Gaston et al. 1997). The

TABLE 7.2. The same as in Table 7.1 but for $N = 100$.

System	K	m	$\mu = 2 * 10^{-4}$	$\mu = 0.5 * 10^{-4}$
1×8	10	0.04	537 (345;697)	1,897 (1,177;3,129)
		0.08	– (3,161;–)	– (–;–)
	30	0.04	313 (249;377)	2,249 (1,337;4,857)
		0.08	1,209 (665;2,169)	– (–;–)
2×8	10	0.04	473 (345;697)	3,801 (1,273;8,505)
		0.08	– (–;–)	– (–,–)
	30	0.04	281 (217;345)	2,201 (1,433;3,353)
		0.08	– (–;–)	– (–;–)
3×8	10	0.04	377 (281;473)	1,209 (729;1,849)
		0.08	– (2,201;–)	– (–;–)
	30	0.04	217 (217;217)	1,433 (921;1,945)
		0.08	921 (473;1,753)	– (–;–)

parameter configurations with $N = 100$ in 2×8 systems describe a population that both occupies twice as many demes and has a local density twice as big as that with $N = 50$ in 1×8 systems. In 24 out of 25 pairwise comparisons, T_m is smaller in the latter case than in the former case ($P < 0.001$, two-sided sign test). This shows that populations with larger range sizes and higher local densities have a smaller chance of speciation than populations with smaller range sizes and local densities. However, if the range size and local density are too small, speciation will not occur because of the lack of necessary genetic variation.

Location of the first break   Gavrilets et al. (2000b) also calculated the number of breaks observed per each connection in Figure 7.1 for all parameter configurations used. In general, in 2×8 and 3×8 systems the vertical connections (such as connections 8 through 15) were broken less frequently than the horizontal ones (such as connections 1 through 7). Also, in 3×8 systems the internal horizontal connections (connections 16 through 22) were broken less frequently than external horizontal connections. First, Gavrilets et al. (2000b) tested whether the distribution of the location of the first break among external horizontal connections deviated from the uniform distribution. The latter is implied in various "broken stick" models (e.g., Sugihara 1980; Nee et al.

1991; Takeshi 1993; Barraclough and Vogler 2000). In most cases, the distribution of the location of the first break significantly deviated from the uniform one. With the G-test (Sokal and Rohlf 1995, Chapter 15), the null hypothesis was rejected at $P < 0.001$ in 74 out of 149 tests performed; the results were not significant in 48 tests.

Next, Gavrilets et al. (2000b) compared the average number of breaks per connection for peripheral and central connections. By *peripheral* connections I mean connections 1 and 7 in 1×8 models, connections 1, 7, 16, and 22 in 2×8 systems, and connections 1, 7, 31, and 37 in 3×8 systems (see Figure 7.1). By *central* connections I mean connections 3, 4, and 5 in 1×8 models, connections 3, 4, 5, 18, 19, and 20 in 2×8 systems, and connections 3, 4, 5, 33, 34, and 35 in 3×8 systems. In general, the average number of breaks per connection for intermediate connections (connections 2 and 6 in 1×8 systems, connections 2, 6, 17 and 21 in 2×8 systems, and connections 2, 6, 32, and 36 in 3×8 systems) was intermediate between those for central and peripheral connections.

To analyze the effects of different parameters on the location of the first break, Gavrilets et al. (2000b) performed a series of pairwise tests. The most general patterns observed concern the effects of μ, K, and N. Increasing the rate of mutation μ from the smallest value to the largest value significantly increases the difference between the average number of breaks per peripheral and central connections in 42 out of 46 pairs for which data are available ($P < 0.0001$, one-sided sign test). Thus, decreasing μ shifts the location of the first break towards central areas, whereas increasing μ shifts the location of the first break towards peripheral areas. This effect is illustrated in Figure 7.2(a). Increasing parameter K from the smallest value to the largest value decreases the difference between the average number of breaks per peripheral and central connections in 38 out of 47 pairs for which data are available ($P < 0.001$, one-sided sign test). Thus, increasing K shifts the location of the first break towards central areas, whereas decreasing K shifts the location of the first break towards peripheral areas. This effect is illustrated in Figure 7.2(b). Increasing subpopulation size from $N = 50$ to $N = 100$ in 2×8 and 3×8 systems significantly increases the ratio of the average number of breaks per peripheral and per central connections. (This ratio increases in 34 out of 35 pairs for which data are available, resulting in $P < 0.0001$, one-sided sign test.) Thus, increasing population size shifts the location of the first break towards

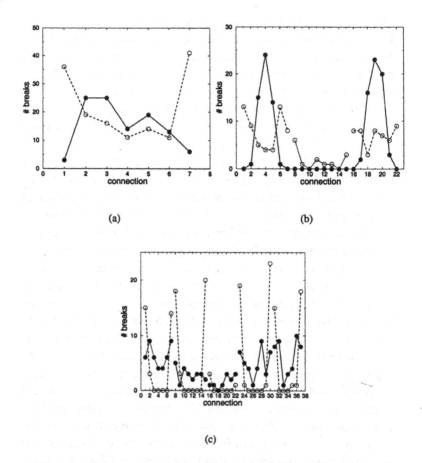

(a)   (b)

(c)

FIGURE 7.2. The number of breaks at different positions observed in 100 runs. (a) Mutation rate: $\mu = 0.00005$ (black circles) and $\mu = 0.0002$ (open circles). An 1×8 system with $N = 50, K = 10$, and $m = 0.06$. (b) Minimum genetic change necessary for RI: $K = 10$ (black circles) and $K = 30$ (open circles). An 2×8 system with $N = 100, m = 0.06$, and $\mu = 0.0002$. (c) Subpopulation size: $N = 50$ (black circles) and $N = 100$ (open circles). An 3×8 system with $K = 20, m = 0.06$, and $\mu = 0.0001$. (From Gavrilets et al. (2000b), fig.2,3,4.)

peripheral areas. This effect is illustrated in Figure 7.2(c). There are no obvious patterns in the effects of migration rate and system size.

Many evolutionary biologists appear to believe that very weak migra-tion, on the order of one individual exchanged between two populations per generation, is sufficient to prevent any genetic differentiation, making speciation impossible. Here, the highest number of migrants still compatible with speciation was $Nm = 8$ migrant gametes per subpopulation per generation. This suggests that rapid speciation by mutation and random drift is possible even when subpopulations exchange several individuals per generation. However, smaller migration rates do make speciation more plausible. In simulations, the time scale for speciation is short (from a few hundred to a few thousand generations), meaning that restrictions on migration between subpopulations do not need to be long-lasting. As already argued (e.g., Palumbi 1992), a relatively brief period of reduced migration (or isolation) may be sufficient for initiating significant genetic divergence and evolution of RI.

Of course, these conclusions should not be interpreted as suggesting that parapatric speciation by random genetic drift and mutation is inevitable or can be accomplished very easily in general. In the simulations, speciation was observed only for specific sets of parameter values (see Tables 7.1 and 7.2). Extrapolating the data, one can conclude that choosing less favorable parameter values (e.g., making N or m significantly larger or μ smaller) will make rapid parapatric speciation practically impossible within the framework of the model used.

7.1.5   Interpretations

Species-level characteristics and speciation rate   There has been extensive discussion in the literature of the relationships between speciation rate and species-level characteristics, such as local abundance, range size, and dispersal ability (e.g., Stanley 1986, 1990; Rosenzweig 1995; Wagner and Erwin 1995; Chown 1997; Gaston 1998). The results presented here allow us to put previous theoretical arguments, which were mostly verbal, on firmer grounds. Geographic range size is usually positively correlated with local abundance and dispersal ability (e.g. Gaston et al. 1997; Gaston 1998). Species with very small range sizes and low abundances may not have enough genetic variation to initiate speciation. Given everything else is equal, increasing population range size (and, resulting from this, greater subdivision) will increase the likelihood of speciation. However, increasing local population size and/or migration rate significantly decreases the probability of parapatric speciation. The positive effects of geographic range size on the likelihood of speciation will be overwhelmed by the negative effects of population density and dispersal ability. Thus, the model substantiates the claims that species with intermediate range sizes should have higher speciation rates.

Asymmetry of range division at speciation   In most cases, the distribution of the location of the first break significantly deviates from a uniform distribution. The asymmetries of range division between sister species can be understood in terms of limiting factors. Any speciation event as considered here involves two necessary steps: increase in genetic variation and breakage of a cohesive group of genotypes into evolutionary independent units. Mutation is a major factor controlling the level of genetic variation. Dispersal of individuals between subpopulations is a major factor preventing the breakage. Central populations are characterized by higher levels of genetic variation than peripheral populations, but the latter are less affected by migration than the former. Thus, if new genetic variation is the limiting factor, then central populations should be the place where most splits (speciation events) take place. This will happen if mutation rate μ is relatively small, or if local abundances N are relatively low, or relatively many genetic changes are required for speciation (K is too high). If there is a sufficient amount of genetic variation, which will take place with high μ, high N, and low K, then migration becomes the major factor controlling speciation. In this case, speciation events are expected to involve mainly peripheral populations.

These results have implications for an old and unsettled dispute regarding the relative importance of central and peripheral populations in speciation (see Chapter 1). In his theory of peripatric speciation, Mayr (1954, 1963) singled out peripheral populations as the primary source of new species. In contrast, in his theory of centrifugal speciation, Brown (1957) argued that the main source of genetic novelty are central populations. The results described above show that both Mayr's (1954; 1963) and Brown's (1957) arguments are sound, but within their own specific domains.

Some data suggest that splits resulting in sister species with similar ranges are more common than splits producing species with very different ranges (Lynch 1989; Chesser and Zink 1994). This is compatible with patterns expected on the basis of a general belief that new genetic variation is a major limiting factor in evolution. However, some caution in interpreting such data is always necessary, because the extent of postspeciational change in geographic range sizes is typically unknown, and extinction of species will strongly influence the distribution of species-range size (e.g., Gaston 1998; Losos and Glor 2003).

Patterns of parapatric speciation can be affected by other factors not considered here. In a heterogeneous environment, the population will most likely split along geographic areas where population densities are low, or migration is restricted, or at which the selection changes significantly (Endler 1977). The importance of spatial heterogeneity of the environment is well recognized (Mayr 1942, 1963; Endler 1977; Rosenzweig 1995). The results discussed here clearly show that genetic factors are important as well.

7.2 DEME-BASED MODELS: SPREAD OF MUTUALLY INCOMPATIBLE NEUTRAL GENES

Unfortunately, individual-based simulations are not practical for systems with a large number of populations and, consequently, a large overall number of individuals. However, important insights about the behavior of such systems can be gained by using deme-based models, to be described next. These models are closely related to the Markov chain models studied in Chapters 5 and 6.

7.2.1 Model

Here, I will consider sexual species with nonoverlapping generations. I will refer to the whole set of populations in the system as a *clade*.

Spatial arrangement. I consider a habitat subdivided into a large but finite number n of discrete patches, arranged on a line in the case of one-dimensional (1D) systems or on a two-dimensional (2D) square lattice. Each patch can support one population of one species. Each patch can receive colonizers from up to two (in the 1D cases) or up to four (in the 2D cases) neighboring patches. The number of neighboring patches will be smaller for patches at the boundary.

Population state. I assume that there is a large number $\mathcal{L}$ of possibly linked diallelic loci, affecting RI or other phenotypic traits (e.g., morphological, behavioral) that differentiate species (e.g., genera, families). Each population is characterized by the genetic sequence of its most common genotype. I will neglect within-population genetic variation. This implies that the size of local populations is relatively small and that the rates of mutation and migration are small as well.

Clade state. The clade state is characterized by the set of states of the populations present. One can imagine a population as a point in the multidimensional genotype space. The clade will be a cloud of points that changes both its structure and location in the genotype space as a consequence of ecological and evolutionary processes.

Fixation of mutations. At each time step, in each population a mutation can be fixed at each locus under consideration, with a very small probability μ. The fixation rate per genotype, $\nu = \mu \mathcal{L}$, is assumed to be small as well. Following the general framework discussed above, I will assume that mutations are nearly neutral. For neutral mutations, the probability of fixation is equal to the mutation rate (Kimura 1983). Thus, for haploid species, μ is equal to the probability of mutation per allele, whereas for diploid species, μ is equal to twice the probability of mutation per allele.

Extinction and recolonization. At each time step, each population may go extinct with a small probability δ. Extinction is rapidly followed by colonization from one of the neighboring patches chosen randomly. Alternatively, one can think of extinction of a local population as being *caused* by successful invasion from one of the neighboring demes. A newly established population rapidly grows to the equilibrium size.

Genetic clusters and species. I will assume that genetic differences lead to genetic or phenotypic incompatibilities, for example as specified in the previous chapter. I will assign different populations to different genetic clusters (e..g, species, families, genera), based on the degree of genetic divergence characterized by genetic distance d. Recall that d is defined as the number of genes that differ between two populations. The maximum divergence allowed within a cluster will be characterized by parameter K (see equation (6.42) on page 182). For assigning populations to clusters, I use the single linkage clustering technique (e.g., Everitt 1993). This means that two populations separated by a distance d larger than the corresponding threshold K may potentially still belong to the same cluster if there is another intermediate population linking them together. For example, if both the genetic distance d_{12} between populations 1 and 2 and the genetic distance d_{23} between populations 2 and 3 are not larger than K, then all three populations 1, 2, and 3 will belong to the same cluster, even if the genetic distance d_{13} between populations 1 and 3 is larger than K. According to this definition, all populations forming a ring species (e.g., Mayr 1942, 1963; Wake 1997; Irwin et al. 2001; see page 99) would belong to the same species. Note that if $K = 0$, each cluster (for example, species) is defined by a unique sequence of genes at the set of loci under consideration. The case of $K = 1$ corresponds to the two-locus BDM model. Alternatively, if new species result from the accumulation of a number of genetic (morphological) differences, then larger values of K are more appropriate. Genetic clusters corresponding to different values of K can also be interpreted as describing different levels of taxonomic classification. For example, let us specify an increasing sequence $K_1 < K_2 < K_3 < \dots$. Then, all populations at a genetic distance not larger than K_1 can be thought of as belonging to the same species, all populations at genetic distances that are larger than K_1 but not larger than K_2 can be thought of as belonging to different species within the same genus, all populations at genetic distances that are larger than K_2 but are not larger than K_3 can be thought of as belonging to different species and genera within the same family, and so on.

Migration into occupied patches. I will assume that migration into occupied demes has no effect on the genetic composition of the resident population, even if immigrants are coming from the same species and are genetically compatible and able to mate with the residents. As a working example, I envision a plant metapopulation where local demes produce a large number of seeds of which only few germinate. In this case, migrant seeds will have an extremely small probability of germinating unless there is an extinction event eliminating all or most resident plants. In a similar way, if there is frequency-dependent selection against immigrants then again one can neglect effects of migration other than bringing colonizers into an empty patch. The assumption of no effects of gene flow is justified only if the rates of immigration are very small or selection against immigrants is very strong.

Dynamic scenario. I wish to identify and understand dynamic features of the process of diversification following colonization of a new environment. As an initial condition I assume that all patches are occupied by populations with exactly the same founder genotype. This implicitly assumes that the spread of the species across the system of patches from the point of its initial invasion happens on a shorter time scale than that of mutation that appears to be reasonable. Initial spread is followed by the diversification phase during which the founder species splits into an increasing number of different clusters. Eventually the system reaches a state of stochastic equilibrium in which the number of clusters and their different characteristics fluctuate around certain values. Note that even after reaching this state, the clade keeps evolving as different species (or clusters) go extinct and their place is taken by new species (clusters).

7.2.2   Parameters and dynamic characteristics

The following is a list of parameters of this model together with their numerical values used in numerical simulations:

• The system size n, that is, the number of patches in the system ($n = 64; 256; 1,024; 4,096$).

• Spatial arrangement of the populations: $8^2 \times 1, 16^2 \times 1, 32^2 \times 1, 64^2 \times 1$ for 1D systems, and $8 \times 8, 16 \times 16, 32 \times 32, 64 \times 64$ for 2D systems.

• Number of loci ($\mathcal{L} = 100$).

• Fixation rate per locus ($\mu = 10^{-6}, 4 \times 10^{-6}, 16 \times 10^{-6}$).

• Extinction-recolonization rate ($\delta = 0.25 \times 10^{-2}, 10^{-2}, 4 \times 10^{-2}$).

• Clustering level and the number of substitutions leading outside the cluster ($K = 0, 1, 3, 7, 15$; note that with this choice of K, the number of substitutions leading outside the cluster, which I denote as $K^* = K + 1$, forms a geometric progression: $K^* = 1, 2, 4, 8, 16$).

The number of runs for each parameter configuration is 40.

The values of μ, $\mathcal{L}$, and K used appear to be biologistically realistic. The number of subpopulations n had to be relatively small because of the computational constraints. The values of δ chosen reflect the interest in the case of relatively low extinction.

I wish to understand the effects of these parameters on the following characteristics:

• The average time to the beginning of radiation, T, defined as the average waiting time until the first split of an initially uniform population into at least two clusters.

• The average duration of radiation, τ, defined as the average waiting time from T to the time when the number of clusters reaches the (stochastic) equilibrium value for the first time.

• The diversity, S, defined as the average number of clusters in the clade at the stochastic equilibrium.

- The average cluster range, $\mathcal{R}$, at the stochastic equilibrium, defined as the average number of populations per cluster.
- The average pairwise genetic distance, $\overline{d}_w$, between the members of the same cluster.
- The cluster diameter, $d_{w,\max}$, defined as the maximum genetic distance between the members of the same cluster.
- The turnover rate, $\mathcal{T}$, defined as the number of new clusters emerging per unit of time divided by the standing diversity S.
- The clade disparity, $\overline{d}$, defined as the average pairwise distance between all populations in the system.
- The average genetic distance of the clade from the founder, $\overline{d}_f$, defined as the average of pairwise distances between all populations and the species-founder.

7.2.3   Results

The analytical approximations for the diversity S, the average cluster range $\mathcal{R}$, and the turnover rate $\mathcal{T}$ will imply that the system size is sufficiently large so that the effects of the boundaries are negligible. For the model under consideration, the characteristic linear size is

$$l_c = \sqrt{\frac{\delta K}{\nu}} \qquad (7.1)$$

(compare with Sawyer 1977a,b, 1979; Bramson et al. 1996; Durrett and Levin 1996). Patches separated by (spatial) distances much larger than l_c demes are expected to behave largely independently. Also, for systems with the linear dimension l larger than l_c, the effects of borders will be small. (In 1D systems $l = n$, whereas in 2D systems $l = \sqrt{n}$.) This implies that for large systems (with $l > l_c$), the diversity S will increase linearly with the number of patches in the system, whereas the range $\mathcal{R}$ will not depend on n. In the 1D numerical examples, the effects of borders will be insignificant except for the smallest system with the largest K. In contrast, in the 2D examples the effects of borders will be important even in the largest system if K is large. Unfortunately, increasing the system size is currently impossible because of computation speed considerations.

Transient dynamics   Figure 7.3 illustrates the transient dynamics of the number of different clusters as well as the clade disparity $\overline{d}$ and the average distance from the founder $\overline{d}_f$.

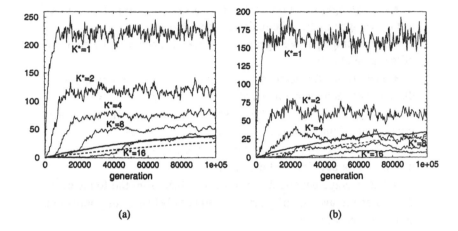

(a)   (b)

FIGURE 7.3. The dynamics of the diversity S at different clustering levels K^* ($= K + 1$), of the clade disparity $\overline{d}$ (bold line), and of the average distance from the founder $\overline{d}_f$ (dashed line). Parameters: $\mu = 4 \times 10^{-6}, \delta = 0.01$. The statistics are computed every 250th generation. (a) $32^2 \times 1$ system. (b) 32×32 system. (From Gavrilets (2004), fig.12.3.)

The dynamics of the average distance from the founder depends only on the fixation rate per locus μ and is approximated by equation

$$\overline{d}_f(t) = \frac{\mathcal{L}}{2} \left[1 - e^{-2\mu t}\right] \tag{7.2}$$

(Gavrilets 1999b). That is, $\overline{d}_f$ asymptotically approaches the distance equal to one-half of the maximum possible distance. This implies that after a sufficiently long time the members of the clade will be different from the founder in half of the genes on average. Moreover, the clade can be equally likely found in any part of the genotype space. Equation (7.2) can be used both to check the constancy of the rate of evolution in time and to estimate its value (see Gavrilets 1999b).

The dynamics of the clade disparity $\overline{d}$ do not seem to depend on the spatial dimensionality. The initial dynamics of $\overline{d}$ are similar to that of $\overline{d}_f$, with the difference that $\overline{d}$ increases twice as fast. A simple explanation of this fact is that while the disparity is computed on the basis of pairs of evolving lineages, in computing $\overline{d}_f$ one lineage in each pair (i.e., the founder) does not change. For a clade with no spatial structure the dynamics of disparity $\overline{d}$ are understood (Gavrilets 1999b). Unfortunately, for spatially explicit systems neither the equilibrium value of $\overline{d}$ nor its dynamics on the intermediate time scales are known. In larger systems, approaching an equilibrium for $\bar{d}$ takes a very long time. Figure 7.3 illustrates the important observation that the diversity S at low taxonomic levels (i.e., at small K) equilibrates faster than the clade disparity $\bar{d}$, whereas the equilibration of the diversity at higher taxonomic levels (i.e., at large K) can take a comparable or longer time. In the latter case, very high values of $\bar{d}$ (relative to its asymptotic equilibrium value) can be observed simultaneously with very low taxonomic diversity. The pattern of elevated disparity early in the history of many clades has been traditionally explained by paleontologists by invoking explanations that postulate temporal changes in the types and/or levels of forces driving divergence (Valentine 1980; Foote 1992, 1999; Erwin 1994; Wagner 1995; Lupia 1999). However, both the previous work (Gavrilets 1999b) and the models described here show that these patterns are perfectly compatible with the null model of time-homogeneous diversification.

Figure 7.4 illustrates the dependence of the time to the beginning of radiation, T, and the duration of radiation, τ, on parameters. The time to the beginning of radiation increases (approximately exponentially) with K. In biological terms, higher taxonomic groups arise later in the clade's history. T decreases with the fixation rate ν (apparently as $1/\nu$). Increasing ν by a certain factor results in a smaller decrease in T than the proportional decrease in K. The time T weakly increases with the extinction/recolonization rate δ and system size n. At $K = 0$, T can be approximated as the inverse of the expected number of mutations per clade, that is, $T \approx 1/(n\nu)$. The duration of radiation τ is not very sensitive to K and δ (weakly increasing with both these parameters), but is much more sensitive to ν. It appears that τ is on the order of $1/\nu$. This feature of the dynamics of radiation is compatible with that for the dynamics of parapatric speciation, where the time interval during which the intermediate forms are present has the order of the reciprocal of the mutation rate (as shown in the two previous chapters). Numerical simulations also show, as expected, that small systems (with small n) reach stochastic equilibrium faster than large systems (with large n). The dynamics of T and τ are similar in 1D and 2D systems with the same n. The dynamics of other characteristics crucially depend on the spatial dimensionality of the system.

Stochastic equilibrium: 1D systems   One can use the theory of coalescing random walks to approximate cluster ranges, diversity, and turnover rates (Gavrilets et al. 2000a; Gavrilets 2004). The formulas given below assume that the fixation rate per genotype is much smaller

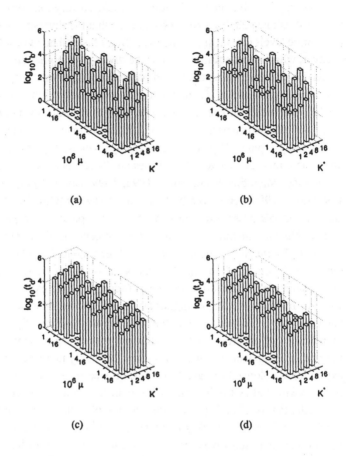

FIGURE 7.4. The average waiting time to the beginning of radiation T (a and b) and the average duration of radiation τ (c and d). Left column: $32^2 \times 1$ system. Right column: 32×32 system. $K^* = K + 1$. Within each figure the three sets of bars correspond to $\delta = 0.0025, 0.01,$ and 0.04 (from left to right). Within each set, individual bars correspond to different fixation rates.(From Gavrilets (2004), fig.12.4.)

than the extinction/colonization rate ($\nu \ll \delta$).

The average range of a cluster can be approximated as

$$\mathcal{R} = \sqrt{\frac{\pi \delta}{2\nu}} \frac{K!}{\Gamma(K + 1/2)}, \tag{7.3a}$$

where Γ is the gamma function (Gradshteyn and Ryzhik 1994). This

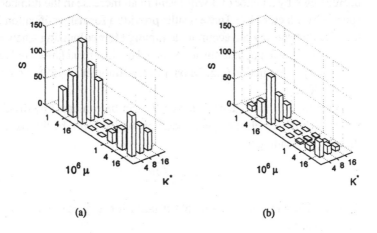

(a)   (b)

FIGURE 7.5. The number of clusters. (a) $32^2 \times 1$ system. (b) 32×32 system. $K^* = K + 1$. Within each figure the two sets of bars correspond to $\delta = 0.01$ (left) and 0.04 (right). The averages are taken over generations 200,000 through 500,000 and over 40 runs. (From Gavrilets (2004), fig.12.5.)

expression simplifies to

$$\mathcal{R} = \sqrt{\frac{\delta}{2\nu}} \qquad (7.3b)$$

for $K = 0$, that is, if a single substitution results in a new cluster. If K is large,

$$\mathcal{R} = \sqrt{\frac{\pi \delta K}{2\nu}} \qquad (7.3c)$$

The above equations assume that each mutation is unique.

The average diversity is just $\mathcal{S} = n/\mathcal{R}$, leading to

$$\mathcal{S} = \sqrt{\frac{2\nu}{\pi\delta}} \frac{\Gamma(K + 1/2)}{K!} \, n. \qquad (7.4)$$

Figure 7.5(a) illustrates the dependence of $\mathcal{S}$ on the parameters of the model observed in simulations. Biological intuition tells one that increasing the rate of fixation of new mutations should increase the rate of speciation, thus increasing the number of species in the system. Decreasing the rate of extinction-colonization should have a similar effect because larger levels of genetic variation will accumulate in the system. Equations (7.3) and (7.4) support these intuitions. For example, decreasing δ by a factor of 4 will result in an increase in the number of species by a factor of 2. These results provide a formal justification for the idea that species can accumulate rapidly after colonizing a new environment if local populations in the novel environment have a reduced probability of extinction (e.g. Mayr 1963; Allmon et al. 1998; Schluter 1998, 2000).

With $K = 0$, the rate of turnover, T, can be evaluated by dividing the number of new clusters per generation, which is νn, by the standing diversity S, leading to

$$T = \sqrt{\frac{\delta \nu}{2}}. \tag{7.5a}$$

The consideration of the time that it takes for a typical cluster to go extinct leads to

$$T = \frac{\nu}{K} \tag{7.5b}$$

for large K (Gavrilets et al. 2000a). The turnover rate weakly depends on δ for $K = 0$ and becomes approximately independent of δ for large K. The latter counterintuitive conclusion is explained by the fact that the increase in the overall extinction rate of species resulting from an increase in δ is exactly balanced by a decrease in the number of species S maintained in the system.

Numerical simulations have confirmed the validity of these approximations (Gavrilets et al. 2000a). In most cases, equation (7.4) underestimates the average number of species by a couple of percents, whereas equation (7.5) overestimates the turnover rate by about 5% to 10%. In the case of the smallest mutation rate, the errors are slightly higher.

Additional simulations were used to analyze the structure of the clade in genotype space. Figures 7.6 shows that although there is plenty of within-cluster genetic variation, typical members of a cluster are at distances that are smaller than K. Effects of ν and δ do not seem to be significant.

Stochastic equilibrium: 2D systems Approximating the average range of clusters $\mathcal{R}$ in the 2D case is much more difficult than in the 1D case. If $K = 0$, then the results of Bramson et al. (1996) on the number of species $\hat{R}$ found within a square embedded within a much larger area give

$$\hat{\mathcal{R}} = \frac{2\pi\delta}{\nu} \frac{1}{[\ln(\delta/\nu)]^2}. \tag{7.6a}$$

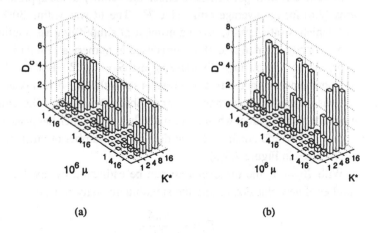

(a)   (b)

FIGURE 7.6. The average pairwise distance $\bar{d}_w$ within clusters. (a) $32^2 \times 1$ system. (b) 32×32 system. Within each figure the three sets of bars correspond to $\delta = 0.0025, 0.01$, and 0.04 (from left to right). The averages are taken over generations 200,000 through 500,000 and over 40 runs. $K^* = K + 1$. (From Gavrilets (2004), fig.12.6.)

This approximation implies that the size of the system is sufficiently large. $\hat{R}$ overestimates R because many species occupying nearby areas may still have few representatives within the sampling square.

One can also estimate the average number of populations $\tilde{R}$ that have the same type as the population from a randomly chosen patch. $\tilde{R}$ is somewhat larger than R. (For example, if there are 5 clusters with 1, 2, 3, 4 and 5 populations, respectively, then $R = (1+2+3+4+5)/5 = 3$, but $\tilde{R} = (1^2 + 2^2 + 3^2 + 4^2 + 5^2)/15 = 3.67$.) If $K = 0$, then range $\tilde{R}$ can be found by integrating the probability of identity of two genotypes at a certain distance apart (found by Sawyer 1977a) over all possible spatial positions. This approach leads to

$$\tilde{R} = \frac{\pi \delta}{4\nu} \frac{1}{\ln(\delta/4\nu) + 2\pi\delta}. \tag{7.6b}$$

(Gavrilets 2004). For large K,

$$\tilde{R} = \frac{\pi \delta K}{2\nu} \Lambda(K, \nu), \tag{7.6c}$$

where $\Lambda(K, \nu)$ is a function that depends only weakly on its arguments For the parameter values used here, $\Lambda(K, \nu)$ stays between 0.1 and 0.2 (Gavrilets 2004).

Simulations were performed to check the validity of the approximations (7.6) for the average range size $\mathcal{R}$. The fit was within 30% to 50% which is satisfactory, given a number of simplifying assumptions involved and the small size of systems used in numerical simulations.

As before, the average diversity is just $S = n/\mathcal{R}$. Both analytical approximations and numerical results show that the diversity in 2D systems is (much) lower than in comparable 1D systems. The differences are most apparent when S is relatively small, which happens with large δ and K and small μ. Some data on the number of clusters are summarized in Figure 7.5(b).

With $K = 1$, the turnover rate can be estimated by dividing the number of new clusters, νn, by the standing diversity S, leading to

$$T = \frac{2\pi\delta}{[\ln(\delta/\nu)]^2} \tag{7.7a}$$

in the case of $\mathcal{R}$ as given by equation (7.6a), and to

$$T = \frac{\pi\delta}{4} \frac{1}{\ln(\delta/4\nu) + 2\pi\delta} \tag{7.7b}$$

in the case of $\tilde{R}$ as given by equation (7.6b).

To approximate the turnover rate in the case of $K > 1$, one can use some intuitive but not rigorous approximations which lead to the conclusion that T is of the order of ν/K (Gavrilets 2004). The latter expression does not depend on the extinction rate, which is similar to the 1D case.

Simulations were also used to analyze the structure of the clade in the genotype space. Figure 7.6b illustrates the average pairwise within-cluster distance, $\overline{d}_w$. As in the 1D case, this figure shows that although there is plenty of within-cluster genetic variation, typical members of a cluster are at a distance that is smaller than K. Effects of ν and δ on $\overline{d}_w$ and $d_{w,\max}$ do not seem to be significant. Typically, populations that belong to the same cluster are also spatially contiguous.

The number of patches in the system n is also of importance. If n is too small, significant diversification will be prevented. For example, consider a case with $\delta = 10^{-2}, \mu = 10^{-6}$, and $K + 1 = 16$. Numerical simulations show that small systems with 8×8 patches fail to diversify and have a single cluster present. In contrast, in 16×16 systems there are on average just over two clusters, whereas in 32×32 systems this number goes to eight.

7.2.4 Interpretations

The above analysis leads to a number of potentially important generalizations.

• The waiting time to the beginning of radiation T increases with decreasing fixation rate per locus μ and increasing the number of genetic changes necessary for speciation K. The local extinction and recolonization rate δ, the dimensionality of the system, and the number of patches n have much smaller effects. In numerical simulation, the waiting time to the beginning of radiation was on the order of 10^3 (for $K = 0$ and 1) to 10^5 generations (for large K).

• The duration of radiation τ depends mostly on the fixation rate μ. Time τ weakly increases with K and system size n and weakly decreases with δ. The duration of radiation is longer in the 1D systems than in the 2D systems. In numerical simulations, the order of τ ranges from 10^4 to 10^5 generations.

• The transient dynamics of the diversity S (that is, the average number of clusters) and the disparity $\overline{d}$ (that is, the average pairwise distance between the populations in the clade) are decoupled to a certain degree. At low taxonomic levels (i.e., with small K), diversity increases faster than disparity, whereas at high taxonomic levels (i.e., with large K) disparity increases faster than diversity. This observation explains the difference between the patterns of diversification as observed in the fossil record (which are usually summarized at higher taxonomic levels, e.g., Valentine 1980; Foote 1992, 1999; Erwin 1994; Wagner 1995; Lupia 1999) and in more recent groups (which are usually summarized at lower taxonomic levels, e.g., Schluter 2000).

• The average genetic distance from the species founder increases monotonically at a constant rate controlled by the fixation rate. Note that, previously, Gavrilets (1999b) used this property to develop a method for testing the constancy of the rate of evolution and for estimating its rate using morphological data.

• The clade as a whole keeps changing genetically, even after the number of species (or other genetic clusters) has approached an equilibrium level.

• The general effects of the model parameters on different equilibrium characteristics of the clade are mostly as suggested by biological intuition. For example, diversity increases with mutation rate and decreases both with the local extinction and recolonization rate and with the number of genetic differences required for speciation. Counter-intuitively, however, the model predicts that the turnover rates do not depend (or weakly depend) on extinction rates and are mostly controlled by parameters ν and K. Intriguingly, the turnover rates apparently did not differ between the two groups of aquatic beetles in the family *Dytis-cidae* associated either with running or stagnant water bodies in spite of the fact that local extinction rates are higher in the latter group (Ribera et al. 2001).

• Diversification requires that the overall number of patches in (or spatial area of) the system exceeds a certain minimum value. This effect may have contributed to the fact that in adaptive radiation of the West Indian *Anolis* lizards, within-island speciation occurred only on bigger islands, in spite of the fact that the degree of spatial heterogeneity does not seem to differ between the islands (Losos 1998). In very large systems the overall diversity increases linearly with the number of patches (or area).

• The results presented here show profound effects of spatial dimen-sionality on the dynamics of diversification and significant differences between 1D systems (such as describing rivers, shores of lakes and oceans, areas at a constant elevation in a mountain range) and 2D sys-tems (such as describing oceans, continental areas).

• In general, the characteristics of 2D systems are (much) more sen-sitive to parameter values than those of 1D systems. For example, increasing the local extinction rate δ by a factor of 25 will typically de-crease species diversity by the same factor in 2D systems. In contrast, in 1D systems the decrease will only be by factor of 5.

• The diversity in 1D systems is predicted to be (much) higher than that in 2D systems. For example, let there be $\mathcal{L} = 100$ loci, let the mutation rate per locus per generation be $\mu = 4 \times 10^{-5}$, let the local extinction/recolonization rate be $\delta = 10^{-2}$ per deme per generation, and let there be $n = 1,024$ local demes. If the demes are arranged in a 32×32 square, numerical simulations show that there are on average 2.6 genetic clusters at the clustering level corresponding to $K = 16$. In con-trast, if the demes are arranged on a line (i.e., in a $32^2 \times 1$ pattern), there are on average 17.2 such clusters. These effects may have contributed to the extraordinary divergence of cichlids in the great lakes of Africa, most species of which inhabit the relatively narrow band along the shore-line and have extremely restricted dispersal abilities (e.g., Markert et al. 1999; Kornfield and Smith 2000).

• Both the average pairwise distance within a cluster $\overline{d}_w$ and the cluster diameter $d_{w,\max}$ are mostly controlled by parameter K and are close to its numerical value. Typically, genetic clusters in 1D systems are denser than in 2D systems.

• It has been argued that species can accumulate rapidly after colonizing a new environment if in the novel environment the species have a reduced probability of extinction. This could happen because reduced extinction can extend the lifetime of a lineage, thus increasing its chance to accumulate enough genetic changes to result in RI (Mayr 1963; Allmon 1992; Schluter 1998, 2000). The same could also happen after developing a key innovation that decreases the extinction rate. The results presented here quantify these arguments. As discussed above, decreasing the extinction rate by a certain factor in a 2D metapopulation will increase the equilibrium diversity by approximately the same factor.

The above conclusions are based on a specific model. Therefore, certain cautiousness is required when trying to apply them to more general and realistic situations. In particular, I note that here the spatial arrangement of demes in a 2D system is (unrealistically) symmetric. Allowing for some demes to be unsuitable is expected to increase the possibilities for differentiation and speciation in both 1D and 2D systems. However, if space were continuous rather than discrete, these possibilities would be significantly reduced in 2D systems. One should also note that the deme-based model considered here does not include the deleterious effects of increasing local densities and migration rates on the possibility of diversification discussed in the previous section.

7.3 DEME-BASED MODELS: SPREAD OF MUTUALLY INCOMPATIBLE ADVANTAGEOUS GENES

In principle, it is possible that the alleles involved in incompatibilities with other alleles have pleiotropic effects on other characters subject to direct natural selection. In the two previous chapters, we have seen that direct selection can have profound effects on the possibility of speciation in systems with two demes. This section considers some recent results on multideme models.

Advantageous alleles have three specific effects on the dynamics of speciation. First, the rate at which advantageous alleles are produced by mutation is much smaller than that for neutral alleles. This decreases the overall probability of speciation. Second, advantageous alleles sweep through the deme much faster than neutral alleles and, thus, have a much smaller probability of being lost as a result of genetic drift or local extinction. This increases the overall probability of speciation. Third, once established in a deme, advantageous alleles will tend to spread through the whole system in a wavelike fashion (Fisher 1937; Kolmogorov et al. 1937). Because the waves of advantageous alleles sweep through the whole species fast, the chance that two (or more) waves of mutually incompatible advantageous alleles will be present simultaneously is reduced. This decreases the overall probability of speciation.

Consider a situation when two mutually incompatible advantageous genes get established in two different demes. The two waves of genes will spread until they become established in a pair of neighboring demes. Assuming that the loss in fitness due to incompatibility is larger than the selective advantage of each allele, the advancing waves will be stopped. As a consequence, a hybrid zone will be formed (Endler 1977) that will represent a genetic barrier to gene flow (Barton 1979; Spirito et al. 1983; Barton and Bengtsson 1986; Gavrilets 1997c; Gavrilets and Cruzan 1998; Piálek and Barton 1997; Box 3.2). Although in a spatially uniform environment, different hybrid zones are expected to form in different locations, stochasticity due to genetic drift, extinction, and colonization is expected to result in an accumulation of hybrid zones in certain locations (Bazykin 1969). Eventually this process is expected to split the species into spatially separated and mutually incompatible populations. The remainder of this chapter considers some simple models aiming to describe this process in situations when genetic incompatibilities are pairwise, as in the original BDM model.

Church and Taylor (2002)   These authors were the first to model the spread of mutually incompatible advantageous genes in a 1D stepping-stone system. They assume that the effective gene flow between neighboring subpopulations is constant and does not depend on the number of incompatibilities separating the subpopulations. No extinction of local populations is allowed. The speciation event is defined as the moment when a pair of demes (not necessarily neighbors) accumulate a certain number Q of incompatibilities (in their numerical simulations, $Q = 3$). Under these assumptions, Church and Taylor have demonstrated the formation of different clusters of demes, with complete compatibility within clusters and some incompatibility between the clusters. The probability of speciation increased with the number of demes. With just two demes, the waiting time to speciation T was shortest in the absence of migration. In contrast, with more than two demes, T was minimized at a small but nonzero migration rate. This happens because with low migration, advantageous alleles that originated in different demes have a chance to migrate into the same deme and, thus, occur there simultaneously as a group that may be incompatible with another group of genes formed in a similar way in a different deme. If the migration rate is too high, advantageous mutations rapidly spread across the whole system, which reduces the probability of differentiation.

Kondrashov (2003) Kondrashov used analytical approximations to estimate the rate of accumulation of incompatibilities in a similar system. There are three crucial parameters in his model: the probability ω that an advantageous mutation is established in a local population per generation, the time B that the spread of the advantageous allele across the whole system takes given the allele has been established locally, and the probability ε that two advantageous alleles are mutually incompatible. In generation t, let $X(t)$ be the number of different advantageous alleles that have not been able to spread through the whole system because of their incompatibility with some other advantageous alleles. Each new allele established in a local population will increase X by one if it is not able to sweep through the whole system. This can happen if it encounters an incompatible allele in another advancing wave or in a previously formed hybrid zone. Because of symmetry, the expected number of alleles potentially encountered on the other side of all hybrid zones is $X/2$. During its spread, an advancing allele will encounter on average ωB other advancing waves, of which $\varepsilon \omega B$ will have incompatible alleles. With probability $\varepsilon \frac{X}{2}$, the advancing wave will be stopped by an existing hybrid zone. Therefore, the rate of change in $X(t)$ can be approximated by a differential equation:

$$\frac{dX}{dt} = \omega \left[\varepsilon \frac{X}{2} + \varepsilon \omega B \right]. \qquad (7.8a)$$

Each collision of two waves and each collision of a wave with a previously formed hybrid zone establishes one new incompatibility in the system. Therefore, the overall number of incompatibilities $J(t)$ in the system changes according to a differential equation:

$$\frac{dJ}{dt} = \omega \left[\varepsilon \frac{X}{2} + \frac{1}{2} \varepsilon(\omega B) \right]. \tag{7.8b}$$

These equations can be used to show that initially both $X(t)$ and $J(t)$ increase linearly in time:

$$X(t) \approx B\omega^2 \varepsilon t, \; J(t) \approx \frac{1}{2} B\omega^2 \varepsilon t. \tag{7.9}$$

Asymptotically (i.e., when $X(t)$ is large), most waves are stopped by previously formed hybrid zones. In this case, $X(t)$ increases linearly in time, whereas $J(t)$ increases quadratically with time, as in the original Orr-Orr model (see equation (6.9) on page 160). This case describes situations in which the species has already split into a number of different clusters separated by a large number of incompatibilities.

The dynamics of hybrid zones   Hybrid zones formed by mutually incompatible alleles can delay the spread of other alleles. Hybrid zones also can move as a result of stochastic factors and pile up leading to stronger and stronger RI between adjacent populations and eventually resulting in parapatric speciation (Bazykin 1969). To illustrate these processes, let us consider a 1D stepping-stone system assuming that mutation, migration, and selection are weak.

Local establishment of advantageous mutant alleles. If each deme (local population) has N diploid individuals and the probability of an advantageous mutation is ν_a per gamete per generation, then there are approximately $2N\nu_a$ new advantageous alleles per deme per generation. If the relative advantage of the new allele over the corresponding resident allele is s, then the probability of local fixation is approximately $2s$ (Haldane 1927). These approximations lead to the rate of local establishment of new advantageous alleles being

$$\omega = 2N \times \nu_a \times 2s = 4Ns\,\nu_a \tag{7.10}$$

per deme per generation.

Local extinction and recolonization. As before, let us assume that each deme is subject to extinction and recolonization from one of its neighbors at rate δ per deme per generation.

Genetic incompatibility. Assume that any two new alleles can be incompatible with probability ε. Here, incompatibility means that organisms (hybrids) carrying both alleles have reduced fitness or that the probability of mating between two organisms carrying incompatible alleles is reduced.

Gene flow. Once established in a local population, an advantageous gene will tend to spread across the whole system. The following simple model for the dynamics of the spread will be used. Assume first that there is a single advantageous gene. Then the population at a border of the range of this gene sends $2N \times (m/2)$ alleles into its neighboring deme, where the allele has not yet been established. Because the probability of fixation is $2s$, the probability that the gene is established in the neighboring population is

$$\beta_0 = 2N \times \frac{m}{2} \times 2s = 2Nsm \qquad (7.11)$$

per generation. Here, subscript zero indicates that there are no incompatibilities between the two populations. Note that because the mutation rate ν_a is typically much smaller than the rate of migration m, ω must be much smaller than β_0.

Next let us consider a hybrid zone formed by two mutually incompatible alleles. The effects of the hybrid zone on the gene flow across it can be quantified using a gene flow factor η (see Box 5.2). Although η can be explicitly found for some specific models, it crucially depends on a number of parameters that are generally unknown. As a simple heuristic approximation, I will assume that the gene flow factor for a hybrid zone formed by c pairwise incompatibilities is

$$\eta = 1 - \frac{c}{Q} \qquad (7.12)$$

if $c \leq Q$, and is 0 otherwise. Here, Q is the number of incompatibilities resulting in complete RI. Recall that the product of migration rate and η gives the effective migration rate (see Box 5.2). Equation (7.12) implies that the effective gene flow reduces linearly with the number of incompatibilities. Then the probability that the deme on the other side of a hybrid zone formed by c incompatibilities fixes a new advantageous allele brought by immigrants is

$$\beta_c = 2N \times \frac{m}{2} \eta \times 2s = \beta_0 \left(1 - \frac{c}{Q}\right) \qquad (7.13)$$

if $c < Q$, and is 0 otherwise.

Summarizing, the parameters of the model are:
- The number of demes in the system n.
- The rate of local establishment of new advantageous mutations ω.

• The rate of local extinction and recolonization δ.

• The probability of pairwise incompatibility ε.

• The rate of local establishment of new alleles brought by immigrants β_0.

• The number of incompatibilities resulting in complete reproductive isolation Q.

A number of numerical simulations of this model have been performed. Figure 7.7 illustrates the observed dynamics. Part (a) shows the dynamics of the number X of alleles that have not spread across the whole system and the overall number J of pairwise incompatibilities. During the first 15,000 to 20,000 generations, the growth of J is linear and becomes quadratic after that, as predicted. However, populations continuously accumulate new alleles and incompatibilities, both before and after the actual acquisition of complete RI. Therefore, understanding the dynamics of J is not sufficient to predict other important characteristics, such as the waiting time to speciation and the number of reproductively isolated clusters (i.e., species) in the system. These two are illustrated in part (b) of the Figure. Both the waiting time to speciation and species diversity are quite compatible with those in the neutral models considered above. As expected, increasing the local extinction-recolonization rate δ decreases the number of species maintained in the system (data not shown).

Part (c) of the Figure describes the dynamics of cluster borders. Specifically, I use shaded pixels, linearly arranged along the abscissa, to represent the number of incompatibilities between the neighboring demes. For example, a pixel in position i represents the number of incompatibilities between demes i and $i + 1$, using five shades of gray (from white at zero incompatibilities to black at four or more incompatibilities). The figure shows random movement of borders as a result of extinction and recolonization accompanied by their collision. Collision of borders causes them to stick together. Different hybrid zones pile up, resulting in stronger and stronger RI. New borders are introduced in the system as a result of mutation and spread. Higher rates of extinction and colonization result in more sporadic movement of the borders.

Relative to the models in which incompatible genes are neutral, this model has more parameters, specific numerical values of these parameters are more difficult to justify (because of the lack of relevant data), the assumptions the model is based upon are heuristic and/or ad hoc,

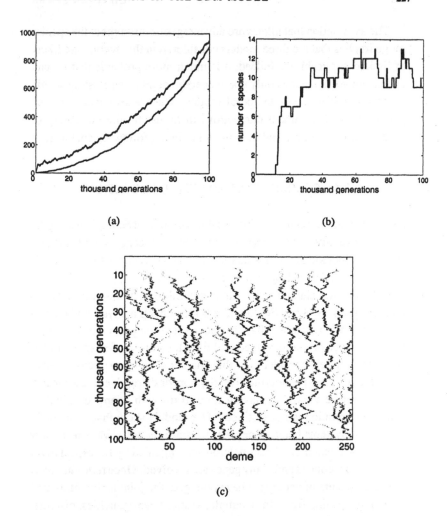

FIGURE 7.7. Spread of mutually incompatible advantageous alleles. Parameters: $n = 256, \omega = 2 \times 10^{-4}, \beta_0 = 0.1, Q = 4,$ and $\delta = 0.01$. (a) Dynamics of the number of segregating alleles X (higher curve) and pairwise incompatibilities I (lower curve). (b) Dynamics of the number of species. (c) Dynamics of cluster borders.

and the simulations are in general slower. For these reasons, a more detailed study of the model was not undertaken. However, it would be very interesting and important to perform an analysis of this model comparable to those in the previous sections.

The assumption that alleles are advantageous throughout the species range implies that the three models considered in this section are likely of limited applicability. Indeed, it is much more probable that mutant alleles are advantageous only under specific conditions existing within a certain spatial and/or temporal range. The consequences of such genotype-by-environment interactions in fitness on the possibility of speciation will be explored in more detail in the following chapters.

7.4 COMMENT ON ADAPTIVE RADIATION

Adaptive radiation can be defined as the evolution of ecological and phenotypic diversity within a rapidly multiplying lineage (Schluter 2000). Examples include the diversification of Darwin's finches on the Galápagos islands, *Anolis* lizards on the Carribean islands, Hawaiian silverswords, and a mainland radiation of columbines, among many others (see Simpson 1953 and Schluter 2000 for general analysis and discussion). It is often assumed that disruptive selection plays the key role in the emergence of multiple species during the course of adaptive radiation. The models considered in this chapter account explicitly neither for ecological and phenotypic diversity nor for disruptive selection and, thus, might be viewed as having little to do with adaptive radiation. This interpretation is not justified. The models show that in spatially structured systems heterogeneity in selection is not always necessary for rapid and extensive diversification accompanied by RI, especially if extinction is reduced and many genes are involved. Once reproductively isolated clusters of genotypes have emerged, the joint action of natural selection (coming from, for example, spatial heterogeneities, discreteness of recources, and interactions between different individuals) and various stochastic factors are expected to result in quick phenotypic and ecological divergence. According to this view, the crucial step of adaptive radiation is the achievement of RI rather than ecological diversification, which is expected to proceed relatively easily soon after RI has been achieved. For example, the achievement of RI preceded as a rule the achievement of ecological segregation in the diversification of birds of Northern Melanesia (Mayr and Diamond 2001). At the same time, heterogeneity in selection is expected to accelerate the process of diversification significantly. However, its effects have not been studied yet using explicit mathematical models of adaptive radiation.

7.5   SUMMARY

• Parapatric speciation in spatially structured systems can follow as a byproduct of a localized spread of mutually incompatible (combinations of) genes.

• Diversification by this mechanism requires the range of the species to be sufficiently large. Therefore, diversification on small islands or in small lakes may be limited.

• Rapid diversification by this mechanism can follow after colonizing a new environment or developing a key innovation.

• With regard to the effects of the incompatible alleles on other components of fitness (such as viability and fertility), these alleles can be nearly neutral, or advantageous locally, or advantageous throughout the species range.

• Dynamics of parapatric speciation in multideme models have been studied using individual-based and deme-based models. The most thoroughly investigated case is that of nearly neutral mutually incompatible genes. The following conclusions have emerged:

- Under favorable conditions (i.e., if the local population size is small, mutation rate is high, and migration rate is low), parapatric speciation can happen on the timescale of a few hundred to a few thousand generations.

- If genetic variation is a major limiting factor, populations will tend to split into reproductively isolated groups in the middle of the species range in accordance with the theory of centrifugal speciation. If genetic variation is abundant, speciation will occur more easily in peripheral populations in accordance with the theory of peripatric speciation.

- There are significant differences in the dynamics of speciation between 1D systems (such as describing rivers, shores of lakes and oceans, and areas at a constant elevation in a mountain range) and 2D systems (such as describing lakes, oceans, and continental areas). The former are more prone to speciation and tend to maintain higher species diversity than the latter. These effects may have contributed to the extraordinary divergence of cichlids in the great lakes of Africa, most species of which inhabit the relatively narrow band along the shoreline and have extremely restricted dispersal abilities.

- Reduced extinction rates (e.g., observed in the absence of predators and competitors and with plenty of recources) are expected to result in much higher diversity relative to the cases of large extinction rates.

- The turnover rates are mostly independent of the local extinction rates and are primarily controlled by the rates of origination of genetic variation.

- The transient dynamics of the diversity (that is, the number of clusters) and the disparity (that is, the average pairwise distance between the populations in the clade) are decoupled. At low taxonomic levels, the diversity increases faster than the disparity, whereas at high taxonomic levels the diversity increases slower than the disparity. This observation explains the difference between the patterns of diversification as observed in the fossil record (which are usually summarized at higher taxonomic levels) and in more recent groups (which are usually summarized at lower taxonomic levels).

- If speciation does occur, multiple new species tend to emerge from different parts of a spatially structured population at approximately the same time. This possibility should be incorporated more explicitly into the methods for reconstructing phylogenies.

• Parapatric speciation can be caused by the spread of mutually incompatible advantageous genes. On the one hand, advantageous mutations are less frequent and tend to spread through the whole system more rapidly. This decreases the possibility of creating geographic variation. On the other hand, advantageous mutations are less likely to disappear from the population as a result of random drift or local extinction. This increases the possibility of creating geographic variation.

7.6   CONCLUSIONS

Populations of many species are spatially structured. Resulting isolation by distance and spatial heterogeneity in selection are important factors controlling the dynamics and patterns of parapatric speciation. Simple mathematical models, which neglect heterogeneity in selection, have uncovered a number of important features of diversification on holey fitness landscapes in the presence of spatial structure. These include the quantitative estimates of the timescale of speciation, its limiting factors, the importance of the size and spatial dimensionality of the habitat, and the plausibility of the simultaneous emergence of multiple species. The most important next step is the development of mathematical models explicitly dealing with spatial heterogeneity in selection in species represented by networks of interconnected populations.

Part III
Speciation via the Joint Action of Disruptive Natural Selection and Nonrandom Mating

CHAPTER EIGHT

Maintenance of genetic variation under disruptive natural selection

The approaches used in the previous part of this book allowed one to neglect genetic details to a certain degree. In this part, I consider models of speciation in which the population genetic state is described explicitly in terms of haplotype or genotype frequencies. These are the models that have received the most attention in speciation research. Explicit consideration of the dynamics of haplotype (or genotype) frequencies allows one to treat frequency-dependent selection and get additional insights into the dynamics of speciation driven by spatially heterogeneous selection. The price for these gains is that only a very small number of loci (one, two, or three) can be handled in analytical models, whereas detailed studies of numerical models are prevented because of a very large number of potentially important parameters, the values of which are generally unknown. In terms of fitness landscapes, the models to be discussed here are characterized by landscapes that are altered as a result of changes in the population genetic structure or in spatial location of individuals.

In considering analytical models below, I will use the standard methods of the dynamical systems theory (e.g., Hofbauer and Sigmund 1988; Glendinning 1994). Although nonequilibrium dynamics such as cycling and chaos are possible in population genetics models (e.g., Hadeler and Liberman 1975; Hastings 1981; Gavrilets and Hastings 1995a), in the models to be considered below, the populations will typically end up at an equilibrium state. If there is only one stable equilibrium, the population will end up there for any initial condition. If there are multiple simultaneously stable equilibria, the outcome of the evolutionary dynamics will depend on initial conditions, typically with the population evolving to the nearest equilibrium. In either case, the approach to an equilibrium usually occurs rapidly (relative to the timescale of speciation in the previous part). For this reason, the main emphasis will be on conditions for existence and stability of different equilibria rather than of the waiting time to and the duration of transition to certain states that were the focus of the previous three chapters. In this part, I start with models describing the maintenance of genetic variation under disruptive selection. Then, I review and generalize some models of nonrandom mating. Finally, I bring together the models for disruptive selection and nonrandom mating to study the dynamics of sympatric and parapatric speciation.

Traditionally, disruptive selection is defined as selection that favors different optima in the population (e.g., Mather 1955). Under such selection intermediate phenotypes (i.e., the phenotypes that fall between the optima) have reduced fitness. I will use a broader notion of disruptive selection that also includes situations when different parts of a population experience selection in different directions or when the population experiences selection in different directions at different times. The latter situations are sometimes described using the term *diversifying selection*. Although earlier studies indicated that disruptive selection is less common than stabilizing selection (Endler 1986), new data suggest that both types of selection are equally common (Kingsolver et al. 2001).

That disruptive selection can maintain genetic variation is well established (e.g., Mather 1955). Disruptive selection maintaining genetic variation creates conditions in which getting rid of low-fitness intermediates is a general way to increase the average fitness of the population. One of the possible ways to accomplish this is the evolution of prezygotic RI between clusters of optimum organisms (to be considered in the next chapter). Therefore, disruptive selection maintaining genetic variation creates a situation favoring *adaptive speciation*, i.e., the situation where speciation is an outcome that increases fitness. One has to be absolutely clear that evolution of RI is but one of several ways to increase the average fitness of the population. Other possibilities include evolution of dominance, epistasis, phenotypic plasticity, sexual dimorphism, and migration (e.g., Sanderson 1989; van Dooren 1999; Matessi et al. 2001; Bolnick and Doebeli 2003). Therefore, the content of this chapter is also relevant for the theories of these other processes.

In this chapter, I discuss several simple models describing the maintenance of genetic variation under disruptive selection. Some of these models have been used as major components of the models of speciation to be considered in Chapter 10. First, I consider models of spatially heterogeneous selection in which different genotypes and phenotypes are favored in different parts of the population range. Then, I describe models of spatially uniform selection in which certain intermediate organisms have reduced fitness independently of environmental conditions. After that, I briefly consider a simple classical model of selection varying in time. Finally, I look at models of frequency-dependent selection. I will describe both a standard phenomenological approach for introducing frequency-dependent selection in population genetics models and more detailed approaches in which fitness emerges from certain ecological interactions. I will also outline a series of novel approximations known as *adaptive dynamics* which are very helpful in studying evolutionary dynamics under frequency-dependent selection.

8.1   SPATIALLY HETEROGENEOUS SELECTION

Spatial heterogeneity of biotic and abiotic environments across a species range is ubiquitous. This heterogeneity is expected and has been repeatedly demonstrated empirically to result in variation in selection acting on individual alleles (and traits) or their combinations (e.g., Hedrick et al. 1976; Endler 1978, 1980; Hedrick 1986; Bell 1997; Mitton 1997; Smith and Skúlason 1996; Stratton and Bennington 1998). Commonly, there are trade-offs, such that the alleles that are advantageous in one environment become deleterious in another (e.g., Futuyma and Moreno 1988; Van Tienderen 1991; Smith and Skúlason 1996; Cooper and Lenski 2000). For example, Schluter (2000) analyzed the published data from reciprocal transplant experiments in which performances of pairs of phenotypically differentiated morphs, populations, or closely related species are measured in both their environments. In 36 out of 42 studies, the native type was superior to the foreign transplant. An important evolutionary implication of these trade-offs is that they create an opportunity for the maintenance of genetic variation in natural populations, especially if migration is limited (e.g., Levene 1953; Dempster 1955; Felsenstein 1976; Hedrick et al. 1976; Karlin 1982; Hedrick 1986).

8.1.1   The Levene model

I will start with a classical Levene model that assumes that individuals are randomly dispersed among the demes at the beginning of each generation, so that no geographic differentiation is maintained. For example, one can think of a plant population inhabiting a highly heterogeneous area with pollen being dispersed over much larger distances than those characterizing local variation in selection.

Following Levene (1953), let us consider a one-locus, two-allele diploid population inhabiting n niches that differ in selection regimes. The three genotypes are **AA, Aa,** and **aa.** Let the genotype fitnesses (viabilities) in niche j be $w_{AA,j}, w_{Aa,j}$, and $w_{aa,j}$. Each niche j contributes a constant proportion c_j to a pool of mating adults ($\sum c_j = 1$). Assume that mating is random in this pool and that offsprings are then dispersed randomly over the niches. Let p and $q = 1 - p$ be the frequencies of alleles **A** and **a** in the pool of offspring. At the beginning of each generation, allele frequencies in all niches are the same. In each niche, selection will change the allele frequencies to some degree. The change in the frequency of **A** in the j-th niche is given by the standard equation (2) in Box 2.1, which I present here for convenience:

$$\Delta_j p = pq \frac{w_{A,j} - w_{a,j}}{\overline{w}_j}. \tag{8.1}$$

Here $w_{A,j} = w_{AA,j}p + w_{Aa,j}q$ and $w_{a,j} = w_{Aa,j}p + w_{aa,j}q$ are the induced fitnesses of allele **A** and **a** in niche j, and $\overline{w}_i = w_{AA,j}p^2 + w_{Aa,j}2pq + w_{aa,j}q^2$ is the mean fitness of individuals in niche j. The local changes will result in a global change in the allele frequency in the pool of mating adults:

$$\Delta p = \sum_{j=1}^{n} c_j \, \Delta_j p. \tag{8.2}$$

This equation can be used to show that the frequency of allele **A** will increase when this allele is rare (i.e., when $p \ll 1$) if

$$\sum_j c_j \frac{w_{Aa,j}}{w_{aa,j}} > 1, \tag{8.3a}$$

and will decrease when this allele is common (i.e., when $q \ll 1$) if

$$\sum_j c_j \frac{w_{Aa,j}}{w_{AA,j}} > 1 \tag{8.3b}$$

(Levene 1953). Together, conditions (8.3) guarantee that neither allele can be lost deterministically and, thus, genetic variation will be protected. Strobeck (1974) showed that conditions for protection of genetic variation remain the same even if mating takes place within the niches or within mating groups defined on the basis of the place of birth rather than in a common mating pool.

The expressions in the left-hand side of equations (8.3) can be interpreted as the average ratios of the fitnesses of heterozygotes and homozygotes. Thus, conditions (8.3) require overdominance at the population level. However, at the level of a single deme, overdominance is not required. For example, let us assume that there are niches of type I and type II, with C_1 and C_2 being the proportions of the niches ($C_1 + C_2 = 1$). Assume that all niches contribute equally and that viabilities of the three genotypes are

$$\begin{array}{cccc} & \textbf{AA} & \textbf{Aa} & \textbf{aa} \\ \text{in niche I} & 1+s & 1 & 1-s \\ \text{in niche II} & 1-s & 1 & 1+s \end{array} \qquad (8.4)$$

Here, the coefficient s measures the strength of additive selection acting within each niche ($0 \le s \le 1$). The direction of selection in the two types of niches are opposite, with allele A being advantageous in the niches of type I and deleterious in the niches of type II. Simplifying conditions (8.3), one finds that genetic variation is protected if $s >$ $|C_1 - C_2|$. That is, stronger local selection (i.e., larger s) is required for more asymmetric fitness regimes (corresponding to larger absolute values of the difference of C_1 and C_2). The equilibrium allele frequency can be written as

$$p^* = \frac{1}{2} + \frac{C_1 - C_2}{2s}.$$

The last equation shows that the allele frequency can be close to $1/2$ if the ratio $|C_1 - C_2|/2s$ is small. In this case, a large level of genetic variation will be maintained. I note that Ludwig (1950) apparently was the first to show that an inferior competitor can survive by using an alternative resource in a somewhat similar model.

Biological systems fitting the scenario of the Levene models are known. For example, the African finch *Pyrenestes* exhibits a polymorphism in bill size. Morphs differ in diet and feeding performance on soft and hard seeds, but breed randomly with respect to bill size. Smith (1993) provided evidence that the polymorphism results from a single genetic factor and is maintained by disruptive selection, with different morphs being superior with regard to feeding performance on seeds differing in hardness. Another example is given by Schmidt and Rand (1999) who studied polymorphism in the *Mpi* locus in the northern acorn barnacle, *Semibalanus balanoides*. Their data indicate that temperature- and/or desiccation-mediated disruptive selection is operating at *Mpi* or a linked locus and that *Mpi* genotypes experience differential mortality in the different intertidal microhabitats.

8.1.2   Two-locus, two-allele haploid version of the Levene model

The model to be considered next was introduced (but not studied explicitly) by Felsenstein (1981) as part of his numerical study of parapatric and sympatric speciation. Following Felsenstein, let us consider a haploid population inhabiting a system of two demes. Assume that viability is controlled by two diallelic loci with alleles **A**, **a** and **B**, **b**. I will use the following parameterization of the genotype fitnesses:

$$
\begin{array}{ccccc}
 & \mathbf{AB} & \mathbf{Ab} & \mathbf{aB} & \mathbf{ab} \\
\text{in niche I} & 1+\sigma & 1-s & 1-s & 1-\sigma \\
\text{in niche II} & 1-\sigma & 1-s & 1-s & 1+\sigma,
\end{array}
\tag{8.5}
$$

where $0 \leq \sigma, s \leq 1$. Here, parameter σ measures the difference in fitness between the two "extreme" genotypes **AB** and **ab** in each niche. The average fitness of these genotypes across the niches is equal to one. The average fitness of "hybrid" genotypes **Ab** and **aB** across the niches is $1 - s$. That is, s measures the relative strength of selection acting on hybrid genotypes. Positive s implies that the population experiences disruptive selection on average. If $s = 0$, the fitnesses are additive. Note that Felsenstein assumed multiplicative selection which corresponds to $\sigma^2 = s(2 - s)$. I will consider the general case.

Assume that offspring disperses randomly over the two niches. In this model, the population evolves to an equilibrium state. The two monomorphic equilibria with a hybrid genotype fixed (**Ab** or **aB**) cannot be stable. The two monomorphic equilibria with an extreme genotype (**AB** or **ab**) fixed are locally stable if

$$
\sigma^2 < \min\left(s, \frac{r}{2-r}\right),
\tag{8.6}
$$

where r is the recombination rate. That is, stability of these equilibria requires that the difference in fitness between the extreme genotypes is not too large.

There is also a polymorphic equilibrium with all allele frequencies equal to $1/2$ and a certain linkage disequilibrium D (defined by equa-

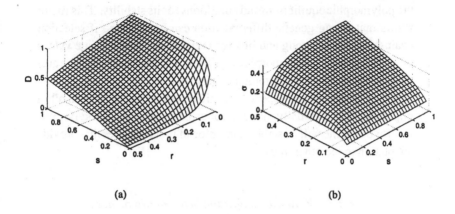

(a)   (b)

FIGURE 8.1. Polymorphic equilibrium in the two-locus, two-allele haploid version of the Levene model. (a) Normalized linkage disequilibrium D is a function of r and s. (b) Conditions for local stability. The equilibrium is stable for parameter values above the surface.

tion 2.7 on page 31). Instead of linkage disequilibrium D, it will often be convenient to use a *normalized* linkage disequilibrium defined as

$$D' = \frac{D}{\sqrt{p_1 q_1 p_2 p_2}}, \tag{8.7}$$

where p_i and q_i are the frequencies of the alternative alleles at the i-th locus. D' ranges from -1 to 1. Below for simplicity of notation, I will drop the prime and will use symbol D for normalized linkage disequilibrium, making sure that it is clear whether linkage disequilibrium is normalized or not. In the model under consideration, the normalized linkage disequilibrium at a polymorphic equilibrium is given by the positive root of quadratic

$$sD^2 + r(2 - s)D - s(1 - r) = 0. \tag{8.8a}$$

Positive D means that hybrid genotypes Ab and aB are in deficiency. If this equilibrium is stable, the population maintains genetic variation under disruptive selection. The condition for local stability is

$$\sigma^2 > \frac{1}{2} \frac{1 - D}{1 + D} s [2 - s(1 - D)] ; \tag{8.8b}$$

that is, the difference in fitness between the extreme genotypes must be large enough. Figure 8.1 illustrates both the value of D maintained at the polymorphic equilibrium and conditions for its stability. This figure shows that strong genetic differentiation can be achieved if selection against hybrids is strong and linkage is tight. For intermediate values of σ, both the two monomorphic equilibria and the polymorphic equilibrium are locally stable simultaneously. In this case, whether genetic variation is maintained depends on initial conditions. If σ is small, a population that is initially monomorphic for an extreme genotype optimal for one niche will be stable to invasion of mutant alleles that provide adaptation to the other niche.

8.1.3   Restricted migration between two niches

Because in the Levene model offspring disperse randomly over the niches, the genotype frequencies are the same in all niches at the beginning of each generation. In contrast, if dispersal is limited (which is true for many organisms), spatially heterogeneous selection can result in substantial genetic divergence between the niches. For example, the south-facing slope and north-facing slope of the "Evolution Canyon," Mount Carmel, Israel, display dramatic physical and biotic contrasts at microscale (Nevo 1997). Higher solar radiation on the south-facing slope makes it warmer, drier, and more spatially and temporally heterogeneous than the north-facing slope, even though the slopes are separated by only 100 m at the bottom and 400 m at the top. Many of some 1,300 species identified there display qualitative and quantitative divergence between the slopes. In particular, the populations of *D. melanogaster* have diverged in body size, heat and desiccation tolerance, oviposition thermal preference, fluctuating asymmetry, rates of migration and recombination, mate preference and microsatellite markers (e.g., Nevo 1997; Korol et al. 2000; Michalak et al. 2001).

I will consider the following model. There are two distinct niches. The life cycle is viability selection in a niche, migration, and reproduction. Let there be a single diallelic locus controlling viability, and let the viabilities of the diploid genotypes be as defined in expressions (8.4). That is, allelic effects on fitness are additive, with allele A being advantageous in the first niche and disadvantageous in the second niche. Let surviving adults stay in the niche in which they were born with probability $1 - m$ and migrate to the other niche with probability m. Mating takes place between adults in the same niche.

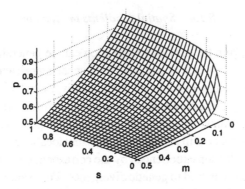

FIGURE 8.2. The equilibrium frequency of a locally advantageous allele in the niche as a function of the migration rate m and the strength of selection s in the two-niche model.

Let p_j and q_j be the frequencies of alleles A and a in the j-th niche ($j = 1, 2$) at the beginning of a generation. The changes in p_j due to selection are given by equations analogous to equation (8.1). The changes in allele frequencies due to migration are $m(p_2' - p_1')$ and $m(p_1' - p_2')$ in the first and second niches, respectively, where p_j' is the corresponding allele frequency *after* selection. It is straightforward to show that the population will quickly approach an equilibrium at which the frequency of a locally advantageous allele in offspring is

$$p^* = \frac{s(1 + 2m) - 2m + \sqrt{s^2 + 4m^2(1 - s^2)}}{2s(1 + 2m)}. \tag{8.9}$$

If $m = 1/2$ (i.e., if dispersal is random as in the Levene model), then $p^* = 1/2$ and the allele frequencies in the niches are the same. However, if migration is limited (i.e., if $m < 1/2$), the difference between the allele frequencies can be substantial. For example, if the rate of migration is very small relative to the strength of selection (i.e., if $m \ll s$), then allele A has a very high frequency in the first niche and a very low frequency in the second niche ($p_1^* \approx 1 - m/s, p_2^* \approx m/s$). Figure 8.2 illustrates the dependence of the equilibrium frequency of a locally advantageous allele on m and s in more detail. This figure shows that this frequency can be close to 1, implying that spatial heterogeneity in selection can induce significant genetic divergence. Similar conclusions hold if the strength of selection differs between the niches.

8.1.4   Spatial gradients in selection

The results from the previous subsection can be generalized for more than two niches. A case of particular biological interest and significance is that of a spatial gradient in selection, that is, when the strength of selection changes monotonically as one moves across a system of demes or across a continuous habitat. Differences in temperature, humidity, or food level between different demes (or spatial locations) can induce differences in selection experienced by the corresponding subpopulations, which, in turn, can lead to genetic divergence. For example, Smith et al. (1997) studied populations of the little greenbul (*Andropadus virens*), a common African rain forest passerine bird. Their results show large morphological differences in fitness-related characters between populations in rain forest and transitional habitats bordering savanna, despite substantial gene flow. Another example is provided by Ogden and Thorpe (2002), who compared variation in microsatellite allele frequencies and morphology across habitat boundaries for the Caribbean lizard, *Anolis roquet*, in northern Martinique. Their results show reductions in gene flow and increased genetic divergence to be concordant with divergent selection for habitat type.

To illustrate the effects of selection gradients, I first consider the well-studied one-locus case (e.g., Endler 1977). Then, I analyze selection on an additive quantitative character.

One-locus clines   Assume that there are n demes arranged on a line. Let fitnesses be additive (i.e., the fitness of heterozygotes is always equal to the average of that of the homozygotes). Let p_j be the frequency of allele A in the j-th deme at the beginning of the generation, and let s_j be a relative selective advantage (or disadvantage) of allele A in deme j $(j = 1, \ldots, n)$. The allele frequencies after selection p'_j can be found using equations analogous to equation (8.1). If we assume migration at rate $m/2$ to the nearest deme, the allele frequencies after migration are

$$p''_j = p'_j + \frac{m}{2}(p'_{j-1} + p'_{j+1} - 2p'_i). \qquad (8.10)$$

If mating is random, reproduction will not change the allele frequencies, so that p''_j will also give the corresponding frequency at the beginning of the next generation.

In this model, the allele frequencies evolve to certain equilibrium values controlled by a balance of selection and migration. For example, Figure 8.3 illustrates the equilibrium clines of an allele frequency ob-

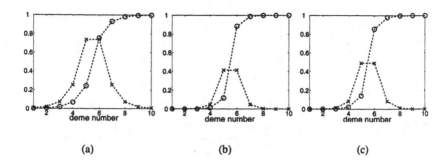

FIGURE 8.3. The equilibrium allele frequency p_i (given by crosses) and the normalized genetic variance $4p_j(1 - p_j)$ (given by circles) in the model with a step change in the selection coefficient in a system with 10 demes. (a) $s = 0.025, m = 0.05$. (b) $s = 0.20, m = 0.05$. (c) $s = 0.025, m = 0.01$.

served in numerical simulations, assuming that allele **A** is deleterious with selective coefficients $s_j = -s < 0$ in the first $n/2$ demes and is advantageous with the same selective coefficient $s_j = s$ in the remaining demes. Increasing the strength of selection or decreasing migration rate makes, as expected, the resulting clines steeper. Similar effects are observed when the coefficient of selection s_j changes linearly from $-s$ in deme 1 to s in deme n (see Figure 8.4). These figures show that steep clines can evolve as a result of a small step change in selection (Figure 8.3) or along smooth environmental gradients (Figure 8.4) if migration is not too high and that spatially heterogeneous selection can maintain high levels of genetic variation in the demes bordering the area where the sign of the selection coefficient changes. Similar conclusions hold if, instead of discrete demes, one considers a continuous habitat (Endler 1977).

Additive quantitative trait   The clines resulting from stabilizing selection on a quantitative character with the optimum changing in space have received much less attention than clines in one-locus models (but see Slatkin 1978; Kirkpatrick and Barton 1997; Barton 1999).

I start with a simple two-locus, two-allele diploid model. I assume that the allelic contributions to the trait value z are equal to 0 and 1 and disregard the environmental effects. In this case, the possible values of z range from 0 to 4, with the midrange value at $z = 2$. As before, there are n demes arranged on a line with migration between neighboring demes

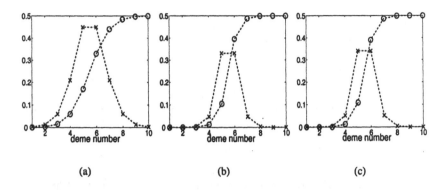

(a)   (b)   (c)

FIGURE 8.4. The equilibrium allele frequency p_i (given by crosses) and the normalized genetic variance $4p_i(1 - p_i)$ (given by circles) in the model with a linear change in the selection coefficient in a system with 10 demes. (a) $s = 0.25, m = 0.05$. (b) $s = 0.5, m = 0.05$. (c) $s = 0.25, m = 0.025$.

at rate $m/2$. Let stabilizing selection acting within the j-th deme be described by a Gaussian fitness function

$$w_j(z) = \exp\left(-\frac{(z - \theta_j)^2}{2V_s}\right), \tag{8.11}$$

where the optimum value θ_j changes linearly from $2 - a$ in deme 1 to $2 + a$ in deme n. Parameter V_s characterizes the strength of stabilizing selection, with larger V_s corresponding to weaker selection. Parameter a characterizes the maximum deviation of the optimum from the midrange value of z.

Figure 8.5 illustrates the properties of the equilibrium states in this model observed in numerical simulations using $a = 1, 2,$ and 3. In these three cases, the range of optimum values is narrower than, coincides with, and is wider than the range of possible phenotypic values. The loci are unlinked. A number of generalizations emerge:

• The clines in allele frequencies can be shifted relative to each other.

• Linkage disequilibrium is negligible and the value of the recombination rate is unimportant (data not shown).

• The average trait values $\bar{z}_j$ are close to the optimum values θ_j.

• Genetic variance $V_{g,j}$ is substantial and nonuniform in space.

• There can be substantial phenotypic variation.

In general, the interaction of spatially heterogeneous selection and restricted migration can result in complex patterns of genetic and phenotypic variation.

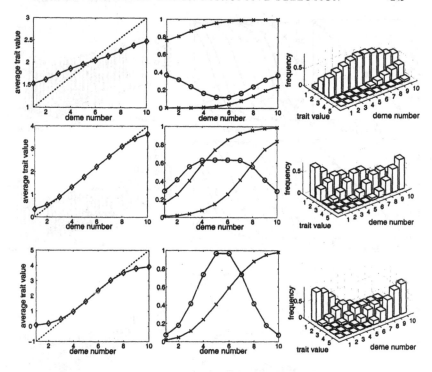

FIGURE 8.5. The two-locus model of stabilizing selection with linearly varying optimum. Left column: the phenotypic optimum (dashed line) and the mean trait value (diamonds). Middle column: the equilibrium allele frequencies $p_{1,j}, p_{2,j}$ (crosses), and the normalized genic variance $V_{g,j} = \sum_{i=1}^{2} 2p_{i,j}(1 - p_{i,j})$ (circles). Right column: the phenotypic distribution. First row: $a = 1$; second row: $a = 2$; third row: $a = 3$; in this case, $p_{1,j} = p_{2,j}$. Other parameters: $m = 0.05$, $V_s = 10$.

Next I assume that there are $\mathcal{L}$ diallelic loci with effects 0 and 1. In this case, the range of possible trait values is from $z = 0$ to $z = 2\mathcal{L}$, with the midvalue at $z = \mathcal{L}$. I will assume that stabilizing selection acting within each deme is described by a Gaussian fitness function (8.11), with the optimum θ changing linearly from $\mathcal{L} - a$ to $\mathcal{L} + a$. I will disregard linkage disequilibrium, which appears to be a safe simplification if selection is not too strong and linkage is loose.

Figure 8.6 illustrates the properties of equilibria observed in this model, which are similar to those in the two-locus model considered above. In this model, there are multiple simultaneously stable equi-

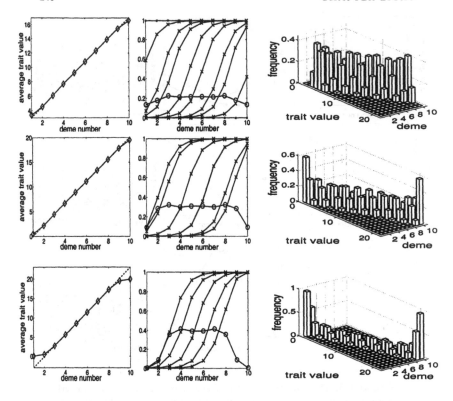

FIGURE 8.6. The ten-locus model of weak stabilizing selection with linearly varying optimum. Left column: the phenotypic optimum (dashed line) and the mean trait value (diamonds). Middle column: the equilibrium allele frequencies (crosses) and the normalized genotypic variance (circles). Right column: the phenotypic distribution. First row: $a = 7$. Only eight clines are visible because one locus is fixed for allele 0 and one locus is fixed for allele 1. Second row: $a = 10$. Only six clines are visible because the two outermost and the two middle curves correspond to two coincident clines each. Third row: $a = 13$. The five visible clines correspond to sets of 3, 1, 2, 1, and 3 loci (from the left to the right). Other parameters: $m = 0.05$, $V_s = 10$. The initial allele frequencies were chosen randomly from the uniform distribution on the interval $[0, 0.05]$.

libria, so that the eventual outcome of the dynamics depends on initial conditions (Barton 1999). Clines at different loci can coincide, which appears to be a general feature (Barton 1999). Numerical simulations also demonstrate the possibility of stable reversed clines (i.e., clines going from high allele frequencies in the demes on the left to low allele frequencies in the demes on the right), as well as nonmonotonic clines. Such clines are observed more frequently as the migration rate becomes smaller. Figure 8.6 shows that steep clines in allele frequencies can result from smooth gradients in the optimum trait value which is analogous to the properties of one-locus clines (Endler 1977).

Asexual population If the population is asexual, the interaction of spatially heterogeneous selection and migration can result in a new regime. To illustrate this, let us consider a one-locus, multiple-allele haploid population inhabiting a system of demes arranged on a line. Assume that individuals differ with respect to a trait z. Let selection be described by the Gaussian function (8.11) with a linearly varying optimum $\theta(x)$, where variable x characterizes the spatial location of demes. Figure 8.7 describes equilibrium clines arising in this model. There are three possible outcomes. If migration is low, substantial local adaptation is possible. In this case, the average trait value is very close to the optimum across the whole system, and within-deme genetic variance is low (Figure 8.7, first row). If migration is high, local adaptation is prevented and the system is genetically uniform across the space (Figure 8.7, second row). If migration is intermediate, the population splits into two genetic clusters that are spatially separated (Figure 8.7, third row). In the middle area where the clusters are sympatric they maintain their genetic separation, and intermediate genotypes are at very low frequencies.

The third outcome is a manifestation of *boundary effects* which are well known in studies of reaction-diffusion systems (e.g., Murray 1989). The intuitive explanation is as follows. Migration brings locally maladapted genes. However, for central demes, the effects of migration from the neighboring demes on the right and on the left practically cancel each other. In contrast, the peripheral demes are subject to the one-sided immigration, which pushes them significantly away from their optimum trait values toward trait values that are optimal in the nearby demes. This results in the emergence of two relatively large and genetically similar groups of demes close to each of the two boundaries. The tails of the corresponding clusters (see the third row of Figure 8.7) are maintained by a balance of immigration from these groups and local selection acting against the immigrants. I note that similar observations and interpretations were made independently by N. H. Barton and J. Polechova (personal communication).

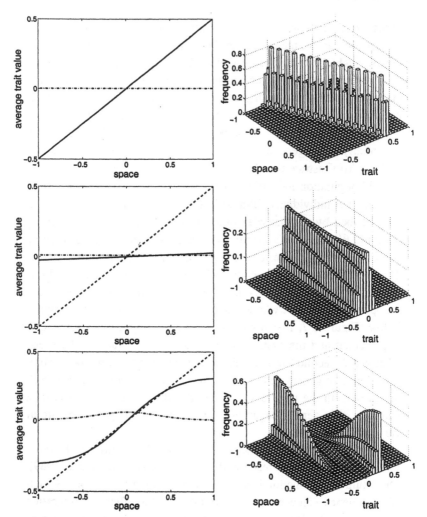

FIGURE 8.7. Equilibrium clines in the asexual one-locus, multiple-allele model of stabilizing selection with linearly varying optimum. Left column: the phenotypic optimum (dashed line), the mean trait value (solid line), and the normalized genotypic variance (dash-dot line). Right columns: the phenotypic distribution. First row: slow migration ($\sigma_m = 0.01$). Second row: fast migration ($\sigma_m = 0.40$). Third row: intermediate migration ($\sigma_m = 0.14$). *Details of numerical simulations*: The trait values were discretized into $\mathcal{L} = 31$ values between -1 and 1. There are 31 demes labeled by variable x which takes 31 equally spaced values between -1 and 1. $\theta(x) = 0.5x, V_s = 1$. Each individual immigrates with probability m and moves a distance taken from a Gaussian distribution with mean zero and variance σ_m^2. Offspring is subject to mutation with probability μ. The deviation of the offspring trait from that of its parent is taken from a Gaussian distribution with zero mean and variance σ_μ^2. In simulations, $m = 1, \mu = 0.001, \sigma_\mu = 0.1$. Boundaries for both x and z are reflective. Results are given for generation 5,000 starting with a random distribution.

8.1.5 Coevolutionary clines

In the models considered so far in this chapter, individual fitness was constant. However, many fitness components depend on interactions with other species inhabiting the same geographic area and, thus, can change as a result of coevolution. Spatial and temporal heterogeneity in selection experienced (and generated) by coevolving populations play a profound role in the geographic mosaic theory of coevolution (Thompson 1994, 1999a,b).

To illustrate the effects of coevolutionary interactions on the maintenance of genetic variation in spatially structured populations, let us consider a model of two haploid species (Nuismer et al. 1999, 2000). Species 1, the parasite or mutualistic symbiont, has alleles A and a, with frequencies at each deme j given by $p_{1,j}$ and $q_{1,j}$. Species 2, the host, has alleles B and b with frequencies $p_{2,j}$ and $q_{2,j}$, respectively. Assume that between-species interactions affecting fitness occur only when individuals carrying matching alleles A and B or a and b meet. The simplest symmetric choice of fitnesses (viabilities) in deme j is

$$w_{A,j} = 1 + c_j p_{2,j}, \quad w_{a,j} = 1 + c_j q_{2,j}, \qquad (8.12a)$$

$$w_{B,j} = 1 + b_j p_{1,j}, \quad w_{b,j} = 1 + b_j q_{1,j}. \qquad (8.12b)$$

Here, c_j is the fitness sensitivity in deme j of species 1 to changes in the frequency of its matching allele in species 2, and b_j is the fitness sensitivity in deme j of species 2 to the changes in the frequency of its matching allele in species 1. When either $c_j > 0, b_j < 0$, or $c_j < 0, b_j > 0$, the interaction is antagonistic, and one species benefits from interacting while the other is harmed. If both b_j and c_j are positive, the interaction is mutualistic. Fitness scheme (8.12) represents a special case of a model with linear frequency-dependent fitnesses introduced and studied by Gavrilets and Hastings (1998) using a differential approximation. The changes in $p_{i,j}$ due to selection are given by equations analogous to equation (8.1). In an isolated deme, both species quickly fix matching alleles if interaction is mutualistic. If interaction is antagonistic, allele frequencies oscillate in both species, with the amplitude of these oscillations increasing through time (Seger 1988).

Let there be a number of demes arranged along a line with migration at a constant rate $m/2$ between neighboring demes. The corresponding changes in allele frequencies due to migration are described by equations analogous to equations (8.10). Assume that the interaction between the

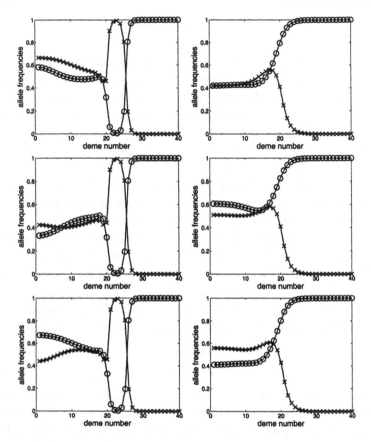

FIGURE 8.8. The dynamic clines in allele frequencies in the symbiont (circles) and host (diamonds). There are 40 demes. $c_j = 0.05$ in all demes; $b_j = -0.05$ in demes 1 through 20 (implying antagonostic interactions in these demes) and $b_j = 0.05$ in demes 21 through 40 (implying mutualistic interactions in these demes). Migration rate $m = 0.02$ (left column) and $m = 0.20$ (right column). The initial allele frequencies are drawn randomly and independently from the interval [0,0.05]. The rows 1 through 3 correspond to generations 4,700, 4,800, and 4,900, respectively.

species varies geographically from antagonism to mutualism. Such a situation may occur when the relative costs and benefits of the interaction change with the differences in local ecological conditions. One example is found in interactions between bacteria and those plasmids that produce drug resistance or allow the use of alternative carbon sources (Duncan et al. 1995). The plasmid is beneficial to its associated bacterium when selection favors the expression of the plasmid-encoded gene, but costly when selection for its function is absent. At the same time, plasmid always benefits from bacterium, independent of local conditions.

To describe such situations, let us assume $b_j = b_l < 0$ for $j = 1, \ldots, n/2$ and $b_j = b_r > 0$ for $j = n/2 + 1, \ldots, n$, with $c_j = c > 0$ in all demes. Numerical studies of this model demonstrate the possibility of both static clines and clines that oscillate in time (see Figure 8.8). The amplitude of oscillations is the highest in the area of antagonistic interactions. Figure 8.8 shows that spatially heterogeneous coevolutionary interactions can lead to both the maintenance of genetic variation and the generation of geographic variation. The left column in Figure 8.8 illustrates the possibility of maintaining genetic variation in the area of mutualism where selection is homogeneous. This happens because random variations in initial allele frequencies get amplified and then are maintained due to a small migration rate. With higher migration (Figure 8.8, right column), spatially heterogeneous solutions in the mutualistic region are not stable.

8.2   SPATIALLY UNIFORM DISRUPTIVE SELECTION

Next, I consider a model where intermediate genotypes have reduced fitness throughout the population range. The main conclusion of this subsection will be that alternative genotypes can coexist in the population by forming spatially separated clusters in which the competing genotypes are excluded to a large degree.

8.2.1   Migration-selection balance: the Karlin-McGregor model

Let genotype fitnesses (viabilities) be $w_{AA} = 1, w_{Aa} = 1 - s, w_{aa} = 1$, with $0 \leq s \leq 1$. This is the model of underdominant selection describing disruptive selection under which "extreme" genotypes AA and aa are favored and intermediate genotypes Aa are selected against.

There are two demes. Following Karlin and McGregor (1972), let p_j be the frequencies of allele A in the j-th deme. The change in p_j due to selection is given by an equation analogous to equation (8.1). Assume that surviving individuals migrate between the demes at a constant rate m. This system has always two locally stable monomorphic equilibria at which $p_1 = p_2 = 0$ or $p_1 = p_2 = 1$. However, it also can have two additional locally stable *polymorphic* equilibria at which allele frequencies differ between the demes. These equilibria are given by equation

$$p_1^* = 1/2 \mp \sqrt{1/4 - \epsilon}, \tag{8.13}$$

with $p_2^* = 1 - p_1^*$, where $\epsilon = m/s$. If $\epsilon \ll 1$, the equilibrium allele frequencies are approximately $1 - \varepsilon$ and ε. The polymorphic equilibria exist (i.e., are feasible) if

$$s > 4m, \tag{8.14a}$$

and are locally stable if

$$s > 6m - 4m^2. \tag{8.14b}$$

Both these conditions require migration to be sufficiently weak relative to selection. The polymorphic equilibria cannot be approached if initially both populations have similar allele frequencies. In this case, the population will evolve to a monomorphic equilibrium losing genetic variation. However, if two isolated populations have somehow fixed alternative alleles, then a secondary contact between them will not lead to the fusion of the populations and genetic variation will be stably maintained provided conditions (8.14) are satisfied.

8.2.2 *Migration-selection balance: the Bazykin model*

Similar conclusions hold if space is treated as continuous rather than discrete. Following Bazykin (1969), let us consider a population inhabiting an infinite one-dimensional area. Let $p(x, t)$ be the frequency of allele A at spatial location x at time t. Assuming that selection is weak ($s \ll 1$), the dynamics of $p(x, t)$ are approximated by a partial differential equation

$$\frac{\partial p}{\partial t} = spq(p - q) + \frac{m}{2} \frac{\partial^2 p}{\partial x^2}. \tag{8.15}$$

Here, the first term in the right-hand side is analogous to the second term in the right-hand side of equation (3.1) with $\overline{w} \approx 1$ (because of the weak selection assumption). The second (diffusion) term can be viewed as an approximation of the second term in the right-hand side of equation (8.10b). It describes random movement of individuals (see Nagylaki 1975 for a rigorous derivation and discussion of underlying

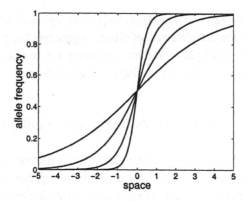

FIGURE 8.9. The clines in allele frequency in the Bazykin model for the cline width $a_w = 1$ (the steepest), 2, 4 and 8 (the shallowest).

assumptions). The coefficient m is the mean square distance between the location where an individual was born and the location where it experienced selection. m corresponds to the migration rate in models with migration between the neighboring demes only.

Equation (8.15) has three spatially homogeneous solutions: two locally stable monomorphic equilibria $p(x, t) = 0$ and $p(x, t) = 1$ and an unstable polymorphic equilibrium $p(x, t) = 0.5$. In addition, there are two families of polymorphic spatially heterogeneous equilibria describing clines in allele frequency:

$$p(x) = \frac{e^{\pm\sqrt{\frac{2s}{m}}(x - x_0)}}{1 + e^{\pm\sqrt{\frac{2s}{m}}(x - x_0)}}. \tag{8.16}$$

The solutions corresponding to the plus sign change from 0 at $x = -\infty$ to 1 at $x = \infty$. The solutions corresponding to the minus sign change from 1 at $x = -\infty$ to 0 at $x = \infty$. Cline width, b, can be defined as the inverse of the gradient dp/dx at the center of the cline. In the Bazykin model, $b = 2\sqrt{2m/s}$. Strong selection (i.e., large s) and weak migration (i.e., small m) result in narrow clines. Figure 8.9 illustrates the shape of the clines observed in the Bazykin model.

In the spatially homogeneous case, the center of the cline x_0 can be at any spatial position. As already mentioned above, Bazykin (1969) argued that introducing spatial heterogeneity in the model would result in the center of the cline getting trapped in the region of the smallest migration rate. If there is not one but a number of underdominant genes, then the clines describing their allele frequencies are expected to assemble in the same region. This can potentially result in formation of two spatially separated clusters of different genotypes (e.g., as modeled in Chapter 7). The hybrids will have strongly reduced fitness, meaning strong RI between the two resulting geographic forms.

8.3   TEMPORAL VARIATION IN SELECTION

Genetic variation can also be maintained if selection varies in time rather than in space. A classical example of a polymorphism maintained by temporal variation in selection is provided by a 10-year long study of a population of the two-spot ladybird beetle, *Adalia bipunctata L.*, which has a black and a red morphs (Timoféeff-Ressovsky and Svirezhev 1966). The color is apparently controlled by a single diallelic locus with the "black gene" being dominant. The black morph has higher fertility during the summer (which roughly corresponds to two generations), whereas the red morph has higher viability during the winter (which roughly corresponds to one generation). The estimates of relative fitnesses support the conclusion that the polymorphism is stable (Timoféeff-Ressovsky and Svirezhev 1966).

Following Haldane and Jayakar (1965), let us consider a one-locus, two-allele model of a diploid randomly mating population with fitnesses varying from generation to generation (see also Nagylaki 1992, Section 4.5). Let the relative fitnesses of the three diploid genotypes in generation t be $w_{AA,t}, w_{Aa,t}$, and $w_{aa,t}$, and let $\tilde{w}_{AA}, \tilde{w}_{Aa}$, and $\tilde{w}_{aa}$ be the corresponding geometric mean fitnesses. (For example, the geometric mean of w_1, w_2, and w_3 is $(w_1 w_2 w_3)^{1/3}$ and that of w_1, w_2, w_3, and w_4 is $(w_1 w_2 w_3 w_4)^{1/4}$.) Then allele **a** will increase in frequency when rare and, thus, cannot be lost if

$$\tilde{w}_{AA} < \tilde{w}_{Aa}, \tag{8.17a}$$

that is, if the geometric mean fitness of **Aa** is larger than that of **AA**. In a similar way, allele **A** cannot be lost if

$$\tilde{w}_{aa} < \tilde{w}_{Aa}. \tag{8.17b}$$

If conditions (8.17) are satisfied simultaneously, genetic variation is protected. Thus, polymorphism is maintained if the geometric mean fitnesses of homozygotes are smaller than that of heterozygotes. This is possible even if the arithmetic mean fitnesses of homozygotes exceed that of heterozygotes, that is, if on average the population experiences disruptive selection.

For example, let us assume that the fitnesses in generation t are $w_{AA} = 1 - s, w_{Aa} = 1, w_{aa} = 1 + s$ with probability $1/2 + \varepsilon$ and $w_{AA} = 1 + s, w_{Aa} = 1, w_{aa} = 1 - s$ with probability $1/2 - \varepsilon$. Under this selection regime, the population experiences directional selection within each generation, but the direction of selection fluctuates. In this case, the inequalities (8.17) simplify to a single condition

$$|\varepsilon| < \frac{1}{2} \frac{\ln[(1 - s)(1 + s)]}{\ln[(1 - s)/(1 + s)]}. \tag{8.18}$$

This inequality implies that genetic variation is protected if the relative frequencies of both types of selection are not too different. For example, if $s = 0.1$ (i.e., selection within each generation is weak), then $|\varepsilon|$ has to be smaller than 0.025. In contrast, if $s = 0.9$ (i.e., selection within each generation is strong), then $|\varepsilon|$ has to be smaller than 0.282.

Bazykin (1965) used an alternative model to argue that selection varying in time can maintain genetic variation and that sympatric speciation may result from the population evolving to increase its fitness in a temporary varying environment. For more results on the maintenance of multilocus variation under fluctuating selection see, for example, Korol et al. (1996); Kondrashov and Yampolsky (1996) and Bürger and Gimelfarb (2002).

8.4 FREQUENCY-DEPENDENT SELECTION IN A SINGLE POPULATION

Frequency-dependent selection is well known for its ability to maintain genetic variation (e.g., Wright 1969; Cockerham et al. 1972; Matessi and Cori 1972; Clarke 1979; Udovic 1980; Asmussen and Basnayake 1990). One example is provided by the orchid *Dactylorhiza sambucina*, which shows a stable and dramatic flower-color polymorphism, with both yellow- and purple-flowered individuals present in natural populations throughout the range of the species in Europe. Gigord et al. (2001) performed an experiment varying the relative frequency of the two color morphs of *D. sambucina* and showed that rare morphs have reproductive advantage through pollinator (bumblebee) preference.

Here, I discuss three simple models of frequency-dependent selection. First, I consider a phenomenological approach in which fitnesses are specified without considering any specific mechanisms leading to frequency dependence. Then, I describe two models in which frequency dependence of fitnesses emerges as a consequence of intraspecific competition for resources.

8.4.1   Phenomenological approach

Let us consider a one-locus, two-allele randomly mating diploid population. Let p and $q = 1 - p$ be the frequencies of alleles A and a. Because mating is random, the genotype frequencies in offspring are in Hardy-Weinberg proportions, and the population state is completely characterized by a single variable p. Let $w_{AA}(p)$, $w_{Aa}(p)$, and $w_{aa}(p)$ be nonnegative, continuous, and differentiable functions of p specifying (frequency-dependent) fitnesses of the three diploid genotypes. The change in p per generation is described by the standard equation (2) in Box 2.1. Then, allele A cannot be lost deterministically if

$$w_{Aa}(0) > w_{aa}(0), \qquad (8.19a)$$

and allele a cannot be lost deterministically if

$$w_{Aa}(1) > w_{AA}(1) \qquad (8.19b)$$

(e.g., Cockerham et al. 1972; Asmussen and Basnayake 1990). If both inequalities are satisfied simultaneously, genetic variation is protected. The biological interpretation of these conditions is straightforward: genetic variation is protected if in a population where one allele is very rare, heterozygotes have higher fitness than common homozygotes. Note that if fitnesses are constant, conditions (8.19) imply overdominance. In general, this model can possess a number of simultaneously stable equilibria and can exhibit nonequilibrium dynamics, including cycles, chaos, transient chaos, and intermittency (Asmussen and Basnayake 1990; Altenberg 1991; Gavrilets and Hastings 1995b).

In Chapter 10, a symmetric version of this model with

$$w_{AA} = f(p-q), \quad w_{Aa} = h(p-q), \quad w_{aa} = f(q-p) \quad (8.20)$$

will play an important role. Here $f(\cdot)$ and $h(\cdot)$ are continuous functions normalized so that $f(0) = 1$. In this model, there always exists a polymorphic equilibrium with equal allele frequencies ($p = q = 1/2$) at which the fitnesses of homozygotes are equal to 1 and the fitness of heterozygotes is $h(0)$, which I will write as $1 - s$. That is, coefficient s is the relative strength of selection against heterozygotes at the equilibrium. A necessary and sufficient condition for local stability of this equilibrium can be written as

$$-2(2 - s) < s + f'(0) < 0 \qquad (8.21)$$

(Udovic 1980). Here, $f'(0)$ is the derivative of the fitness function of homozygotes AA evaluated at $p = 1/2$. The left inequality in (8.21) guarantees the absence of growing oscillations in p in a neighborhood of $p = 1/2$. This condition will always be satisfied in the applications to be considered below. Then the right inequality in (8.21) shows that the polymorphic equilibrium can be locally stable, even if heterozygotes have reduced fitness at equilibrium (i.e., if $s > 0$), so that selection is disruptive at equilibrium. This requires that the fitness of the homozygotes decreases sufficiently rapidly as the corresponding allele frequency p exceeds $1/2$. The biological interpretation of the latter condition is that rare genotypes have sufficient selective advantage over common genotypes.

8.4.2   Intraspecific competition

Interactions between individuals in a population can induce frequency-dependent selection. In particular, competitive interactions can result in disruptive selection when individuals with extreme or rare phenotypes that consume an underutilized resource gain selective advantage because of reduced competition. For example, Hori (1993) demonstrated that the direction of mouth-opening (either left-handed or right-handed) in scale-eating cichlid fish *Perissodus Microlepis* of Lake Tanganyika is determined by a one-locus, two-allele system, in which dextrality is dominant over sinistrality. Attacking from behind, right-handed individuals snatched scales from the prey's left flank and left-handed ones from the right flank. Within a given population, the frequencies of the two phenotypes remain approximately equal. This phenomenon was effected through frequency-dependent selection exerted by the prey's alertness. Thus, individuals of the rare phenotype had more success as predators than those of the more common phenotype (Hori 1993). Another example is provided by crossbills (Benkman 1996). In crossbills *Loxia*, the lower mandible crosses with equal frequency to the left and to the right in several crossbill populations. A crossbill always orients towards closed conifer cones so that its lower mandible is directed towards the cone axis. Thus, only part of the cone can be reached easily when crossbills have few perch sites and the cone cannot be removed from the branch or otherwise turned around. Since crossbills may visit cones that have previously been foraged on by other individuals, the rarer morph has selective advantage. Experimental data support this hypothesis (Benkman 1996).

Let us consider a population of size N where individuals differ with respect to a certain metric trait z. Assume that trait z controls the type of resource utilized by the individuals. Let function $K(z)$ characterize the abundance of the resource used by individuals with trait z. A common way is to interpret $K(z)$ as the carrying capacity, that is, the number of individuals with trait z that the environment can support if no individuals with other trait values are present. One usually assumes that $K(z)$ is a unimodal function of z. Alternatively, one can think of $K(z)$ as a measure of the frequency-independent component of viability of organisms with trait z. If $K(z)$ is a unimodal function of z, then $K(z)$ can be thought of as a fitness function describing stabilizing selection.

Let $\phi(z)$ be the distribution of z in the population. A standard way to introduce competition into the modeling framework is to specify a certain *competition function* $\xi(x,y)$ that measures the degree of competition between a pair of individuals with trait values x and y (e.g., Roughgarden 1972, 1979; see Figure 3 in Roughgarden 1972 for an example of function $\xi(x,y)$ for *Anolis* lizards). Then function $C(z)$, defined as

$$C(z) = \sum_y \xi(z,y)\phi(y), \qquad (8.22)$$

measures the average competition experienced by organisms with trait z. If the distribution of z is continuous, the sum in the above definition of $C(z)$ is replaced by an appropriate integral. One usually assumes that $\xi(x,y)$ is a unimodal function of the difference $x - y$, that is, $\xi(x,y) = \xi(x-y)$, with a maximum at $x - y = 0$. The last assumption reflects the idea that competition is strongest for similar organisms.

Functions $K(z)$ and $C(z)$ control the expected fitness $w(z)$ of organisms with trait z. From biological considerations, fitness $w(z)$ is supposed to increase with carrying capacity $K(z)$ and decrease with the average strength of competition $C(z)$. A number of different functional forms have been used in the literature, for example,

$$w(z) = \left[\rho - \frac{NC(z)}{\kappa}\right] K(z), \qquad (8.23a)$$

where ρ and κ are constants (Bulmer 1974, 1980; Bürger 2002a,b);

$$w(z) = 1 + r\frac{N - NC(z)}{K(z)}, \qquad (8.23b)$$

where r is a constant (Pimm 1979);

$$w(z) = 1 + r\left(1 - \frac{NC(z)}{K(z)}\right), \qquad (8.23c)$$

where r is a constant (Roughgarden 1972, 1979; Slatkin 1979; Dieckmann and Doebeli 1999);

$$w(z) = 1 + V[K(z) - NC(z)], \qquad (8.23d)$$

where V is a constant (Christiansen and Loeschcke 1980; Loeschcke and Christiansen 1984);

$$w(z) = \frac{\lambda}{1 + a\left[\frac{NC(z)}{K(z)}\right]^b}, \qquad (8.23e)$$

where λ, a, and b are constant (Doebeli 1996); and

$$w(z) = \frac{1 + r}{1 + r\frac{NC(z)}{K(z)}}, \qquad (8.23f)$$

where r is a constant (Drossel and McKane 2000). A related approach commonly used in individual-based simulations (e.g., Dieckmann and Doebeli 1999; van Doorn and Weissing 2001; Doebeli and Dieckmann 2003) is to assume that the death rate of individuals is proportional to $NC(z)/K(z)$. Note that fitness functions (8.23b,c,e,f) imply that selection is absent at low densities, whereas fitness functions (8.23a,d) allow for selection even at low population densities.

The Christiansen and Loeschcke model   To illustrate the properties of evolutionary dynamics expected under intraspecific competition let us first consider the classical one-locus, multi-allele model of Christiansen and Loeschcke (1980). This model uses the fitness function (8.23d), which is the simplest from the mathematical point of view. Specifically, let us assume that the competition function is

$$\xi(x - y) = \exp\left[-\frac{(x - y)^2}{2V_C}\right]. \qquad (8.24a)$$

The strength of competition is controlled by parameter V_C (larger V_C means weaker competition). Competition for resource results in disruptive selection, where extreme phenotypes are favored because they experience weaker competition. Assume also that the carrying capacity for phenotype z is given by function

$$K(z) = K_{\max} \exp\left[-\frac{z^2}{2V_K}\right], \qquad (8.24b)$$

where $K_{\max}$ is the maximum possible carrying capacity. The assumption that $K(z)$ has a maximum at zero effectively means that the population experiences phenotypic stabilizing selection with an optimum value of z at zero. The strength of stabilizing selection is controlled by parameter V_K (larger V_K means weaker stabilizing selection).

Christiansen and Loeschcke assume that character z is controlled by a diploid locus with n alleles A_i having additive allelic contributions a_i $(i = 1, \ldots, n)$. That is, the character value for genotype $A_i A_j$ is $z_{ij} = a_i + a_j$. Assume for definiteness that the alleles are labeled in the order of their effects, so that $a_1 < a_2 \cdots < a_n$. Let the allelic contributions be small (namely, $a_i^2 \ll V_K, V_C$). In this case the population evolves to a single globally stable equilibrium (Christiansen and Loeschcke 1980), at which the population size is maximized (Christiansen 1988). This equilibrium is either monomorphic or dimorphic. That is, no more than two alleles can be present simultaneously at a stable equilibrium. Let

$$\kappa = \frac{V_C}{V_K}, \qquad (8.25)$$

where κ characterizes the strength of stabilizing selection relative to that of selection resulting from competition. Which alleles will be maintained can be summarized as follows:

Strong competition ($\kappa < 1$). If the contributions of the extreme alleles A_1 and A_n are of opposite signs, they both will be maintained. If the contributions of the extreme alleles A_1 and A_n are of the same sign, then the extreme allele with the smallest absolute contribution (i.e., allele A_1 if $0 < a_1 < a_n$ or allele A_n if $a_1 < a_n < 0$) always persists. In addition, the extreme allele with the largest absolute contribution will be maintained if its contribution is sufficiently large, specifically, if

$$|a_{\max}| > \frac{\kappa + 3}{\kappa - 1} |a_{\min}|, \qquad (8.26)$$

where $|a_{\min}|$ and $|a_{\max}|$ are the smallest and the largest allelic effects. Christiansen (1988) showed that if genetic variation is maintained and competion is strong enough, then heterozygotes will have smaller fitness than both homozygotes. In this case, one will observe disruptive selection at the population level.

Weak competition ($\kappa > 1$). Then, in most cases, only the allele with the smallest absolute contribution $|a_i|$ will be maintained. Two alleles $\mathbf{A}_i$ and $\mathbf{A}_{i+1}$ with the smallest absolute contributions can be maintained simultaneously only if their contributions are of opposite signs (i.e., $a_i \leq 0 \leq a_{i+1}$) and are not too different, specifically, if

$$|a_i| \frac{1 - \kappa}{\kappa + 3} < a_{i+1} < |a_i| \frac{\kappa + 3}{1 - \kappa}. \tag{8.27}$$

These results imply that if competion is strong, a very high level of genetic variation can be maintained with two extreme alleles segregating in the population. In contrast, if competition is weak, either all genetic variation is eliminated by stabilizing selection, or a low level of variation is maintained with two alleles with the smallest contributions segregating in the population.

To illustrate the implications of these results, let us consider the following scenario. Assume that initially the most frequent alleles in the population have allelic contributions of the same sign. In this case, selection will be directional, favoring alleles with the contributions maximizing carrying capacity K. As a result, the phenotypic distribution will shift toward $z = 0$. However as the population density increases, competition intensifies and new alleles with extreme contributions become favored because their carriers experience reduced competition. If competition is strong enough, selection becomes disruptive. This leads to the emergence of two phenotypically diverged groups of genotypes (i.e., genotypes $\mathbf{A}_1\mathbf{A}_1$ and $\mathbf{A}_n\mathbf{A}_n$), the hybrids between which (i.e., genotypes $\mathbf{A}_1\mathbf{A}_n$) have reduced fitness. This process can be very important in creating conditions favoring sympatric speciation, as first hypothesized by Rosenzweig (1978) and Pimm (1979) and first demonstrated by Pimm (1979) using a mathematical model. This process will be explored in detail below.

The Bürger model   Following Bürger (2002a), let us also consider a simple symmetric model of a two-locus, two-allele haploid population of sexual organisms. Assume that the alleles contribute additively to a trait z. Assuming equal effects of the loci and neglecting the effects of environment, the trait values for genotypes ab, Ab, aB, and AB are $z_1 = 0, z_2 = z_3 = 1$, and $z_4 = 2$, respectively. The overall fitness function is given by expression (8.23a). Assume that the coefficients

ξ_{ij} measuring the strength of competition between genotypes i and j are given by a quadratic function of the difference in trait values $z_i - z_j$:

$$\xi_{ij} = 1 - \frac{(z_i - z_j)^2}{2V_C}. \tag{8.28a}$$

Assume also that the fitness component coming from stabilizing selection is given by a quadratic function:

$$K(z_i) = 1 - \frac{(z_i - \theta)^2}{2V_K}. \tag{8.28b}$$

Note that quadratic functions (8.28) can be viewed as approximations of the corresponding Gaussian functions (8.24) when the coefficients V_C and V_K are large. In the Bürger model, the overall fitness can be rewritten in a normalized form as

$$w_i = \left[1 + c \sum_j (z_i - z_j)^2 x_j\right] \left[1 - s(z_i - \theta)^2\right], \tag{8.29}$$

where positive coefficients $s = 1/(2V_K)$ and $c = 1/[2V_C(\rho\kappa - 1)]$ measure the strength of stabilizing selection and competition, respectively (Bürger 2002a,b). Let us choose the optimum θ to be at the middle of the trait range, that is, $\theta = 1$. To guarantee that fitness is nonnegative, the coefficient s has to be smaller than 1. Bürger (2002a) showed that:

• If $c \leq s/(1 - s)$ (weak competition), genetic variation is not maintained and the population evolves to a state with a hybrid genotype (**Ab** or **aB**) fixed.

• If $s/(1 - s) < c < s/(1 - 3s/2)$ (moderate competition), the population evolves to one of the two polymorphic states at which allele frequencies satisfy the equality $p_A + p_B = 1$, but are different, and linkage disequilibrium is absent ($D = 0$).

• If $c \geq s/(1 - 3s/2)$ (strong competition), the population evolves to a globally stable polymorphic equilibrium with allele frequencies equal to 1/2 and a positive linkage disequilibrium D.

The positivity of D implies the deficiency of hybrid genotypes **aB** and **Ab** in the population. Figure 8.10 illustrates the level of linkage disequilibrium D maintained at the symmetric polymorphic equilibrium. As expected, smaller r and s and larger c increase D.

Loeschcke and Christiansen (1984) used numerical simulations to study a two-locus, two-allele version of this model. Asmussen (1983)

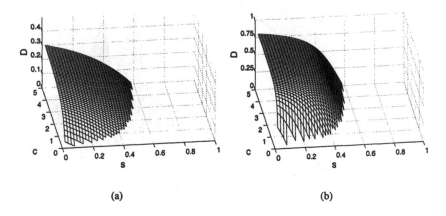

(a)   (b)

FIGURE 8.10. Normalized linkage disequilibrium in the Bürger model.
(a) $r = 0.5$. (b) $r = 0.1$.

and Wilson and Turelli (1986) used analytical methods to study alternative one-locus, two-allele models. Bürger (2002a,b) performed a
thorough analytical investigation of a two-locus, two-allele model using fitness function (8.23a). The results of these authors support the
conclusion reached previously by Roughgarden (1972), Bulmer (1974,
1980), Pimm (1979), Slatkin (1979), and Christiansen and Loeschcke
(1980) that sufficiently strong competition can maintain genetic variation. At polymorphic equilibria, the intermediate genotypes often have
reduced fitness. In addition, the two-locus results of Loeschcke and
Christiansen (1984) and Bürger (2002a,b) show that the equilibrium
structure of the model can be rather complex with multiple simultaneously stable equilibria present, so that the outcome of the evolutionary
dynamics crucially depends on the initial conditions.

8.4.3 Spatially heterogeneous selection and competition

What happens if spatially heterogeneous selection and selection resulting from competition act simultaneously? Figure 8.11 illustrates some
of the possible patterns arising in an asexual one-locus, multiple-allele
model. Numerical simulations used to produce this figure utilized the

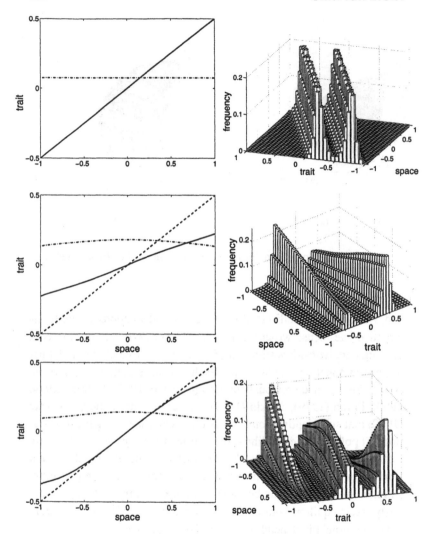

FIGURE 8.11. Equilibrium clines in the asexual one-locus, multiple-allele model of stabilizing selection with linearly varying optimum and competion. See the legend to Figure 8.7. $V_K = 1.0, V_C = 0.75$.

approach similar to the one described in the legend of Figure 8.7. The difference was that the overall fitness function was given by a version of function (8.23a), with Gaussian functions (8.24a) and (8.24b). Coefficients N, ϱ, and κ were chosen so that the maximum possible fitness reduction from competition was 50%. Figure 8.11 shows that the joint action of strong competition and spatial heterogeneity can result in the formation of complex genetic and spatial clustering, which is not surprising given that each of these two factors can cause clustering alone. Endler (1977) considered a sexual one-locus diallelic population and showed that competition steepens clines.

8.4.4   Adaptive dynamics approach

From the mathematical point of view, the population genetics models of frequency-dependent selection are much more complex than those of constant selection. The mathematical complexity becomes almost prohibiting if one allows for multiple alleles and mutation. Therefore, it is desirable to have a set of approximations that ideally are both general and simple enough to allow for significant analytical progress. One such set of approximations, which has emerged and become very popular in the past several years, is provided by a theoretical framework loosely known as *adaptive dynamics* (AD) (e.g., Metz et al. 1996; Geritz et al. 1997, 1998; Dieckmann 1997; Kisdi 1999; Kisdi and Geritz 1999; Geritz and Kisdi 2000). The AD is important because it can be used to find conditions for the maintenance of genetic variation and indicate directions of evolutionary dynamics. AD has provided a conceptual basis for several recent models of sympatric speciation which I will discuss later in the book.

The AD framework has grown from two major sources. One involves the attempts to generalize the approach based on the notion of evolutionarily stable strategies (ESS; e.g., Maynard Smith 1982) by explicitly considering the process of population convergence to an ESS. Another involves the population genetics studies of the dynamics of invasion of new alleles (and genotypes) in a resident population. For additional discussions of the relationships between AD and alternative approaches, see Abrams (2001) and Waxman and Gavrilets (2004). In describing the AD approach, I will follow Metz et al. (1996) and Geritz et al. (1997, 1998), adapting their approach to the population genetics methodology.

The AD approach is based on the following major assumptions, the importance of which will be discussed at the end of this section: (i) reproduction is asexual, (ii) individuals differ with respect to a single trait that can vary continuously, (iii) mutations have very small effects, (iv) mutations are very rare.

Invasion fitness   Let us consider a population of asexual organisms that are different with respect to a trait z. Let $f(z)$ be the distribution of z in the population. In the most general case, frequency-dependent selection acting on the trait can be formally specified using a fitness function (or functional) $w(x; f(\cdot))$, which gives the fitness of an organism with trait $z = x$ in a population in which trait z has distribution $f(z)$. Rather than using this general fitness function, the AD approach concentrates on its special case – the so-called *invasion fitness*. This is defined as the fitness of a rare (mutant) organism with trait value $z = y$ in a (resident) population with a certain distribution of z. In most applications, the resident population is assumed to be monomorphic or dimorphic. The standard notation for invasion fitness is $s_x(y)$ for the case of a monomorphic resident population with $z = x$, and $s_{x_1,x_2}(y)$ for the case of a dimorphic population with trait values $z = x_1$ and $z = x_2$, respectively. The fact that variables $x, x_1,$ and x_2 are written as subscripts is merely to make clear the distinction between residents and mutants. Otherwise, these variables are treated as the independent variables of function s in the same way as variable y.

Invasion in a monomorphic population   Assume that there is a monomorphic population of residents with trait value x. Let function $s_x(y)$ give the (invasion) fitness of a rare mutant y. What $s_x(y)$ exactly looks like depends on the specific biological situation. Expanding $s_x(y)$ in a Taylor series at $y = x$ gives

$$s_x(y) \approx s_x(x) + D(x)(y - x) + \ldots, \qquad (8.30)$$

where the fitness gradient $D(x)$ is defined as

$$D(x) = \left[\frac{\partial s_x(y)}{\partial y} \right]_{y=x}. \qquad (8.31)$$

If y is close to x and $D(x) \neq 0$, the term linear in $y - x$ dominates the higher order terms. Therefore, the fitness of rare mutants, $s_x(y)$, is higher than that of residents, $s_x(x)$, and, thus, mutants will increase in frequency if $D(x)$ and $y - x$ have the same sign. The sign of $D(x)$ determines what mutants can invade. If $D(x) > 0$, then mutants with $y > x$ will increase in numbers (and frequency) when rare (because they have a higher fitness than the residents). If $D(x) < 0$, then only mutants with $y < x$ will increase in numbers (and frequency) when rare. This behavior is analogous to those predicted by Wright's equation (2.2) (see page 23) and Lande's equation (2.12a) (see page 44).

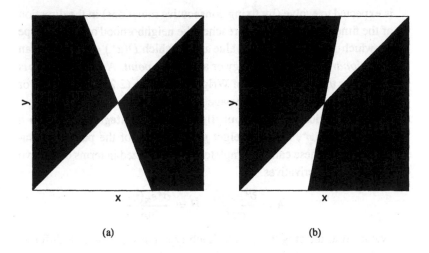

(a) (b)

FIGURE 8.12. Pairwise invasibility plot (PIP). (a) The PIP in a neighborhood of an evolutionarily stable strategy. (b) The PIP in a neighborhood of a branching point. White areas: no invasion. Dark areas: successful invasion.

The outcome of invasion in a monomorphic population can be analyzed graphically using a *pairwise invasibility plot* introduced by Christiansen and Loeschcke (1980). This plot gives the sign of $s_x(y) - s_x(x)$ as a function of x and y (see Figure 8.12). To see which mutants will increase in frequency when rare given a resident population with trait x, one looks along a vertical line through a point on the x-axis corresponding to x. The parts of this line inside a dark region denote trait values of mutants that can invade. The parts of this line inside a white region denote trait values of mutants that cannot invade. On the principal diagonal, $s_x(y) = s_x(x)$ by definition. This equality can be true for other combinations of y and x. In Figure 8.12 it is assumed that there is another line on which $s_x(y) = s_x(x)$, which intersects with the diagonal at point x^*. For all x values excluding x^*, the regions above and below the principal diagonal are marked by opposite colors. In contrast, the regions above and below x^* must have the same sign. This is only possible if $D(x^*) = 0$, because in this case $s_x(y) - s_x(x)$ will be proportional to $(x - y)^2$ (see equation (8.30)).

The assumptions initially made guarantee that generically any mutant phenotype that increases when rare will actually oust the resident phenotype (Geritz et al. 2002; Jacobs, unpub.). Therefore, the population is expected to evolve (by fixing consecutive mutations) in the direction of the fitness gradient until it reaches the neighborhood of a phenotype for which $D(x) = 0$. A trait value x^* for which $D(x^*) = 0$ is called an *evolutionarily singular strategy* or a *singular point*. A singular point is analogous to an equilibrium of Wright's equation (2.2) (see page 23) or Lande's equation (2.12a) (see page 44).

Classification of evolutionarily singular strategies Close to a singular point x^*, there are eight generic types of the pairwise invasibility plot. These can be completely characterized in terms of the two second-order derivatives

$$A = \frac{\partial^2 s_x(y)}{\partial x^2}, \ B = \frac{\partial^2 s_x(y)}{\partial y^2}$$

evaluated at the singular point. Each type corresponds to a different evolutionary scenario that can be interpreted in terms of four properties: (i) invasibility, (ii) the ability to invade a population when rare, (iii) convergence stability, and (iv) the possibility of protected polymorphism (e.g., Dieckmann 1997).

(i) A singular strategy x^* is *not invasible*, if a resident population consisting solely of x^* individuals cannot be invaded by any nearby mutant. This is the case if $s_{x^*}(x^*) > s_{x^*}(y)$ for all y close to x^*. Geometrically, this implies that at x^* the invasion fitness $s_x(y)$ has a local maximum with respect to y. Mathematically, this corresponds to

$$B < 0. \tag{8.32}$$

(ii) A singular strategy x^* invades a population with trait x if $s_x(x^*) > s_x(x)$ for all x close to x^*. Geometrically, this implies that at the singular point the invasion fitness $s_x(y)$ has a local minimum with respect to x (Geritz et al. 1997). Mathematically, this corresponds to

$$A > 0. \tag{8.33}$$

(iii) A singular strategy is *convergence stable* if a population of a nearby phenotype x can only be invaded by mutants with a phenotype closer to x^* than x is itself. This is the case if the fitness gradient, $D(x)$, is positive for $x < x^*$ and negative for $x > x^*$. This implies that at the singular point the fitness gradient $D(x)$ is a decreasing function of x. Mathematically, this corresponds to

$$A - B > 0. \tag{8.34}$$

(iv) A *protected polymorphism* arises if two strategies, say y_1 and y_2, making up the polymorphism can mutually invade, i.e., if $s_{y_1}(y_2) > s_{y_1}(y_1)$ and $s_{y_2}(y_1) > s_{y_2}(y_2)$. The set of all pairs of mutually invasible traits is given by the overlapping parts of the dark regions in the pairwise invasibility plot and its mirror image taken along the main diagonal. Geometrically, these conditions imply that along the secondary diagonal defined by equality $y - x^* = -(x - x^*)$ the invasion fitness $s_x(y)$ has a local minimum at x^*. Mathematically, this corresponds to

$$A + B > 0. \tag{8.35}$$

Locally stable singular points (that is, singular points satisfying condition (8.32)) are the subject of the ESS framework (Maynard Smith 1982). Being locally stable, however, does not guarantee that the population will actually get there. For this to happen, the condition of convergence stability (8.34) has to be satisfied. A singular strategy that is both locally stable and convergence stable is called *continuously evolutionarily stable* (Eshel 1983, 1997). If conditions (8.32) and (8.34) are satisfied, then the population reaches a (monomorphic) state from which no further evolutionary change is possible. Figure 8.13(a) gives a typical example of evolution towards an evolutionary stable strategy.

Evolutionary branching *Branching point* is defined as a singular strategy that is convergence stable ($A - B > 0$) and at the same time unstable to invasions ($B > 0$). These two conditions also imply that the singular strategy can invade ($A > 0$) and that nearby polymorphisms are protected ($A + B > 0$). What happens in the vicinity of a branching point? Assume that a monomorphic resident population has trait x that is already close to the branching point x^*. For definiteness, let $x < x^*$. Because x^* is convergence stable, a mutant with $y < x$ will not be able to invade, whereas a mutant with $x < y < x^*$ will be able to invade. Such a mutant will necessarily oust the resident allele. A mutant with $y > x^*$ sufficiently close to x^* will also be able to invade. However, its invasion will not be followed by the extinction of trait x because the dimorphism is protected. Thus, starting in a neighborhood of a branching point, an initially monomorphic population evolves by fixing mutations toward the branching point, where it inevitably becomes polymorphic.

Assume next that the population has reached a stable dimorphic equilibrium. Let us rename the trait values present in the population (that is, former x and y) to x_1 and x_2 ($x_1 < x^* < x_2$). Let function $s_{x_1,x_2}(y)$ give the invasion fitness of a rare mutant y introduced in a stable di-

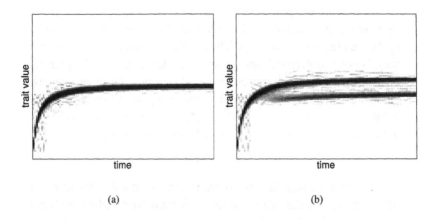

FIGURE 8.13. Two examples of the evolutionary dynamics. (a) Evolution towards an ESS. (b) Evolution towards a branching point with subsequent branching. The singular point is exactly at the middle of the visible range of trait values. The color maps the frequency of the corresponding trait values (from white for zero frequency to black for the highest frequency).

morphic population with alleles x_1 and x_2. The assumption that the dimorphic state is stable implies that the fitnesses of both residents are equal, implying that $s_{x_1,x_2}(x_1) = s_{x_1,x_2}(x_2)$. By expanding the invasion fitness $s_{x_1,x_2}(y)$ in a Taylor series for y, x_1, and x_2 close to the branching point x^*, Metz et al. (1996) and Geritz et al. (1998) show that

$$s_{x_1,x_2}(y) \approx s_{x_1,x_2}(x_1) + B(y \quad x_1)(y \quad x_2) + \dots \qquad (8.36)$$

Invasion is possible only if the mutant fitness $s_{x_1,x_2}(y)$ is higher than that of the residents $s_{x_1,x_2}(x_1)$. Because $B > 0$ at the branching point, the mutants that invade must have a trait value outside the two resident types (that is, with $y < x_1$ or $y > x_2$). For definiteness let $y < x_1$. Now if somehow allele x_1 reaches a low frequency, it will be ousted from the population (because in this case x_1 will be between the two residents y and x_2 and, thus, will not be protected). If this happens, the two phenotypes remaining in the population will be more divergent than before the invasion of y took place. Therefore, with each such step, the two remaining trait values will become more and more distinct. This process, in which an initially monomorphic population evolves to a branching point and subsequently undergoes disruptive selection and splits up into two phenotypically diverging lineages, is referred to as *evolutionary branching*. Figure 8.13(b) gives a typical example of evolutionary branching.

Examples   As an illustration of the AD methodology, let us consider three simple models.

The Christiansen and Loeschcke model.   Let fitness be given by function (8.23d). Assume that both the competition function and the carrying capacity are given by Gaussian functions (8.24a) and (8.24b), respectively. In this case, the invasion fitness is

$$s_x(y) = 1 + V\left[K(y) - N\xi(y - x)\right],$$

where $N = K(x)$. The local fitness gradient is

$$D(x) = -x\frac{VK(x)}{V_K}.$$

The above equation means that the only singular point is $x^* = 0$. One finds that at the singular point the relevant second-order derivatives are

$$A = \frac{VK_0}{V_C V_K}(V_C + V_K), \quad B = \frac{VK_0}{V_C V_K}(V_K - V_C).$$

It follows from the biological meaning of the parameters entering the above equalities that $A > 0$ always, $A - B > 0$ always, and $B > 0$ or $B < 0$ depending on whether $V_K - V_C$ is positive or negative. We conclude that $x^* = 0$ is a branching point if

$$V_K > V_C, \tag{8.37}$$

that is, if competition is stronger than stabilizing selection. In this case, the population first evolves towards the point $x = 0$, where it undergoes splitting into two clusters as in Figure 8.13(b). If inequality (8.37) is reversed, then $x = 0$ is an evolutionarily stable strategy. In this case, the trait values in the population approach $x = 0$ and stay there as in Figure 8.13(a). Condition (8.37) was first found by Roughgarden (1972).

The Kisdi and Geritz (1999) model.   Let us consider a symmetric two-niche version of the Levene model (see Section 8.1.1). Following Kisdi and Geritz (1999) and Geritz and Kisdi (2000), let us assume that the diallelic locus under consideration controls an additive trait z. Assume that fitness $w_j(z)$ in niche j is given by the Gaussian function (8.11), where the optimum phenotypes differ between the niches and are scaled to be $\theta_1 = -\theta_2 = \theta$. With two alleles with effects x and y, the three diploid phenotypes are $2x$, $x + y$, and $2y$, respectively. Their fitnesses can be found by taking the corresponding averages of $w_j(z)$ over the niches. Kisdi and Geritz (1999, equation 4) show that the invasion fitness of a mutant allele y in a resident population x can be written as

$$s_x(y) = \frac{1}{2} \frac{w_1(x+y)}{w_1(2x)} + \frac{1}{2} \frac{w_2(x+y)}{w_2(2x)}. \qquad (8.38)$$

One finds that the fitness gradient is

$$D(x) = -\frac{x}{2V_s},$$

which implies that $z = 0$ is the only singular point. Proceeding as above, one finds that the relevant second order derivatives are

$$A = \frac{\theta^2 + 3V_s}{V_s^2}, \quad B = \frac{\theta^2 - V_s}{V_s^2}.$$

It follows that $A > 0$ and $A > B$. If

$$\theta^2 > V_s, \qquad (8.39)$$

then $B > 0$, which means that $z = 0$ is a branching point. If the inequality (8.39) is reversed, then $B < 0$ and $z = 0$ is convergence stable. In the former case, the population will maintain genetic variation. In the latter case, genetic variation will be lost as the population evolves to a monomorphic state.

The Gavrilets and Waxman (2002) model. The original model developed by Gavrilets and Waxman (2002) describes competition among males for access to females, but I will formulate it in terms of competition among exploiters for a renewable resource. Assume that exploiters are different with respect to phenotypic trait z whose distribution in the population is $\phi(z)$. Let the resource type be characterized by variable R, with $f(R)$ the density distribution of R. Assume that the efficiency at which exploiters with phenotype z can utilize resource R is characterized by a function $U(z, R)$, which depends only on the difference between y and R and has a Gaussian form

$$U(z, R) = \exp\left(-\frac{(z-R)^2}{2V_U}\right). \qquad (8.40)$$

This implies that for each exploiter trait z, there is an optimum resource type $R = z$. Parameter V_U characterizes the efficiency of utilization of nonoptimal resources, which is reduced because of competition with other types. Therefore, V_U implicitly characterizes the strength of competition between different exploiters, with smaller values of V_U implying stronger competition.

I will assume that the distribution of resources $f(R)$ does not change and is given by a Gaussian distribution with zero mean and variance V_R. The assumption about unimodal distribution of resources imposes kind of a stabilizing selection on exploiters, with parameter V_R characterizing the strength of selection. The average efficiency at which resource R is utilized by exploiters is

$$\overline{U}(R) = \int U(z, R)\phi(z)dz, \qquad (8.41)$$

where the integral is taken across the range of possible values of y. The ratio $U(z, R)/\overline{U}(R)$ gives the relative efficiency of exploiter z in utilizing resource R. Therefore, the relative fitness of exploiter z can be defined as its average efficiency over the whole spectrum of available resources:

$$w(z) = \int \frac{U(z, R)}{\overline{U}(R)} F(R)dR. \qquad (8.42)$$

If trait values are discrete, the integral in the above equation is substituted by an appropriate sum.

Let us consider the invasion of a mutant trait y in a resident population that is monomorphic for trait x. In this case, $\overline{U}(R) = U(x, R)$, and a straightforward integration yields the invasion fitness in the form

$$s_x(y) = \exp\left[-\frac{(y - x)}{2V_U}\left(x + y - (y - x)\frac{V_R}{V_U}\right)\right]. \qquad (8.43)$$

The fitness gradient is

$$D(x) = -\frac{x}{V_U}.$$

The above equation implies that the only singular point is $x^* = 0$. One finds that the relevant second-order derivatives are

$$A = \frac{V_F + V_U}{V_U^2}, \quad B = \frac{V_F - V_U}{V_U^2}.$$

It follows from the biological meaning of V_F and V_U that $A > 0$ and $A - B > 0$ always. Furthermore, $B > 0$ or $B < 0$, depending on whether $V_F - V_U$ is positive or negative, respectively. We conclude that $x^* = 0$ is the branching point if

$$V_F > V_U, \qquad (8.44)$$

that is, if competition is stronger than stabilizing selection. In this case, the population of exploiters is predicted to evolve first toward the point of the highest resource abundance $x = 0$, where it undergoes splitting into two clusters as in Figure 8.13(b). If inequality (8.44) is reversed, then $x = 0$ is an evolutionarily stable strategy. In this case, the trait values in the population approach $x = 0$ and stay there, as in Figure 8.13(a).

Limitations of the AD approach The simplicity of the AD approach relative to the methods of population genetics must have some price. Here, I discuss the consequences of violations of the major assumptions underlying this approach listed on page 265.

Reproduction is asexual. In sexual populations, organisms are characterized by a number of genes that are reshuffled by recombination and segregation and can interact nonlinearly in controlling the phenotypic value(s). In general, these features make AD inapplicable to sexual populations. However, there are exceptions. For example, if (i) mating is random and (ii) selection is in the form of the differences in viability, then the state of a one-locus, multiallele diploid population is uniquely defined by the set of allele frequencies. Assume also that (iii) each allele can be assigned a certain allelic contribution z_i, so that the diploid phenotype z_{ij} is uniquely defined by the contributions z_i and z_j of the two corresponding alleles. In this case, the diploid model of frequency-dependent selection is mathematically equivalent to the haploid model in which the fitness of a haploid organism carrying a given allele is played by the induced fitness of the allele in a diploid organism. Then AD results will be applicable to the diploid case. Note that assumption (iii) basically means that the degree of dominance is assumed to be fixed. For example, it will be satisfied if the trait is additive, so that $z_{ij} = z_i + z_j$. This assumption was explicitly made by Christiansen and Loeschcke (1980), Kisdi (1999), Geritz and Kisdi (2000), and Matessi et al. (2001) in their studies of diploid populations. However, if dominance is allowed to evolve, the two variables z_i and z_j are not sufficient to uniquely specify the three diploid phenotypes z_{ii}, z_{ij}, and z_{jj}, and the methods considered above will not apply. Van Dooren (2003) arrives at similar conclusions from considering invasion fitnesses.

Individuals differ with respect to a single trait that can vary continuously. Apparently, not much has been done so far with regard to generalizing AD to the case on multiple traits.

Mutations have very small phenotypic effects. This is a crucial assumption, which is very easily violated biologically. Indeed, mutations with nonnegligible effects are well documented (e.g., Griffiths et al. 1996; Futuyma 1998), and most mutations of specific interest to biologists have observable effects. The consequences of the violation of this assumption can be severe. Indeed, the AD conclusions that (i) successful invasion in a monomorphic population away from a singular point always results in the extinction of the resident allele, and (ii) successful invasion in a dimorphic population always results in the extinction of the allele with an intermediate effect both hinge on this assumption. Conclusion (i) justifies the claim about convergence to a singular point. Conclusion (ii) justifies the claim about the formation of two discrete "branches".

The comparison of the exact results for the Christiansen and Loeschcke model described on page 260 with those based on AD is illuminating. For this model, AD predicts the convergence of the trait value to $z = 0$, with alleles with smaller absolute contributions always ousting the alleles with larger absolute contributions. However, the exact analysis demonstatrates the possibility of a stable polymorphism of two alleles under certain conditions (given by inequalities (8.26)). If competition is weak (i.e., if $V_C > V_K$), the AD predicts that polymorphism cannot be maintained. However, the exact analysis shows that under certain conditions (given by inequalities (8.27)) the maintenance of genetic variation is possible. Moreover, from analyses of simple models of frequency-dependent selection, we know that these models can have multiple simultaneously stable equilibria and also exhibit nonequilibrium dynamics, including cycles, chaos, transient chaos and intermittency (e.g., Cockerham et al. 1972; Asmussen and Basnayake 1990; Altenberg 1991; Gavrilets and Hastings 1995b). The existence of these behaviors implies that AD conclusions can be rather misleading if phenotypic effects of mutations are not too small. The important question of how small mutational effects have to be for the AD approach to work cannot be answered a priori and requires a careful examination in each specific case.

Mutations are very rare. This is another crucial assumption that can be easily violated biologically with severe consequences. The majority of AD results are based on consideration of models with just two, or sometimes three, alleles present in the population simultaneously. The current AD theory has basically not much to say about what happens if the number of alleles is larger than two or three. Note that natural populations typically have many more segregating alleles, even in the loci experiencing selection (e.g., Li 1997). In principle, it is possible that both the dynamics and the equilibrium states with just two alleles

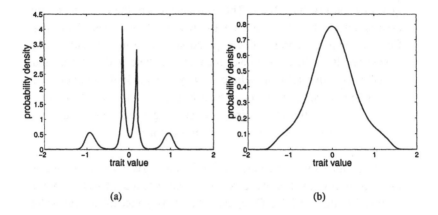

FIGURE 8.14. Effects of initial conditions on the evolutionary dynamics in the Gavrilets and Waxman (2002) model. In both cases, $V_U = 1 < V_F = 1.3$, so that the AD predicts branching. Mutation rate $\mu = 10^{-5}$ per generation, the standard deviation of mutational effects is $\sigma_m = 0.05$, the trait values are discretized with interval 0.025. (a) Initial distribution is uniform on the interval $[-0.15, 0.20]$; that is, the initial distribution is in a neighborhood of a branching point at $x = 0$. (b) Initial distribution is uniform on the interval $[3.75, 4.05]$; that is, the initial distribution is away from the branching point. The distributions are shown after 100,000 generations; however, significant changes were observed only within the first few thousand generations.

and with many alleles are completely different (e.g., Seger 1988). In particular, the alternative equilibria with a different number of alleles present can be stable simultaneously, so that the eventual outcome of evolutionary dynamics will depend on initial conditions.

Figure 8.14 provides an illustration of the different dynamics observed in a neighborhood of a branching point, depending on initial conditions in the Gavrilets-Waxman model considered above. For the parameter values used in this figure, a unimodal distribution presented in Figure 8.14(b) is a much more probable outcome than a distribution with two distinct branches predicted by the AD approximation. In this model, one can show that if $V_F > V_U$, a Gaussian distribution of $f(z)$ with zero mean and variance V_G equal to $V_F - V_U$ represents an equilibrium. It is interesting that in both cases shown in Figure 8.14, V_G is very close to $V_f - V_\psi$ in spite of the fact that the distribution of z is not Gaussian. The main message of Figure 8.14 is that models of frequency-dependent selection can have multiple simultaneously stable equilibria that can be easily missed by the AD approximation.

8.5 SUMMARY

• Disruptive selection can be defined as selection that favors different optima in the population as a whole, or in different parts of the population, or at different times.

• Spatially heterogeneous selection induced by abiotic or biotic factors is very effective in both maintaining genetic heterogeneity and creating spatial differentiation. Weak differences in selection can produce significant spatial differentiation if migration is not too strong. The interaction of spatial gradients in selection with limited migration can result in complex clusters both in geographic and genetic spaces.

• Sharp hybrid zones can be formed by weak gradients in selection.

• Spatially uniform disruptive selection cannot create but can maintain genetic heterogeneity in space and prevent the fusion of diverged populations after their secondary contact. This requires migration rates to be sufficiently small relative to the strength of selection.

• Selection varying in time can both increase and maintain genetic variation in populations.

• Frequency-dependent selection is very effective in both creating and maintaining genetic variation in populations. Frequency-dependent selection can lead to the formation of clearly differentiated genetic clusters within populations.

• Commonly, the dynamic equations describing evolution under disruptive selection have multiple simultaneously stable equations. In this case, the eventual outcome of evolutionary dynamics depends on initial conditions. Therefore, choosing appropriate initial conditions is very important. To investigate the processes of the origin of genetic variation, the most appropriate initial conditions are those describing (almost) monomorphic population. The initial conditions specifying highly polymorphic populations are most appropriate for studies on the maintenance of variation and prevention of fusion of populations that have diverged substantially prior to their secondary contact.

• The AD approach provides a very useful set of simple approximations for analyzing and understanding evolutionary consequences of frequency-dependent selection. As with any approximation, the simplicity gained by using it has certain costs. In particular, the assumptions about the extreme rarity of mutations and the smallness of their effects are crucial for the conclusions of AD to be justified. These assumptions are easily violated in natural populations.

• Evolutionary branching is a process of the origin and amplification of genetic heterogeneity within populations experiencing frequency-dependent selection. The word *branching* implies that only very well defined and differentiated trait values are observed. However, if mutations are not extremely rare or have appreciable effects, evolutionary branching as defined mathematically by AD is quite compatible with multimodal but continuous or unimodal continuous distributions of the trait value. Therefore, the term *branching* is somewhat misleading. When the AD assumptions are not met or their validity cannot be checked, the appropriate way to interpret the conditions for evolutionary branching is as conditions for the maintenance of genetic variation.

• Disruptive selection maintaining genetic variation creates conditions in which getting rid of low-fitness intermediates is a general way to increase the average fitness of the population. One of the possible ways to accomplish this is the evolution of prezygotic RI between clusters of optimum organisms (to be considered in the next chapter). Therefore, disruptive selection maintaining genetic variation creates a situation where speciation is an outcome that increases fitness. Disruptive selection is a crucial component of most models of parapatric and sympatric speciation (to be considered in Chapter 10), which was our main motivation for devoting this chapter to the models of disruptive selection.

8.6   CONCLUSIONS

Disruptive selection can arise from a number of factors, including spatial gradients in selection, selection varying randomly in space or time, and frequency-dependent interactions within the population and between species. Disruptive selection can be modeled using fitness landscapes that change with the population state or in space. Mathematical models have identified a variety of conditions under which disruptive selection maintains genetic variation and leads to the formation of genetic clusters between which some postzygotic RI exists. Disruptive selection maintaining genetic variation creates situations favoring the evolution of prezygotic RI.

CHAPTER NINE

Evolution of nonrandom mating and fertilization

Mating is *random* if each female in a population has an equal chance of mating with any male in the population. Random mating implies that the overall frequency of matings between females of a certain type and males of a certain type is equal to the product of the frequencies of these types in the population. Random mating does not change allele frequencies. Mating is *selective* if the frequency of a certain type among mated individuals is different from its frequency in the population at large (Lewontin et al. 1966). If mating is selective, there are genetically controlled differences in the probabilities of mating. Mating can be selective for male alleles (or traits) only, for female alleles (or traits) only, or for both. Selective mating changes allele frequencies (unless the population is at an equilibrium). In most cases considered below, selective mating results in a loss of genetic variation. Under certain conditions (e.g., if rare phenotypes have a mating advantage), selective mating maintains high levels of genetic variation. By definition, random mating cannot be selective. However, nonrandom mating (NM) is not necessarily selective. The term *assortative mating* is usually used to describe NM when there is a correlation (positive or negative) between genotypes (or phenotypes) of mating individuals.[1] The term *preferential mating* is used to describe NM when there are genetically controlled differences in the probabilities of accepting individuals of the opposite sex as mates. Note that preferential mating is not necessarily assortative. For example, if all females have a preference for a specific male phenotype, mating is preferential but not assortative. In

---

[1]Lewontin et al. (1966) used a more restrictive definition of assortative mating that implied it was also nonselective.

general, mating can be random with respect to a certain trait, selective with respect to another trait, and assortative or preferential with respect to other traits.

I consider models of NM due to genetic factors (rather than NM due to, say, spatial subdivision). In general, mating will be nonrandom if there are genetically controlled differences in the probabilities of (i) encounter of different individuals, or (ii) actual mating given that an encouter has occurred. Most models of NM due to genetic factors used for studying speciation consider only one of these scenarios. Below I will interpret *mating* in a general way that, in particular, will include interactions between gametes. The latter is important both for organisms with external fertilization and for organisms with internal fertilization. Most of the results on NM given below can also be interpreted in terms of preferential or assortative fertilization.

There are many different types of NM that require different mathematical models. Moreover, theoretical conclusions are often model-specific. Therefore, to assure the generality of theoretical conclusions, it is desirable not only to study a specific mathematical model in detail but also demonstrate that similar conclusions emerge from related models. For this reason, in this chapter I consider a large number of different models of NM. I start by describing a theoretical framework for specifying mating probabilities that helps in exposing more clearly the relationships between different models of NM. Then, I review specific models of NM, outline their relationsips, and compare them with regard to (i) the possibility of the maintenance of genetic variation and (ii) the levels of within-population genetic clustering and RI. Some of these models describe allopatric, parapatric, or sympatric speciation by NM. Others will be used in the next chapter as components of more complex models incorporating both NM and disruptive selection. At the end, I consider a simple model of NM controlled by a culturally transmitted trait.

9.1 A GENERAL FRAMEWORK FOR MODELING NONRANDOM MATING AND FERTILIZATION

I will consider populations in which there is a set of different female phenotypes, with $\Phi_{f,i}$ being the frequency of female phenotype i among adults, and a set of different male phenotypes, with $\Phi_{m,j}$ being the frequency of male phenotype j among adults ($\sum_i \Phi_{f,i} = 1$, $\sum_j \Phi_{m,j} = 1$). The overall frequency of $i \times j$ matings among all matings taking place will be denoted as $P(i \times j)$. I will define the *relative mating success* of female phenotype i as

$$w_{f,i} = \frac{\sum_j P(i \times j)}{\Phi_{f,i}}. \tag{9.1a}$$

Here, the numerator gives the overall contribution of female phenotype i to matings. In a similar way, the relative mating success of male phenotype j will be defined as

$$w_{m,j} = \frac{\sum_i P(i \times j)}{\Phi_{m,j}}. \tag{9.1b}$$

The relative mating success of a certain phenotype is equal to one if the overall contribution of this phenotype to matings is equal to its frequency. Mating is selective for females if $w_{f,i}$ vary between different female phenotypes. In a similar way, mating is selective for males if $w_{m,j}$ vary between different male phenotypes. The relative mating success of a phenotype in models of NM as defined by equations (9.1) is analogous to relative viability in models of viability selection. Note that in models of constant viability selection, the ratio of the relative viabilities of any two genotypes is a constant. In contrast, in models of NM to be considered below, the ratio of the relative mating successes, say $w_{m,i}/w_{m,k}$, will depend on the population genetic state. In this regard, the models of NM are similar to models of frequency-dependent viability selection. Mating success as defined above represents an important component of *sexual selection*, by which one usually means differential reproductive success (e.g., Darwin 1859; Andersson 1994). Existing data show that there is genetic variation in female and male mate preferences (e.g., Kirkpatrick 1987; Andersson 1994; Bakker and Pomiankowski 1995) and suggest that, on average, selection on mating success is stronger than on survival (Kingsolver et al. 2001).

In principle, it is possible that not all individuals (or gametes) of one sex in the population can potentially encounter all individuals (or gametes) of the other sex during the mating period (e.g., if there is some structuring in the population with respect to location or timing of mating). To formalize this idea, I will assume that mating takes place within different *mating pools*. Individuals that belong to the same mating pool encounter each other randomly (i.e., with equal probability). Individuals that join different mating pools do not encounter each other during the mating period.

9.1.1 Random mating within mating pools joined preferentially

First, I consider the case when individuals have "preferences" for certain mating pools, but mating within the pools is random. There are a number of biological scenarios leading to this kind of NM. One is when there are genetically based differences in host preference and mating occurs exclusively on or near a certain host. For example, this is the case for the hawthorn- and apple-infesting populations of *Rhagoletis pomonella*, the apple maggot fly. In the mark-and-recapture experiments, the relative preference of hawthorn-origin flies for hawthorn over apple trees was 91.2%, and the relative preference of apple-origin flies for apple over hawthorn trees was 55.2% (Feder et al. 1994). Another factor, which also was shown to be important in *Rhagoletis*, is seasonal isolation (allochrony) due to host-associated differences in the timing of adult eclosion. Apple-infesting flies eclose (i.e., emerge from their pupal cases) significantly earlier than hawthorn-infesting flies, and this difference is genetically based (Prokopy et al. 1988). Because adult longevity is limited and the fruit of apple varieties favored by *R. pomonella* ripen approximately three weeks earlier than those of hawthorns, eclosion time differences may limit the opportunity for the flies to use alternative host species. NM can also result from differences in flowering time. For example, the populations of the grass *Agrostis tenuis* growing on pastures contaminated with heavy metals from mining activities and on surrounding grassland have diverged in flowering time (McNeily and Antonovics 1968). Plants on the mine flower about a week earlier than those on the adjacent pasture, and these differences are genetically determined. Yet another example is NM mediated via pollinator preference. For example, the monkeyflower species *Mimulus lewisii* is bumbblebee-pollinated, whereas a closely related species *M. cardinalis* is pollinated by hummingbirds. The two species have diverged in flower color, corolla form, nectar volume and concentration, and other characters (Bradshaw et al. 1995). Despite these differences, artificially produced hybrids are both viable and fertile. NM can also occur because of *flower constancy*, i.e., the tendency of insects to work a single kind of flower for a period of time. For example, flower constancy in bees is an effective mechanism preventing cross-fertilization between two species of snapdragon (*Antirrhinum majus* and *A. glutinosum*) grown in experimental plots (Mather 1947).

Organisms preferring a specific host, flowering during a specific time interval, or being pollinated by a specific pollinator can be viewed as forming a separate mating pool. Assume that females with phenotype i join the k-th mating pool with probability $\pi_{i,k}$. Let $\pi'_{j,k}$ be the corresponding probability for males with phenotype j. Then the overall fractions of females and males joining mating pool k are $\Pi_k = \sum_i \pi_{i,k} \Phi_{f,i}$ and $\Pi'_k = \sum_j \pi'_{j,k} \Phi_{m,j}$. Correspondingly, the frequencies of female and male phenotypes in the k-th mating pool are $(\pi_{i,k}/\Pi_k)\Phi_{f,i}$ and $(\pi'_{j,k}/\Pi'_k)\Phi_{m,j}$. Let c_k be the relative contribution of the k-th mating pool to offspring ($0 \le c_k \le 1, \sum c_k = 1$). From biological considerations, c_k can depend on the overall frequency of females that join the pool and do mate there. I assume that mating within the mating pools is random. Therefore the contribution of $i \times j$ matings in the k-th mating pool to the overall frequency of $i \times j$ matings in the population is proportional to the product of the frequencies of the corresponding male and female phenotypes in the pool. The overall frequency of matings between females i and males j among all the matings taking place can be written as

$$P(i \times j) = \left[\sum_k c_k \frac{\pi_{i,k}}{\Pi_k} \frac{\pi'_{j,k}}{\Pi'_k} \right] \Phi_{f,i}\Phi_{m,j}. \qquad (9.2)$$

The ratios inside the sum characterize the relative preferences of female type i and male type j for the k-th mating pool. Note that if individuals have no specific preferences with regard to mating pools (that is, if $\pi_{i,k} = \pi_k, \pi'_{j,k} = \pi'_k$), then the genotype frequencies in the mating pools are the same as in the whole population (implying that $\Pi_k = \pi_k, \Pi'_k = \pi'_k$), and expression (9.2) simplifies to $P(i \times j) = \Phi_{f,i}\Phi_{m,j}$. This is, obviously, the case of random mating.

Using expression (9.2) and definitions (9.1), the relative mating success of female phenotype i and of male phenotype j are

$$w_{f,i} = \sum_k c_k \frac{\pi_{i,k}}{\Pi_k}, \qquad (9.3a)$$

$$w_{m,j} = \sum_k c_k \frac{\pi'_{j,k}}{\Pi'_k}. \qquad (9.3b)$$

The relative mating success depends both on the probabilities $\pi_{i,k}, \pi'_{j,k}$ of joining the mating pools and on the relative contributions c_k of the pools to offspring. Under certain conditions, phenotypes have equal mating success. For example, let the trait controlling mating behavior be equally expressed in both sexes (implying that $\pi_{i,k} = \pi'_{i,k}$ and $\Pi_k = \Pi'_k$). Assume also that the relative contribution of each mating pool is equal to the proportion of the population that enters it (implying that $c_k = \Pi_k = \Pi'_k$). Moreover, assume that the phenotypes do not differ with regard to the overall proportion of individuals entering mating pools (implying that the sum $\sum_k \pi_{i,k}$ is the same for all i). Then $w_{f,i}$ and $w_{m,j}$ are the same for all phenotypes and mating is nonselective.

9.1.2 Preferential mating within mating pools joined randomly

Next, I consider the case when individuals encounter each other randomly, but whether mating occurs or not depends on their mating preferences. This kind of NM is relevant for many biological situations arising within the context of sexual selection. Females usually have certain preferences and often accept only males that possess certain characters. For example, male courtship songs play a dominant role in female discrimination in Darwin's finches, grasshoppers, frogs, and *Drosophila* (e.g., Ryan and Wilczynski 1988; Grant and Grant 1997; Grant 1999; Doi et al. 2001; Panhuis et al. 2001). Female mating preferences are associated with male color patterns in cichlids (e.g., Markert et al. 1999; Kornfield and Smith 2000), *Heliconius* butterflies (Naisbit et al. 2001), and many birds (e.g., Andersson 1994; Greene et al. 2000). Preferential fertilization by cospecific sperm and pollen underlies RI between species of *Drosophila*, ground crickets, ladybirds, grasshoppers, Louisiana iris, sunflowers, red flour beetles, sea urchins, and abalone (Howard 1999). Sexual preferences can be strong not only in between-species interactions but also in within-population interactions. For example, laboratory tests show that about a third of mature and healthy courting males of the Hawaiian species *D. silvestris* are repeatedly rejected by females in favor of a minority of males (Carson 2002).

To set up a framework for modeling these situations, let us consider the case of a single mating pool assuming that individuals have equal chance of joining it. Obviously, the phenotype frequencies in the pool are the same as in the whole population. (Equivalently, one can treat the whole population as one mating pool.) I will assume that encounters of individuals within the mating pool are random. Let ψ_{ij} be a genetically controlled *preference* of females with phenotype i for males with phenotype j ($0 \leq \psi_{ij} \leq 1$), in the sense that the probability that a given female i mates with one of the males with phenotype j is proportional to $\psi_{ij} \, \Phi_{m,j}$, with a coefficient of proportionality being independent of j. Let us also define the average preference of female i for males in the mating pool:

$$\overline{\psi}_i = \sum_j \psi_{ij} \, \Phi_{m,j}. \qquad (9.4)$$

In principle, not all females necessarily mate. Let us write the overall probability f_i that a female with phenotype i does mate as a function of the strength of her preference for accessible males: $f_i = f(\overline{\psi}_i)$. Finally, let us define the probability that an average female does mate:

$$\overline{f} = \sum_i f(\overline{\psi}_i) \Phi_{f,i}. \qquad (9.5)$$

Using these variable, the overall frequency of matings between females i and males j among all the matings taking place can be written as

$$P(i \times j) = \left[\frac{f(\overline{\psi}_i)}{\overline{f}} \frac{\psi_{ij}}{\overline{\psi}_i} \right] \Phi_{f,i} \Phi_{m,j}. \qquad (9.6)$$

The first ratio in the right-hand side of equation (9.6) is the relative probability of mating for female phenotype i. The second ratio is the relative preference of female type i for male type j. If females have no specific preferences (i.e., if $\psi_{ij} = $ const), then the average female preferences are equal (i.e., $\overline{\psi}_i = $ const), all females have equal probabilities of mating (i.e., $f(\overline{\psi}_i) = \overline{f}$), and expression (9.6) simplifies to $P(i \times j) = \Phi_{f,i} \Phi_{m,j}$. This is, obviously, the case of random mating.

Using expression (9.6) and definitions (9.1), the relative mating success of female phenotype i and of male phenotype j are

$$w_{f,i} = \frac{f(\overline{\psi}_i)}{\overline{f}}, \qquad (9.7a)$$

$$w_{m,j} = \sum_i \frac{f(\overline{\psi}_i)}{\overline{f}} \frac{\psi_{ij}}{\overline{\psi}_i} \Phi_{f,i}. \qquad (9.7b)$$

Obviously, if females have equal mating success (i.e., if $w_{f,i} = $ const), mating is nonselective for females. However, if females differ in preferences, different males in general will have different mating success, and mating will be selective in males even if all females have equal mating success. Selection in males will cause indirect selection (via linkage disequilibrium and pleiotropy) in females, affecting the corresponding allele frequencies.

Few additional comments are in order. If there is not one but a number of different mating pools that individuals join randomly, then within each of the pools the frequencies of matings will be given by expression (9.6). Therefore, the overall frequencies of specific matings in the whole population will still be given by expression (9.6), even if the mating pool sizes differ. The derivations above implicitly assumed that the mating pools are very large. Alternatively, a mating pool can be formed by a few individuals randomly sampled from the whole population (e.g., Seger 1985b). In this case, different mating pools may have different genetic composition due to stochastic factors, even if there are no genetically controlled differences in the probabilities of joining them. Such a case can be treated by modifying our approach accordingly. If the phenotypic distributions in the population are continuous rather than discrete, the sums in the above equations should be substituted by appropriate integrals. Preference function ψ can be interpreted as a measure of fitness of mating pairs. Therefore, ψ defines a fitness landscape for mating pairs as was already illustrated in Chapters 2 and 5. Finally, I note that, in principle, it is possible that not only one sex (here, females) but both sexes have certain mating preferences (e.g., Seehausen et al. 1999; Bergstrom and Real 2000). The approach described above can be modified accordingly to handle this case. For an example of a speciation model incorporating mating preferences by both sexes, see Lande et al. (2001). More theoretical and empirical work on this topic is necessary.

In discussing the models of NM, I will identify several major components: (i) the nature of mating pools and the probabilities of joining them, (ii) the contributions of specific mating pools to the overall production of offspring, (iii) the mating preferences of different phenotypes, and (iv) female and male mating successes and whether mating is selective. I will classify the models into two general classes, depending on whether mating is controlled by a single trait (or a single set of loci) expressed in both sexes or by two different sex-linked traits (or sets of loci). The former type of NM will be referred to as *similarity-based*, whereas the latter type will be referred to as *matching-based*.

9.2   SIMILARITY-BASED NONRANDOM MATING

Here, I will assume that the genes controlling NM are equally expressed in both sexes. In this case, the sets of male and female phenotypes are identical and the frequencies of different phenotypes in the sexes are identical as well.

9.2.1   Single locus

One-locus models are a natural starting point for theoretical approaches to many evolutionary phenomena. These simple models have some empirical support. For example, single-locus determination of mating preferences appears to be important in discrimination against homospecific males in mole rats (Beiles et al. 1984), in the control of flowering time in the grass *Agrostis tenuis* (Macnair and Christie 1983), in the control of male courting songs in *Drosophila* (Doi et al. 2001), and in the control of the species-specific timing of mating behavior in *Drosophila* (Tauber et al. 2003). Genes with major effects underly sexual isolation in some other groups as well (Ritchie and Phillips 1998).

In this subsection, I consider one-locus diallelic populations with three diploid genotypes **AA, Aa**, and **aa**. The frequencies of the three genotypes will be denoted as Φ_{AA}, Φ_{Aa}, and Φ_{aa}; the frequencies of alleles **A** and **a** will be denoted as $p\,(= \Phi_{AA} + \Phi_{Aa}/2)$ and $q\,(= \Phi_{aa} + \Phi_{Aa}/2)$. The population will evolve to a state where heterozygotes are in deficiency relative to what is expected under random mating at the same allele frequencies. This deficiency reflects the evolution of (partial) prezygotic RI within the population. The degree of RI can be characterized using a heterozygote deficiency index

$$I = 1 - \frac{\Phi_{Aa}}{2pq}, \tag{9.8}$$

which varies between 0 and 1. If the population is in Hardy-Weinberg proportions (i.e., $\Phi_{Aa} = 2pq$), which is expected if mating is random, then RI is absent: $I = 0$. If the population has split into two homozygous groups that do not mate and hybrids (i.e., heterozygotes) are completely absent (i.e., $\Phi_{Aa} = 0$), then RI is complete: $I = 1$.

The O'Donald model   The most commonly used models of assortative mating were originally formulated and studied in a classical paper by O'Donald (1960).

Assortative mating without complete dominance: symmetric case.
Assume that all three genotypes possess distinct phenotypes. In a simple
symmetric version of the O'Donald model, a fixed proportion α of
phenotypes mate with the like phenotypes, while the remainder $1 - \alpha$
of the population mates at random.

In terms of our general framework, here one can think of four different
mating pools: three one-phenotype pools joined by a proportion α of
the corresponding phenotypes, and one three-phenotype pool joined by
the remainder organisms. The sets of male and female phenotypes are
identical, and therefore the probabilities of joining the mating pools
are the same for males and females ($\pi_{i,k} = \pi'_{i,k}$); also, the genotype
frequencies in mating pools are the same for males and females ($\Pi_k =
\Pi'_k$, with $\Pi_1 = \alpha x, \Pi_2 = \alpha y, \Pi_3 = \alpha z, \Pi_4 = 1-\alpha$). Within the pools,
mating is random and each organism has an equal chance of mating.
The contribution of each pool to offspring is equal to the proportion of
the population entering the group ($c_k = \Pi_k$). Therefore, the overall
probability of $i \times j$ mating is given by equation (9.2). This is a model
of assortative nonselective mating.

Therefore, the gene frequencies p and q remain constant. The equation for the frequency Φ_{Aa} of heterozygotes in the next generation, in
which for simplicity of notation I drop the subscript, can be written as

$$\Phi' = \frac{1}{2}\alpha\, \Phi + 2(1 - \alpha)pq. \tag{9.9}$$

If $\alpha = 0$ (i.e., all mating is random), then of course $\Phi' = 2pq$; that is,
the Hardy-Weinberg equilibrium is reached in one generation. If $\alpha > 0$
(i.e., mating is assortative), then the frequency of heterozygotes evolves
towards an equilibrium value

$$\Phi^* = 2pq\,\frac{1 - \alpha}{1 - \alpha/2}.$$

This equilibrium is achieved very rapidly. At equilibrium, the heterozygote deficiency index (9.8) is

$$I = \frac{\alpha}{2 - \alpha}, \tag{9.10}$$

which is always positive, meaning that there always is a deficiency
of heterozygotes relative to the case of random mating. Note that I
is smaller than α unless $\alpha = 1$ (see Figure 9.1a). If $\alpha = 1$ (i.e.,
mating is completely assortative), then $I = 1$. This means that all

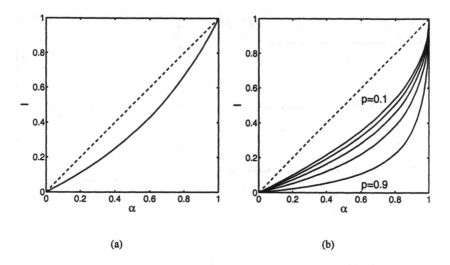

FIGURE 9.1. Isolation index I in the O'Donald model as a function of α. The dashed line shows the diagonal $I = \alpha$. (a) Three distinct phenotypes. I does not depend on the allele frequency p. (b) One allele is completely dominant. Different curves correspond to $p = 0.1$ (closest to the diagonal), $0.3, 0.5, 0.7$, and 0.9 (farthest from the diagonal), where p is the frequency of the dominant allele.

heterozygotes are eventually eliminated and the population splits into two reproductively isolated groups with genotypes AA and aa.

Assortative mating with complete dominance. Similar conclusions hold if one of the alleles is completely dominant. For definiteness, let genotypes AA and Aa not be recognizably different. Assume that a constant proportion α of organisms mate with like phenotypes whereas the remainder mate at random. In this case, one can think of three mating pools: one formed by proportion α of genotypes aa, another formed by proportions α of genotypes AA and Aa, and a third formed by the remainder of the population. As before, mating within the pools is random and individuals have equal chance of mating (that is, mating is assortative and nonselective).

In this model, the allele frequencies do not change, whereas the frequency of heterozygotes in the next generation is

$$\Phi' = \alpha \, \frac{2p\Phi}{2p + \Phi} + 2(1 - \alpha)pq. \tag{9.11}$$

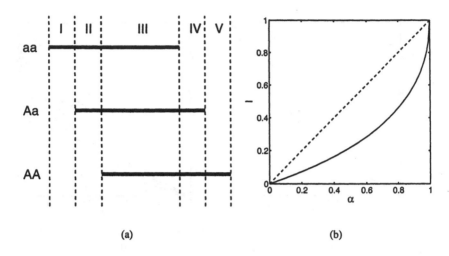

(a)   (b)

FIGURE 9.2. The Crosby model. (a) Mating areas (horizontal bars) of the three genotypes. (b) Isolation index as a function of α for $p = q = 1/2$.

The equilibrium value of Φ is

$$\Phi^* = p\sqrt{(1 - \alpha)[p^2(1 - \alpha) + 4q]} - p^2(1 - \alpha).$$

As before, if $\alpha = 1$ (i.e., mating is completely assortative), then heterozygotes are completely eliminated ($\Phi^* = 0$). Figure 9.1(b) illustrates the dependence of the isolation index (9.8) at equilibrium on α and p in this model. One can see that the isolation index decreases with decreasing the frequency of the dominant allele.

Other similar models. O'Donald (1960) also considered a model of assortative mating without complete dominance, where each genotype is characterized by a separate probability of assortative mating. Overall conclusions were similar to those for the two models discussed above. A number of other models of assortative mating were treated by O'Donald in his subsequent publications, culminating in his 1980 book.

The Crosby model   The model of assortative mating to be described next was first proposed by Crosby (1970) as part of his model for sympatric and parapatric speciation via divergence in flowering time. Crosby's original model was a multilocus one. A one-locus version was numerically studied by Stam (1983) in his study of environmentally triggered divergence in flowering time. An analytical treatment was given by Spirito (1987) and Spirito and Sampogna (1995), whose approach will be followed here.

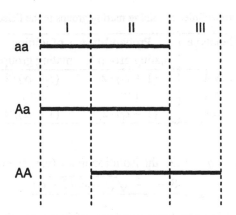

FIGURE 9.3. The Crosby model with dominance.

Consider a plant population. Let the flowering times of different genotypes be treated as one-dimensional intervals that have a unit length, but are displaced one from another by a constant value $\alpha/2$ with $\alpha < 1$ (see Figure 9.2a). Here, one can think of a locus controlling the onset of flowering, with $\alpha/2$ being an additive allelic effect. Assume that a flowering plant can be fertilized by any other plant that is flowering at the same time.

In terms of our general framework, the five subareas in Figure 9.2(a) can be thought of as different mating pools. Individuals aa join the mating pools I through III in proportions $\alpha/2{:}\alpha/2{:}1{-}\alpha$. Individuals Aa join mating pools II through IV in proportions $\alpha/2{:}1{-}\alpha{:}\alpha/2$. Individuals AA join the mating pools III through V in proportions $\alpha/2{:}\alpha/2{:}1{-}\alpha$. As in the O'Donald model, within the pools mating is random, plants have equal chance of being fertilized, and the contribution of each pool to offspring is equal to the proportion of the population entering the pool (i.e., $c_k = \Pi_k$). The probability of $i \times j$ mating is given by equation (9.2). This is again a model of assortative nonselective mating.

In this model, as shown by Spirito (1987), the frequency of heterozygotes changes according to equation

$$\Phi' = 2pq(1 - \alpha) + \alpha\frac{\Phi}{2}\left(\frac{p}{p + \Phi/2} + \frac{q}{q + \Phi/2}\right). \qquad (9.12)$$

The equilibrium value of Φ can be easily found from the above equation. At equilibrium, there is a deficiency of heterozygotes. In the symmetric

TABLE 9.1. Probabilities of joining mating groups in the Felsenstein model.

Genotype	Probability $\pi_{i,k}$ of joining	
	mating group 1	mating group 2
AA	$(1+\alpha)/2$	$(1-\alpha)/2$
Aa	$1/2$	$1/2$
aa	$(1-\alpha)/2$	$(1+\alpha)/2$

case (i.e., if $p = q = 1/2$), the isolation index (9.8) at equilibrium is

$$I = \frac{3 - \alpha - \sqrt{(1-\alpha)(9-\alpha)}}{2}. \tag{9.13}$$

Figure 9.2(b) illustrates the dependence of I on α in this model.

One can also consider a version of the Crosby model with complete dominance (see Figure 9.3). In this case there are three mating pools. Individuals aa and Aa join the mating pools I and II in proportions 1-α:α. Individuals AA join mating pools II and III in proportions α:1-α. It should be clear that the Crosby model with dominance is identical to the O'Donald model with dominance.

The Felsenstein model   The model to be considered next was used (but not explicitly studied) by Felsenstein (1981) as part of his numerical study of sympatric and parapatric speciation. Felsenstein explicitly assumed that mating is random but that it takes place within two mating groups, which individuals join on the basis of genetically determined preferences. Mating groups can correspond to a specific location (e.g., to a specific host) or timing (e.g., in the morning or in the evening) of mating. In the Felsenstein model, the sets of male and female phenotypes coincide. The probabilities $\pi_{i,k} = \pi'_{i,k}$ of joining these groups are given by Table 9.1. Individuals have equal mating success and the contributions of the mating pools to offspring are equal to the proportions of the population entering the corresponding mating pool (i.e., $c_k = \Pi_k$). Therefore, the probability of $i \times j$ mating is given by equation (9.2). This is a model of assortative nonselective mating.

Let Φ_{AA}, Φ_{Aa}, and Φ_{aa} be the genotype frequencies, and p and q the allele frequencies before joining the mating groups. The frequencies of allele A in the first and the second mating pools are

$$p_1 = \frac{p + \alpha\Phi_{AA}}{2\Pi_1}, \quad p_2 = \frac{p - \alpha\Phi_{AA}}{2\Pi_2},$$

where $\Pi_1 = [1 + \alpha(\Phi_{AA} - \Phi_{aa})]/2$ and $\Pi_2 = [1 - \alpha(\Phi_{AA} - \Phi_{aa})]/2$ are the proportions of the population entering the first and second mating

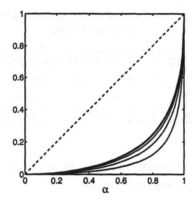

FIGURE 9.4. Isolation index in the Felsenstein model. The dashed line shows the diagonal. Different curves correspond to the deviation of p from $1/2$ equal to 0.0 (closest to the diagonal), $0.1, 0.2, 0.3$, and 0.4 (farthest from the diagonal).

pools. Then, in the next generation, the frequency of heterozygotes is

$$\Phi' = 2p_1q_1\Pi_1 + 2p_2q_2\Pi_2, \qquad (9.14)$$

where $q_i = 1 - p_i$ $(i = 1, 2)$. From this equation, one can show that asymptotically Φ approaches an equilibrium value that is given by the positive solution of quadratic

$$\alpha^2\Phi^2 + 2(1 - \alpha^2)\Phi - 4pq(1 - \alpha^2) = 0. \qquad (9.15)$$

If the allele frequencies are equal ($p = q = 1/2$),

$$\Phi^* = \frac{\alpha^2 - 1 + \sqrt{1 - \alpha^2}}{\alpha^2}. \qquad (9.16)$$

As expected, this value approaches $1/2$ for $\alpha \to 0$ and is equal to 0 if $\alpha = 1$. Figure 9.4 illustrates the dependence of the heterozygote deficiency index (9.8) at equilibrium on α and p. The lack of heterozygotes results in genotypes AA dominating in mating group 1 and genotypes aa dominating in mating group 2. For example, if $p = 1/2$ and $\alpha = 0.5$, then the equilibrium frequency of heterozygotes is 0.46 and the frequencies of the common and rare homozygotes in each mating group are 0.40 and 0.14. If $\alpha = 0.9$, the frequency of heterozygotes is 0.30 and those of homozygotes are 0.66 and 0.04.

Haploid version. Felsenstein (1981) also introduced a haploid version of this model, in which the probabilities $\pi_{i,k}$ of joining two different mating groups are given by Table 9.2. Let p and q be the frequencies of

TABLE 9.2. Probabilities of joining mating groups in a haploid version of the Felsenstein model.

Genotype	Probability $\pi_{i,k}$ of joining	
	mating group 1	mating group 2
A	$(1 + \alpha)/2$	$(1 - \alpha)/2$
a	$(1 - \alpha)/2$	$(1 + \alpha)/2$

genotypes A and a before joining the mating pools. Then the frequencies of A in the mating pools will be

$$p_1 = \frac{(1 + \alpha)p}{2\Pi_1}, \quad p_2 = \frac{(1 - \alpha)p}{2\Pi_2},$$

where $\Pi_1 = [1 + \alpha(p - q)]/2$ and $\Pi_2 = [1 - \alpha(p - q)]/2$ are the proportions of the population entering the first and the second mating pools, respectively. The frequency of A in the next generation will be

$$p' = p_1 c_1 + p_2 c_2, \tag{9.17}$$

where c_1 and c_2 are the relative contributions of the mating pools to the offsping generation. If the contribution of a mating pool to offspring is equal to the proportion of organisms entering it (i.e, $c_i = \Pi_i$), then $p' = p$; that is, the genotype frequencies do not change.

Diehl and Bush (1989) considered a model in which the contributions c_i were constant and equal ($c_1 = c_2 = 1/2$). In this case, rare genotypes have a selective advantage. Then, in the next generation,

$$p' = \frac{1 - \alpha^2(p - q)}{1 - \alpha^2(p - q)^2} \, p. \tag{9.18}$$

The last equation implies that asymptotically the population reaches a state with $p = q = 1/2$ and $p_1 = (1 + \alpha)/2, p_2 = (1 - \alpha)/2$. The overall frequency of heterogametic matings (that is, mating between A and a genotypes) is $P(A \times a) = p_1 q_1 + p_2 q_2$. This frequency plays the role of the heterozygote frequency in the diploid models. Therefore the isolation index in the haploid model can be defined as

$$I = 1 - \frac{P(A \times a)}{2pq}. \tag{9.19}$$

At equilibrium, $P(A \times a) = (1 - \alpha^2)/2$, and

$$I = \alpha^2. \tag{9.20}$$

Kawecki (1996, 1997) studied a three-allele version of this model.

TABLE 9.3. Mating preferences ψ_{ij} in the Moore model.

Female	Male genotype		
genotype	AA	Aa	aa
AA	1	$1-\alpha$	$1-\beta$
Aa	$1-\alpha$	1	$1-\alpha$
aa	$1-\beta$	$1-\alpha$	1

In all models considered above, genetic variation is maintained. The isolation index is the highest in the O'Donald model and is the smallest in the Felsenstein model. Overall, unless the coefficient α is close to one, the level of within-population RI is relatively small.

The Moore model   In the O'Donald, Crosby, and Felsenstein models, all organisms have equal mating success, no matter how choosy they are and how rare their preferred mates are. Moore (1979) introduced his model in particular to illustrate the importance of the costs of being choosy in NM. For example, minority flower variants can be at a disadvantage because odd floral types fall outside the acceptable search image of a specific pollinator, odd plants that flower out of phase will have a reduced chance for cross-pollination, and choosy females of *Drosophila* that spend too much time in courtship can be exposed to higher death risk during this interval. This disadvantage can be high especially if fertilization opportunity is limited, for example, because of limited pollen flow (Koenig and Ashley 2003).

Assume that individuals encounter each other randomly. In terms of the mating pools, there is a single mating pool and individuals join it with equal probabilities (i.e., $\pi_i = \pi_i' = $ const). Given that two individuals have met, they mate with probability $1, 1 - \alpha$, or $1 - \beta$, depending on whether they share 2, 1, or 0 identical alleles. This model implies that preferences ψ_{ij} are given by Table 9.2.1. Moore's equations for the frequencies of different matings mean that the overall probability of mating of females i is implicitly assumed to be equal to the average preference of the female for males:

$$f(\overline{\psi}_i) = \overline{\psi}_i. \tag{9.21}$$

That is, females that have higher preference $\overline{\psi}_i$ for the males in the population have a higher probability of leaving offspring. This implies that mating is selective and must result in changes in allele frequencies. In the Moore model, the probability of $i \times j$ mating given by equation (9.6) simplifies to

$$P(i \times j) = \frac{\psi_{ij}}{\overline{\psi}} \Phi_{f,i} \Phi_{m,j}, \qquad (9.22)$$

where $\overline{\psi} = \sum_{ij} \psi_{ij} \Phi_{f,i} \Phi_{m,j}$ is the average preference.

Moore (1979) used numerical iterations to study the model. However, his model is mathematically equivalent to the model of fertility selection (Bodmer 1965; Hadeler and Liberman 1975; Gavrilets 1997d; Hofbauer and Sigmund 1998) and to a model of parental selection (Gavrilets 1998), for which various analytical results are known. In particular, these results tell us that the only polymorphic equilibrium that can exist in this model is unstable. Therefore, in the Moore model genetic variation is quickly lost (as was already obvious from Moore's simulations). In this model, rare genotypes are at a mating disadvantage, and the loss of genetic variation happens because of direct (disruptive) selection on the allele frequency p, pushing it towards zero or one.

An important question is how much viability selection is needed to overcome the centrifugal force of selective mating (Moore 1979). For the Moore model, the answer is very simple. If the homozygote fitness (viability) is normalized to be 1 ($v_{AA} = 1$), then the deterministic loss of allele A is prevented if the heterozygote fitness, v_{Aa}, satisfies to the inequality

$$v_{Aa} > \frac{1}{1 - \alpha}. \qquad (9.23)$$

Therefore, viability selection for heterozygotes must be stronger than sexual selection against them. For example, if preference of AA for heterozygote Aa is half of that for AA, then the viability of heterozygotes must be more than twice that of AA.

The Spirito model   As was discussed above, an explicit assumption in the Crosby model is that all individuals have equal mating success. Spirito (1987) and Spirito and Sampogna (1995) introduced a modification of the Crosby model in which the probability of mating in each mating pool is equal to the density of individuals joining the pool. This assumption formalizes the idea that the probability of successful fertilization should be proportional to the number of potential mates and uses the simplest functional form – linear – for this dependence (which is similar to the Moore model).

TABLE 9.4. Mating preferences ψ_{ij} in the Spirito model.

Female genotype	Male genotype		
	AA	Aa	aa
AA	1	$1 - \alpha/2$	$1 - \alpha$
Aa	$1 - \alpha/2$	1	$1 - \alpha/2$
aa	$1 - \alpha$	$1 - \alpha/2$	1

In terms of our general framework for NM, let Π_k be the proportion of the population that enters the k-th mating pool, and let l_k be the relative length of the k-th flowering interval (that is, $l_k = \alpha/2, \alpha/2, 1 - \alpha, \alpha/2$, or $\alpha/2$ for $k = 1, 2, 3, 4$, or 5, respectively; see Figure 9.2). In the Spirito model, the overall probability of mating for individuals entering the k-th mating pool is Π_k/l_k. The overall contribution of the k-th mating pool to offspring is equal to the proportion of females that joined the pool and mated there: $c_k = \Pi_k (\Pi_k/l_k)$. These assumptions result in the probabilities of mating being proportional to

$$P(i \times j) \sim \left[\sum_k \frac{\pi_{i,k} \pi_{j,k}}{l_k} \right] \Phi_{f,i} \Phi_{m,j}. \qquad (9.24a)$$

The only nonzero contributions in the sum in the above equation come from the mating pools where phenotypes i and j are present simultaneously. Let us denote the set of such mating pools $i \cap j$. For mating pools in $i \cap j$, $\pi_{i,k} = \pi_{j,k} = l_k$. Therefore, equation (9.2) simplifies to

$$P(i \times j) \sim \left[\sum l_k \right] \Phi_{f,i} \Phi_{m,j}, \qquad (9.24b)$$

where the sum is over all mating pools with both genotypes present (i.e., over k that belong to $i \cap j$). Comparing the last equation with equation (9.22) immediately shows that the Spirito model is identical to the Moore model with preferences ψ_{ij} given by Table 9.4. Therefore, all qualitative conclusions of the Moore model are applicable to the Spirito model. In particular, the costs of being rare and choosy impose selection against rare phenotypes, which quickly eliminates genetic variation from the population.

The Gavrilets-Boake model A model of NM introduced by Gavrilets and Boake (1998) has a parameter that tunes the degree of selection for mating success in females. Let ψ_{ij} be the probability of mating between female i and male j, given that they have encountered each other. Assume that males and females encounter each other randomly.

TABLE 9.5. Mating probabilities in the Gavrilets-Boake model.

Female genotype	Male genotype		
	AA	Aa	aa
AA	a	b	c
Aa	b	a	b
aa	c	b	a

Let females mate only once. The probability that female i mates with the first male she encounters is $\overline{\psi}_i = \sum_j \psi_{ij} \Phi_{m,j}$, which is identical to the average preference of female i defined by equation (9.4). Assume that if the female does not mate with a given male, she may mate with a male she encounters later. If a female has not mated after the n-th encounter, she remains unmated. The latter assumption reflects the idea that the time interval during which females can mate (or, more generally, any other resource necessary for mating, e.g., energy) is limited. I note that the assumption that the maximum number of encounters per female is limited is explicit in many models of sympatric speciation using individual-based simulations that will be discussed later.

In this model, the frequency of $i \times j$ matings among all the matings taking place is given by equation (9.6), with the overall probability $f(\overline{\psi}_i)$ that female i mates being

$$f(\overline{\psi}_i) = 1 - (1 - \overline{\psi}_i)^n \qquad (9.25)$$

If $n = 1$ (that is, mating is a once-in-a-lifetime opportunity), then $f(\overline{\psi}_i) = \overline{\psi}_i$ and the mating frequencies in this model are the same as in the Moore model. If $n = \infty$ (that is, females can afford to wait for the right male effectively indefinitely), then every female eventually mates ($f = \overline{f} = 1$) and the mating frequencies are

$$P(i \times j) = \frac{\psi_{ij}}{\overline{\psi}_i} \Phi_{f,i} \Phi_{m,j}. \qquad (9.26)$$

Note that although in this case, mating is nonselective in females, males still differ in their mating success (defined by equation (9.1b)).

Gavrilets and Boake (1998) studied a one-locus, two-allele version of this model. Specifically, they assumed that the probabilities ψ_{ij} that an encounter between a female and a male results in mating is given by Table 9.5, where $a > b \geq c \geq 0$. Loss of genetic variation is the only possible stable outcome in this model, as was the case in the Moore model and in the Spirito model.

Although the value of n does not influence the final outcome, it affects the rate of convergence toward the monomorphic equilibrium. With large n, rare individuals have a higher chance of finding appropriate mates. This delays the loss of genetic variation, especially when selection for mating success is strong (see Figure 1 in Gavrilets and Boake 1998). Coefficient n is also important if the population size is finite. Specifically, Gavrilets and Boake (1998) showed that in a population undergoing rapid increase in size (e.g., after a founder event), heterozygotes can be lost as a result of random genetic drift. Then, if mating preferences for similar genotypes are much higher than those for different genotypes (i.e., if $a \gg b, c$) and n is large, the population splits into two reproductively isolated groups represented by different homozygotes. In general, this event describing sympatric speciation has a small probability.

9.2.2 Multiple loci

Here, I consider situations when mating preferences depend on more than one loci or on a single quantitative character controlled by $\mathcal{L}$ loci with additive effects.

Neglecting environmental effects, we can write the value of a quantitative trait z as

$$z = \sum_{i=1}^{\mathcal{L}} (a_i + a'_i),$$

where a_i and a'_i are the effects of the paternally and maternally inherited alleles at the i-th locus (Bulmer 1980; Lynch and Walsh 1998). Assuming that the distributions of a_i and a'_i are the same, the variance of z can be written as

$$V_G = V_g + V_{LD} + \hat{V}_g + \hat{V}_{LD}, \qquad (9.27a)$$

where

$$V_g = 2 \sum \text{var}(a_i) \qquad (9.27b)$$

is the genic variance,

$$V_{LD} = 2 \sum_{i \neq j} \text{cov}(a_i, a_j) \qquad (9.27c)$$

is the contribution of linkage disequilibrium, and

$$\hat{V}_g = 2 \sum_i \text{cov}(a_i, a_i'), \qquad (9.27d)$$

$$\hat{V}_{LD} = 2 \sum_{i \neq j} \text{cov}(a_i, a_j'), \qquad (9.27e)$$

are the contributions to the variance arising from correlations between genes from different gametes at the same and at different loci, respectively (Bulmer 1980; Lynch and Walsh 1998). Here, $\text{var}(\cdot)$ and $\text{cov}(\cdot, \cdot)$ denote the corresponding variances and covariances of the locus effects. In a very large population, if mating is random and there is no selection, $\hat{V}_g = \hat{V}_{LD} = 0$, V_{LD} approaches zero asymptotically as recombination randomizes alleles in gametes, whereas V_g does not change. If mating is positive assortative, so that individuals with similar trait values are more likely to mate, positive correlations of individual loci effects will be induced, resulting in an increase in V_G. Therefore, with positive assortative mating based on a quantitative character, one indicator of the progress in the evolution of RI is the increase in V_G. Another indicator is the formation of gaps in the distribution of traits in the population.

To illustrate the consequences of assortative mating based on an additive trait, let us consider a simple case of two diallelic loci with effect 0 for alleles a and b and 1 for alleles **A** and **B** (Crow and Kimura 1970, Chapter 4). The range of possible phenotypes is from 0 to 4. Assume that the alleles are at equal frequency 0.5 and that initially the population is at linkage equilibrium. Then $V_G = V_g = 1$. Let us assume that mating is completely assortative; that is, the trait values of mates are equal. In this case, the frequencies of "extreme" homozygotes aa/bb with phenotype 0 and **AA/BB** with phenotype 4 can only increase as they mate among themselves. In contrast, the frequency of intermediate phenotypes will decrease each generation because some of their offspring produced by recombination and segregation are the extreme homozygotes. Eventually, the population approaches a state where only these homozygotes, which are reproductively isolated, remain. In this state, the phenotypic variance is equal to $V_G = 4$, which is the maximum possible value. In terms of the different contributions to the genotypic variance V_G, at this state all covariances in equation (9.27) have the maximum possible value of 1/4, so that $V_g = V_{LD} = \hat{V}_g = \hat{V}_{LD} = 1$. At this state, there is a gap in the distribution of z in the population, with all intermediate phenotypes being absent.

The Wright model Next, I consider a more general situation. Following Wright (1921) and Crow and Felsenstein (1968), let us assume that assortative mating in a diploid population is based on similarity for an additive character controlled by $\mathcal{L}$ diallelic loci. Let ρ be the coefficient of correlation between the phenotypes of mates. Assume that all individuals have equal mating success (that is, mating is nonselective). In this case, allele frequencies will not change, but assortative mating will result in the buildup of correlations between effects of different loci, which will increase the genotypic variance. Assuming the absence of environmental effects on the trait value and the equality of allelic effects and allele frequencies across the loci, Wright (1921) and Crow and Felsenstein (1968) showed that the equilibrium value of the genetic variance is

$$V_G = \frac{V_g}{1 - \rho(1 - \frac{1}{2\mathcal{L}})}. \tag{9.28}$$

Note that Bulmer (1980, Chapter 8) derived equation (9.28) using a different set of approximations. If $\rho < 1$ and the number of loci $\mathcal{L}$ is very large, then $V_G \approx V_g/(1 - \rho)$. This shows that assortative mating can result in a substantial increase in the genotypic variance. For example, if $\rho = 0.9$, then V_G is 10 times larger than that expected when mating is random (i.e., if $\rho = 0$). The maximum possible increase is observed if $\rho = 1$. In this case, $V_G = 2\mathcal{L}V_g$. Variance V_G grows with increasing the number of loci. At equilibrium

$$\hat{V}_g = V_g\gamma, \tag{9.29}$$

$$V_{LD} = \hat{V}_{LD} = V_g(\mathcal{L} - 1)\gamma, \tag{9.30}$$

where

$$\gamma = \frac{1}{2\mathcal{L}} \frac{\rho}{1 - \rho(1 - \frac{1}{2\mathcal{L}})}. \tag{9.31}$$

Because with large $\mathcal{L}$ the terms V_{LD} and $\hat{V}_{LD}$ are much larger than the term $\hat{V}_g$, most of the increase in the equilibrium genotypic variance V_G is due to the buildup of the correlations between alleles at different loci.

If there are multiple alleles and the gene effects and allele frequencies are not equal, then equation (9.28) still holds, but $\mathcal{L}$ has to be substituted by the effective number of loci

$$\mathcal{L}_e = \frac{\sum_{ij} r_{ij}\sigma_i\sigma_j}{\sum_i r_{ii}\sigma_i^2} \tag{9.32}$$

(Crow and Kimura 1970, Chapter 4; Nagylaki 1992, Chapter 10). Here, r_{ij} is the correlation of the effects of the i-th and j-th loci in different individuals, and $\sigma_i^2 = \mathrm{var}(a_i)$. The effective number $\mathcal{L}_e$ will be equal to $\mathcal{L}$ if there is free recombination and all loci contribute equally to the variance; otherwise it will be smaller (Nagylaki 1992).

How is this model related to the two approaches outlined at the beginning of this chapter? Assume that with probability α an individual mates with another individual that has exactly the same trait value, and with probability $1 - \alpha$ the individual mates randomly. Then the average correlation between the traits of the mating partners is $\rho = \alpha \times 1 + (1 - \alpha) \times 0 = \alpha$. This shows that in the case of one diallelic locus, the Wright model is equivalent to the symmetric version of the O'Donald model. Thus, the Wright model can be interpreted in terms of different mating pools joined preferentially, with random mating within the mating pools.

The results on the Wright model presented above concentrate exclusively on the genetic variance V_G and its components. It would be very interesting to know the whole distribution of trait z in the population and, in particular, at what frequencies the intermediate phenotypes are present. Some approaches to this question will be considered below in the section dealing with the Rice and Kondrashov-Shpak models. But first we consider two simple models in which the equilibrium distributions can be found explicitly. These models will be used in the next chapter in analyzing conditions for sympatric speciation.

In both these models there are two diallelic loci with effects 0 and 1 on an additive trait z controlling mating in a sexual haploid population. Let ϕ_1, ϕ_2, ϕ_3, and ϕ_4 be the frequencies of genotypes ab, Ab, aB, and AB. Neglecting environmental effects, the trait values for genotypes ab, aB, Ab, and AB are $z_1 = 0, z_2 = z_3 = 1$, and $z_4 = 2$, respectively. The level of within-population RI can be characterized by the hybrid deficiency index

$$I = 1 - \frac{f_{obs}}{f_{exp}}, \tag{9.33}$$

where f_{obs} is the observed frequency of hybrids (that is, genotypes Ab and aB), and f_{exp} is the frequency of hybrids expected at the same allele frequencies if mating were random. In the model under consideration, $f_{obs} = p_1 q_2 + q_1 p_2 - 2D$ and $f_{exp} = p_1 q_2 + q_1 p_2$, where p_1, q_1, p_2, and q_2 are the frequencies of alleles A, a, B, and b, respectively, and D is the linkage disequilibrium.

TABLE 9.6. Mating preferences in the two-locus, two-allele haploid version of the Moore model.

Female	Male genotype		
genotype	ab	Ab, aB	AB
ab	1	$1 - \alpha$	$1 - \beta$
Ab, aB	$1 - \alpha$	1	$1 - \alpha$
AB	$1 - \beta$	$1 - \alpha$	1

Two-locus, two-allele haploid version of the Wright model   Assume that the correlation of the trait values in mating pairs is α, or, equivalently, that individuals mate assortatively or randomly with probabilities α and $1 - \alpha$. In this model, the allele frequencies do not change. Linkage disequilibrium D changes according to equation

$$\Delta D = r \frac{(2 - \alpha)D^2 - (p_1 q_2 + q_1 p_2)D + \alpha p_1 q_1 p_2 p_2}{p_1 q_2 + q_1 p_2 - 2D}, \qquad (9.34)$$

where r is the recombination rate. D evolves to a positive value given by the solution of the quadratic in the numerator of equation (9.34). The recombination rate r does not affect the equilibrium values of D, but larger values of r result in faster approach to the equilibrium. If all allele frequencies are equal to $1/2$, then at equilibrium the isolation index is

$$I = \frac{\alpha}{2 - \alpha}, \qquad (9.35)$$

which is the same as in the O'Donald model.

Two-locus, two-allele haploid version of the Moore model   The Wright model considered above is a model of nonselective mating. What happens if one introduces a cost of being choosy into models of NM based on a quantitative character? To see intuitively into this question, let us consider the following model. Assume that the probability of mating between genotypes i and j is proportional to $\psi_{ij} \phi_i \phi_j$, where preferences ψ_{ij} are given by Table 9.6.

Note that if the preference declines as a quadratic function of the phenotypic difference, that is, if

$$\psi_{ij} = 1 - \frac{(z_i - z_j)^2}{2V_\psi}, \qquad (9.36a)$$

where σ_α^2 is a positive parameter measuring the strength of preferences, then $\alpha = 1/(2V_\psi)$ and $\beta = 4\alpha$. If the preference declines as a Gaussian function of the phenotypic difference, that is, if

$$\psi_{ij} = \exp\left[-\frac{(z_i - z_j)^2}{2V_\psi}\right], \qquad (9.36b)$$

then $1 - \alpha = \exp[-1/(2V_\psi)]$ and $1 - \beta = (1 - \alpha)^4$ (e.g., Kirkpatrick and Ravigné 2002).

Following Moore (1979), assume that the overall probability of mating f_i for a female genotype i is equal to her average preference $\overline{\psi}_i = \sum_i \psi_{ij}\phi_j$ for the males in the population. The resulting system of dynamics equations is simple enough to allow for analytical determination of all equilibria and finding conditions for stability for some of them. What follows is a brief summary of these conditions.

Monomorphic equilibria. There are always four monomorphic equilibria with one genotype frequency equal to one and three remaining genotype frequencies equal to zero. These equilibria are locally stable provided $\alpha > 0$ and $\beta(1 - r) + r > 0$, which is always true if mating between similar phenotypes is more probable than between different phenotypes.

Singly polymorphic equilibria. The system always has four singly polymorphic equilibria at which one locus is fixed for an allele, whereas the frequencies of the alleles at the other locus are equal to one half. These equilibria are never stable.

Doubly polymorphic equilibria. There can be up to three symmetric polymorphic equilibria with $p_1 = p_2 = 1/2$, and up to four asymmetric polymorphic equilibria (two equilibria at which $\phi_1 = \phi_4$, and two other equilibria at which $\phi_2 = \phi_3$). One can show that if the loci are unlinked, the symmetric polymorphic equilibria are locally unstable. Numerical simulations suggest that the system always evolves to one of the monomorphic equilibria.

From these results I conclude that introducing costs of being choosy leads to a quick loss of genetic variation, as was the case in the Moore, Spirito, and Gavrilets-Boake models considered above.

The Higgs-Derrida model   Assume that individuals are haploid and differ with respect to a very large number $\mathcal{L}$ of unlinked diallelic loci. Genetic difference between two individuals i and j can be characterized by the genetic distance d_{ij} defined as the number of loci at which they differ (see equation (2.4) on page 26). Assume that each organism chooses a mate among genotypes within genetic distance K from itself.

This is a model of assortative nonselective mating, with preferences ψ_{ij} defined according to a simple rule:

$$\psi_{ij} = \begin{cases} 1 \text{ if } d_{ij} < K, \\ 0 \text{ if } d_{ij} \geq K. \end{cases} \tag{9.37}$$

In an infinite population, a suitable mate will always be found. In a finite population, mutation and recombination may occasionally produce individuals significantly deviating from the rest of the population. Such odd individuals will leave no offspring, introducing a weak selective component in mating.

This model was first introduced[2] and numerically studied by Higgs and Derrida (1991, 1992). Their numerical simulations show that if the threshold value K is sufficiently small and the mutation rate per genotype ν and population size N are sufficiently large, then the evolutionary dynamics are characterized by a continuous splitting of the population into different reproductively isolated genetic clusters, which subsequently go extinct as a result of random genetic drift.

These results can be understood as follows. Classical models of neutral molecular evolution tell us that with no RI, the population will evolve to a state where both the average and variance of the pairwise genetic distance are equal to $\theta = 2N\nu$, which is twice the expected number of new mutations in the population per generation (see Box 2.3 in Chapter 2). Therefore, in the Higgs-Derrida model, if K is sufficiently large relative to θ, the population will not feel the mating threshold because most mating pairs will be at distances smaller than K. However, if K is close to θ, then a proportion of all pairs will be reproductively isolated. This creates an unstable situation where different reproductively isolated clusters will be formed if random genetic drift results in the extinction of intermediate genotypes. Because mating is nonselective, smaller clusters will persist for some time before extinction.

Higgs and Derrida argued that, in their model, *sympatric speciation* occurs if $d > K$ or, equivalently,

$$2N\nu > K. \tag{9.38}$$

Therefore, sympatric speciation is more plausible in larger populations with higher mutation rates when a smaller number of genetic changes

---

[2]In their original publications, Higgs and Derrida (1991, 1992) expressed the model in terms of the overlap $q = (\mathcal{L}/2)(1 - d)$ rather than genetic distance d.

is sufficient for RI. For example, if $\mathcal{L} = 100, \mu = 10^{-5}$ (meaning $\nu = 10^{-3}$), and $K = 5$, then the population size must be larger than 2500.

Manzo and Peliti (1994) demonstrated that introducing spatial subdivision in the Higgs-Derrida model can significantly promote speciation. To understand this effect, it is illuminating again to start with the neutral case. Let us consider the n-island, model assuming the balance of mutation, migration, and drift. In this model the average pairwise distance between a pair of individuals from different islands is

$$d_b = 2N_T\nu + (n-1)\frac{\nu}{m},$$

where m is the migration rate and N_T is the overall population size (Li 1976; Slatkin 1987; Strobeck 1987). The populations will start feeling the mating threshold and split into different clusters if $d_b > K$. Therefore, in this model one expects *parapatric speciation* if

$$2N_T\nu + (n-1)\frac{\nu}{m} > K. \tag{9.39}$$

Strong spatial subdivision makes speciation more plausible. For example, if there are 20 islands, $m = 0.01$, and all other parameters are the same as in the previous example, then parapatric speciation takes place if the overall population size is larger than $1,500$.

The conclusions above are for the case of nonselective mating. The individual-based simulations discussed in Chapter 7 correspond to a version of the Higgs-Derrida model with selective mating. Recall that in these simulations both potential mates were disregarded if the distance between them exceeded a threshold. This implicitly assumes that for each organism the probability of mating $f(\overline{\psi}_i)$ is equal to its average preference $\overline{\psi}_i$ for the members of the population, as in the Moore model. In this model of selective mating, sympatric speciation is impossible, while conditions for parapatric speciation are very restrictive (see Chapter 7).

Other multilocus models   Several studies used numerical simulations to analyze the dynamics of RI with preferences based on a quantitative character. However, in most of these studies (e.g., Crosby 1970; Stam 1983) several other factors were present simultaneously, making it very difficult to evaluate the effects of NM alone. Some information can be extracted from Rice's (1984) numerical simulations. In his model, the preference ψ was a monotonic function of the difference

$$\Delta_{ij} = |z_i - z_j| \tag{9.40}$$

in the values of an additive trait controlled by four diallelic loci. Function ψ decreased from 1 for $\Delta_{ij} = 0$ to 0 for $\Delta_{ij} = 8$ (which is the largest possible difference with four loci with effects 0 and 1). The main focus of Rice's paper was on the consequences of disruptive selection acting on a trait that was subject to assortative mating. However, he also performed numerical simulations for a situation where disruptive selection was absent. In these simulations, assortative mating had little effect when function ψ was linear. However, it did split the population into two reproductively isolated groups with no (or very few) intermediates in his case IV, where ψ quickly dropped to approximately 0.1 for $\Delta_{ij} = 1$ (which corresponds to a single allele substitution). In Rice's model, mating was selective. However, in his simulations, to avoid the loss of genetic variation, one half of the mated females was constrained to have trait values smaller than or equal to the mean trait value in the population, and the other half of the mated females was constrained to have trait values larger or equal to the mean trait value in the population.

Kondrashov and Shpak (1998) introduced and studied two classes of models of assortative nonselective mating based on an additive trait controlled by unlinked diallelic loci with equal effects (0 and 1). In their *interval mode* model, an organism with phenotype z_i can only mate with organisms with phenotype z_j such that $\Delta_{ij} < K$, where K is a parameter of the model. This is a model of assortative nonselective mating with preferences ψ_{ij}, defined according to a simple rule:

$$\psi_{ij} = \begin{cases} 1 \text{ if } \Delta_{ij} < K, \\ 0 \text{ if } \Delta_{ij} \geq K. \end{cases} \tag{9.41}$$

This is similar to the threshold function used by Higgs and Derrida (1991, 1992), with the difference that the distance between mating partners is phenotypic (as given by equation (9.40)) rather than genetic (d). Note that the case of one-locus diploid organisms with $K = 1$ (only organisms with identical phenotypes can mate) corresponds to the symmetric version of the O'Donald model. With uniform allele frequencies across the loci and $K = 1$ or $K = 2$, all intermediate genotypes eventually disappear, leaving only the extreme genotypes 0...0 and 1...1 present (Wright 1921; Kondrashov and Shpak 1998). Thus, in these cases, assortative mating alone leads to sympatric speciation. Kondrashov and Shpak used numerical methods to study situations when $K > 2$, treating haploid and diploid cases separately. Their results show that there is a critical value K_{crit} such that for $K \leq K_{crit}$ intermediate genotypes disappear even if they are in abundance initially. If $K > K_{crit}$ the population stays at a state where all possible gametes are at equal frequencies. The value of K_{crit} grows approximately as $\sqrt{\mathcal{L}}$ with the number of loci.

In the *threshold mode* model, mating is impossible if $i < K_1$ and $j > K_2$, or if $i > K_2$ and $j < K_1$, where K_1 and K_2 are parameters of the model. This is a model of assortative nonselective mating with preferences ψ_{ij} defined according to a rule:

$$\psi_{ij} = \begin{cases} 1 \text{ if } i < K_1 \text{ and } j > K_2 \text{ or } i > K_2 \text{ and } j < K_1, \\ 0 \text{ otherwise.} \end{cases} \quad (9.42)$$

Kondrashov and Shpak proved that if $K_1 = K_2$, sympatric speciation always occurs, although the process takes a very long time. Their numerical results suggest that if $K_1 \neq K_2$, speciation is impossible.

Both Rice (1984) and Kondrashov and Shpak (1998) used a common assumption in numerical studies of sympatric speciation, namely, that the initial genetic variation is at a maximum possible level with allele frequencies at one half at all loci. Such initial conditions could arise if the initial population was produced as a result of hybridization of two populations completely diverged in allopatry. Analyzing the sympatric origin of new species per se requires one to consider the evolutionary dynamics starting with a low level of genetic variation. With realistically small mutation rates and with random genetic drift removing genetic variation, the level of genetic variation maintained in the population is expected to be much smaller than the maximum possible level. Therefore, sympatric speciation will be very difficult.

Kawata and Yoshimura (2000) performed detailed individual-based simulations assuming that all females mate and using the threshold form (9.41) of female preference function. These authors varied the initial degree of genetic divergence of two hybridizing populations and compared a number of strategies used by females if no preferred males were found. Their results show that given sufficient initial divergence, the two populations can rapidly (within 200 generations) diverge further and evolve strong RI with a significant probability.

I note that in terms of fitness landscapes for mating pairs, the Higgs-Derrida, Rice, Kondrashov-Shpak, and Kawata-Yoshimura models all belong to the class of holey landscapes discussed in Part 2.

9.2.3 General conclusions on similarity-based nonrandom mating

If NM is selective and individuals pay some costs for being choosy, genetic variation is not maintained and the formation of stable genetic clusters does not occur. In contrast, nonselective NM or selective NM giving advantage to rare phenotypes can lead to substantial genetic clustering within the population. The degree of clustering increases with initial genetic variation and the number of loci (and traits) underlying NM. [3] Overall, biological mechanisms resulting in similarity-based NM (e.g., habitat choice, divergence in flowering time, divergence in pollinators) can be effective in causing the sympatric origin of strong RI.

9.3 MATCHING-BASED NONRANDOM MATING

In this section I consider models of NM in which the traits underlying mating preferences in males and females are controlled by different genes. As discussed above, this is a very general type of genetic architecture underlying many forms of sexual selection.

Sexual selection is generally viewed as an important engine of speciation. Some data show positive correlation between species richness (i.e., the number of species in a clade) and some measures of the strength of sexual selection. For example, species richness correlates with the proportion of sexually dichromatic species within taxa of passerine birds (Barraclough et al. 1995). In birds, taxa with promiscuous mating systems tend to be more species-rich than their nonpromiscuous sister taxa (Mitra et al. 1996). In insects, groups where females mate with many males exhibited speciation rates four times as high as in related groups where females mate only once (Arnqvist et al. 2000). However, in a study performed by Gage et al. (2002), the degree of male-biased sexual size dimorphism (across 480 mammalian genera, 105 butterfly genera,

---

[3]These observations explain the ease with which very strong RI emerged sympatrically in the experiments with *Drosophila* (Rice 1985; Rice and Salt 1988, 1990). In these experiments, initial genetic variation was maximized, organisms were selected on the basis of their preference for 2 out of 16 habitats differing in four characteristics (position, lighting, odor, and development time), and the experimental protocol, under which the contributions of both habitats to offspring were equal promoted the maintenance of genetic variation.

and 148 spider genera) and degree of polyandry (measured as relative testis size in mammals (72 genera) and mean spermatophore count in female butterflies (54 genera)) showed no associations with the variance in the species richness. Morrow et al. (2003) analyzed data for testis size as an index of postmating sexual selection, and sexual size dimorphism and sexual dichromatism as indices of premating sexual selection obtained for 1,031 species representing 467 genera. None of the variables investigated explained the patterns of species richness.

A growing amount of experimental data coming from studies of sperm or pollen competition between closely related species (e.g., Howard et al. 1998, Arnold et al. 1993; Wade et al. 1994; Rieseberg 1995; Howard 1999) as well as from molecular studies of fertilization proteins (e.g., Palumbi 1998; Aguade et al. 1992; Vacquier and Lee 1993; Metz et al. 1996; Howard 1999) indicates extremely rapid evolution of traits and proteins related to fertilization in many diverging taxa. In *Drosophila*, prezygotic RI between sympatric pairs of species appears to evolve more than 10 times faster than postzygotic RI (Coyne and Orr 1989, 1997). The data on the genetics of sexual isolation are less detailed than those on postzygotic isolation. Some studies indicate a single or just a few genes with major effects (e.g., Beiles et al. 1984; Doi et al. 2001; Ritchie and Phillips 1998; Kyriacou 2002). For example, in the Bananaquit (*Coereba flaveola*) on the Caribbean island of Grenada, a plumage polymorphism is governed by a single, diallelic, autosomal locus with melanic plumage, which is completely dominant over yellow plumage (Grant and Grant 1997). However, polygenic inheritance of plumage traits is more common than single-gene inheritance (Grant and Grant 1997). Large number of loci with minor effects is suggested by other studies of prezygotic RI as well (e.g., Ting et al. 2001; Wu 2001; Ritchie and Phillips 1998). In some situations, prezygotic RI is based on cultural rather than genetic traits. For example, in Darwin's finches, females mate with males that sing the same species song as their fathers. Males sing a single song that is sung unchanged throughout life and acquired by an imprinting-like learning process, usually from their fathers (Grant and Grant 1997).

In this section I consider models of matching-based NM that have been used for speciation modeling. Other types of sexual selection models, such as "good gene models" (e.g., Andersson 1994), will not be discussed here. In models of matching-based NM, mating is always selective in males.

TABLE 9.7. Relative mating preferences ψ_{ij} in the Nei-Kirkpatrick model.

Female genotype	Male genotype	
	AA, Aa	aa
BB, Bb	1	$1 - \alpha_1$
bb	$1 - \alpha_2$	1

9.3.1   Two loci

I start with models in which one locus controls female mating preference and another locus controls the male trait.

The Nei-Kirkpatrick model   Let us consider a diploid population in which diallelic loci A and B control male-limited and female-limited morphological, physiological, or behavioral characters. Assume that alleles **A** and **B** are dominant over **a** and **b**. The model of ethological isolation proposed by Nei (1976) can be specified by Table 9.7.

In this model, all females have equal chances to mate, but females **BB** and **Bb** choose males **AA** and **Aa** more frequently than **aa** males. There is no direct selection on female alleles. However, female preferences result in direct selection on male alleles. Consequently, the frequency of female alleles can change as a correlated response to selection in males if there is linkage disequilibrium between the loci. A special case of this model corresponding to $\alpha_1 = 1, \alpha_2 = 0$ was previously studied by O'Donald (1963). Nei (1976) briefly discussed the dynamics of a haploid version of this model, assuming that the two loci are in linkage equilibrium. The complete analytical study of the haploid model incorporating viability selection in males was done by Kirkpatrick (1982). The following is a summary of his results corresponding to the case when viability selection in males is absent.

Let $p_1, q_1, p_2,$ and q_2 be the frequencies of alleles **A, a, B,** and **b**, and let D be the linkage disequilibrium. Kirkpatrick showed that allele frequencies evolve to a curve of equilibria defined by

$$
q_1 = \begin{cases} 0 & \text{for } q_2 < q^*, \\ \frac{q_2 - q^*}{q^{**} - q^*} & \text{for } q^* \leq q_2 \leq q^{**}, \\ 1 & \text{for } q_2 > q^{**}, \end{cases} \tag{9.43}
$$

where the two threshold values q^* and q^{**} are

$$
q^* = \frac{(1 - \alpha_2)\alpha_1}{\alpha_1 + \alpha_2 - \alpha_1\alpha_2}, \quad q^{**} = \frac{\alpha_1}{\alpha_1 + \alpha_2 - \alpha_1\alpha_2}.
$$

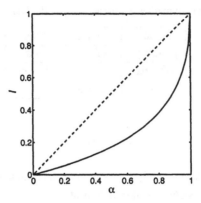

FIGURE 9.5. Hybrid deficiency index as a function of α in the haploid Nei-Kirkpatrick model for $p_a = p_b = 1/2$ and $\alpha_2 = \alpha_1 = \alpha$. The dashed line shows the diagonal $I = \alpha$.

This shows that if one of the female alleles is rare (that is, if $q_2 < q^*$ or $q_2 > q^{**}$), then the corresponding preferred male allele is lost. However if both female alleles are at high frequencies (that is, if $q^* \leq q_2 \leq q^{**}$), then both male alleles will be maintained. In this case the population will achieve a stable level of positive linkage disequilibrium D, with gametes **AB** and **ab** at higher frequencies than expected if alleles are combined into gametes randomly. The value of D at equilibrium is given by the smaller of the two roots of the quadratic

$$D^2 + \left(p_1 p_2 + q_1 q_2 - \frac{2 - \alpha_2 + q_1(\alpha_2 - \alpha_1)}{\alpha_1 + \alpha_2 - \alpha_1 \alpha_2} \right) D + p_1 q_1 p_2 q_2 = 0. \tag{9.44}$$

Note that the equilibrium values of D do not depend on the recombination rate r. The resulting level of within-population RI can be characterized by the hybrid deficiency index I defined by equation (9.33) on page 302. In the symmetric case, when $\alpha_1 = \alpha_2 = \alpha$ and $q_1 = q_2 = 1/2$,

$$I = \frac{2 - \alpha - 2\sqrt{1 - \alpha}}{\alpha}. \tag{9.45}$$

Figure 9.5 illustrates the dependence of I on α for this case.

Note that if females **B** mate indiscriminately (that is, $\alpha_1 = 0$), then allele **a** is fixed for any initial frequency $q_2 > 0$ (because $q^{**} = 0$). In a similar way, if $\alpha_2 = 0$ (that is, females **b** mate indiscriminately), then $q^* = 1$ and allele **A** is fixed for any initial frequency $q_2 < 1$.

In terms of fitness landscapes for mating pairs, specified by the preference function ψ, this is a model of disruptive selection with two peaks. In contrast to the models of disruptive viability selection, here the population can maintain genetic variation under certain conditions because mating is nonselective in females.

With no mutation, within-population RI will evolve only if initially there is substantial genetic variation in female alleles (that is, if $p_1 \leq p_b \leq p_2$). Mutation, however small, will produce a qualitatively different behavior (which is expected because the dynamic systems possessing lines of equilibria are structurally unstable). Mutation will drive the frequency of female alleles toward one half (assuming equality of forward and backward mutation rates). This is subsequently expected to result in the evolution of within-population RI to the maximum level as specified by equation (9.45). However, allowing for random genetic drift can prevent mutation from increasing genetic variation, which in turn will prevent the establishment of within-population RI.

Seger (1985a) introduced a modification of the Nei-Kirkpatrick model ("a best of N" model) in which each female chooses a mate out of a group (e.g., a lek) of N males randomly chosen from the whole population. The case of $N = 2$ corresponds to three different types of mating pools: one with two males B present, another with two males b present, and a third one with one male B and one male b present. The original Nei-Kirkpatrick model corresponds to the case $N = \infty$. Seger showed that for small N, the line of equilibria, which is always stable in the Nei-Kirkpatrick model, becomes unstable. In this case, genetic variation in male alleles is not maintained. Adding mutation to this model is expected to result in very low genetic variation maintained by mutation-selection balance. Reducing the number of males to choose from creates strong selection against rare male phenotypes, which acts even if all females mate. Adding costs of female choice in the form of reduced viability of choosy females has effects similar to those of selection for mating success, with genetic variation for choosiness being quickly removed (Pomiankowski 1987).

The Nei-Maruyama-Wu model   Here, I return to a model that was already mentioned in Chapter 6. Nei et al. (1983) and Wu (1985) consider a two-locus, multiallele haploid model of speciation utilizing the stepwise mutation model of Ohta and Kimura (1973). Specifically they assumed that there is an ordered sequence of alleles ..., A_{-2}, A_{-1},

TABLE 9.8. Relative mating preferences ψ_{ij} in the Nei-Maruyama-Wu model.

Female	Male allele				
allele	A_{i-2}	A_{i-1}	A_i	A_{i+1}	A_{i+2}
B_i	0	$1-\alpha$	1	$1-\alpha$	0

$A_0, A_1, A_2, \ldots$ in the male locus and an ordered sequence of alleles $\ldots, B_{-2}, B_{-1}, B_0, B_1, B_2, \ldots$ in the female locus. The mating preferences ψ_{ij} in the Nei-Maruyama-Wu model are specified in Table 9.8. Note that here ψ_{ij} depend on the number of mutational steps separating two loci rather than on the number of mutational steps separating two sequences (as in the Higgs-Derrida model) or on the phenotypic distance (as in the Rice, Kondrashov-Shpak, and Kawata-Yoshimura models). This is a model in which the corresponding fitness landscape for mating pairs is holey (see Figure 2.4(a) on page 29). Nei et al. (1983) studied the case of $\alpha = 0$. Wu (1985) used two different values of α: 0 and $1/2$. The assumptions about the mating process made by Nei et al. (1983) correspond to those in the Moore model. That is, in their paper, mating is selective and both males and females experience direct selection for mating success. In contrast, Wu (1985) assumed that all females have equal mating success, corresponding to no direct selection in females. Both papers used numerical techniques to simulate the dynamics of allele frequencies under the joint action of NM, mutation, and random genetic drift, starting from a monomorphic population. I will consider this model in detail because it is both simple and captures many important features of the dynamics of NM.

Let us assume that the population size is very big, so that random genetic drift can be neglected. Using the framework developed by Gavrilets and Waxman (2002), one can show that the Nei-Maruyama-Wu model is characterized by the presence of lines and planes of equilibria (rather than by isolated equilibria). Different equilibria differ not only with regard to their position along the ridge of high fitness values (see Figure 2.4(a) on page 29), but in some other features as well. An important question is which equilibria will be stable for generic initial conditions. To obtain some intuition, let us consider the results of numerical simulations presented in Figures 9.6 and 9.7. Both these figures show the distributions of genotype frequencies observed in a certain generation assuming that all females have equal mating success. Figure 9.6 corresponds to a Gaussian preference function

$$\psi_{ij} = \exp\left(-\frac{(i-j)^2}{2V_\psi}\right).$$ (9.46)

Figure 9.7 corresponds to the threshold preference

$$\psi_{ij} = \begin{cases} 1 \text{ if } |i-j| < K, \\ 0 \text{ if } |i-j| \geq K. \end{cases}$$ (9.47)

Positive parameters V_ψ and K specify the strength of mating preferences. Small V_ψ and K imply that females have strong preferences for males with matching alleles, whereas large V_ψ and K imply that preferences are weak.

The first two columns in each figure correspond to the case of no mutation and a very high level of initial genetic variation. Specifically, initial distributions were uniform (slightly perturbed to avoid artifacts of symmetry). The first column in each figure describes the case of unlinked loci ($r = 0.5$). In both cases, female alleles frequencies did not change significantly and stayed practically uniform, which is expected given the absence of direct selection in females. The slight observed changes are due to very weak indirect selection via effects of linkage disequilibrium. In contrast, the frequencies of male alleles changed dramatically. If female preferences were not too strong (see the cases of $V_\psi = 8$ in Figure 9.6 and $K = 6$ in Figure 9.7), the distribution of male allele frequencies becomes monomorphic. As preferences become stronger, first the distribution of male allele frequencies becomes dimorphic (the cases of $V_\psi = 4$ and 2 in Figure 9.6 and $K = 5$ and 4 in Figure 9.7). Then, as the strength of preferences increase further, the distribution becomes trimorphic (the cases of $V_\psi = 1$ in Figure 9.6 and $K = 3$ in Figure 9.7). The second column in each figure describes the case of tightly linked loci ($r = 0.005$). The recombination rate does not seem to play any important role (for example, compare the left and right columns in both these figures). The only apparent effect of tight linkage is some smothering of the female distributions.

These results can be understood in the following intuitive terms which parallel those first given by van Doorn and Weissing (2001) for a different model. In the population there is a strong competition between genetically similar males for mating opportunity. The strength of competition is characterized by the width of the preference function. Females play the role of a resource. If competition is weak relative to

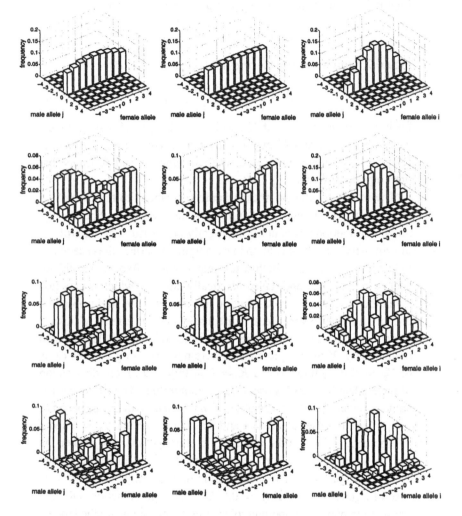

FIGURE 9.6. The distribution of genotype frequencies in the Nei-Maruyama-Wu model with exponential preferences and no selection in females. $V_\psi = 8, 4, 2$, and 1 (from the top row to the bottom row). Left column: slightly perturbed uniform initial distribution, unlinked loci ($r = 0.5$), no mutation, 10^4 generations. Middle column: slightly perturbed uniform initial distribution, tight linkage ($r = 0.005$), no mutation, 10^4 generations. Right column: slightly perturbed monomorphic initial distribution, unlinked loci ($r = 0.5$), mutation rate 10^{-5}, 10^6 generations.

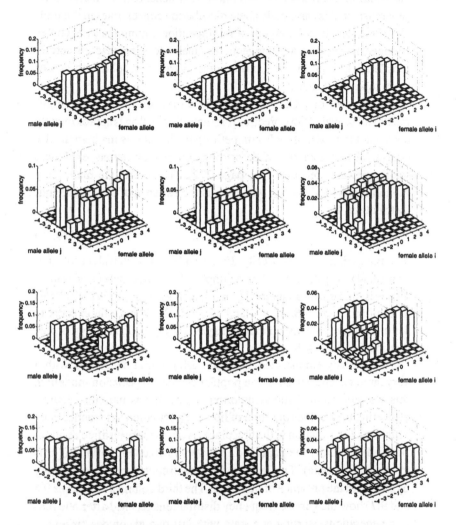

FIGURE 9.7. The distribution of genotype frequencies in the Nei-Maruyama-Wu model with threshold preferences and no selection in females. $K = 6, 5, 4,$ and 3 (from the top row to the bottom row). Left column: slightly perturbed uniform initial distribution, unlinked loci ($r = 0.5$), no mutation, 10^4 generations. Middle column: slightly perturbed uniform initial distribution, tight linkage ($r = 0.005$), no mutation, 10^4 generations. Right column: slightly perturbed monomorphic initial distribution, unlinked loci ($r = 0.5$), mutation rate 10^{-5}, 10^6 generations.

the width of the resource distribution, it is beneficial for males to be generalist. In this case, male allele distribution concentrates in the middle of the female distribution. As competition among males becomes stronger, it becomes advantageous for males to specialize on certain female alleles. In this case, male allele distribution splits into two or more separate clusters.

For the case of a Gaussian preference function (9.46), these arguments can be made more precise. If female distribution is approximately constant, the model of male competition for females is mathematically equivalent to the model of describing competition among exploiters for a renewable recourse (see page 272). Therefore, the males stay monomorphic if $V_\psi > V_x$, where V_x is the variance of the female trait distribution. If

$$V_\psi < V_x, \tag{9.48}$$

then males diversify genetically. Equation (9.48) is equivalent to equation (8.44) on page 273. For example, if the distribution of female alleles is uniform on an interval $[-a, a]$, then $V_x = a^2/3$. With 9 alleles as used in Figure 9.6, $a = 4$ and $V_x \approx 5.33$. Thus, one expects to see a monomorphic distribution in males if $V_\psi \geq 6$, and a polymorphic distribution if $V_\psi \leq 5$.

As I already emphasized, although using initial conditions with a very large genetic variation is a popular choice in speciation modeling, this is a rather unrealistic assumption. Using this assumption requires answering the obvious question of how the population has reached such a high level of variation in the first place. A much more justified assumption is that initially the population has very low genetic variation (or is even monomorphic). This leads to the need to consider the evolutionary dynamics in the presence of mutation. The third columns in Figures 9.6 and 9.7 show the gamete frequency distributions reached after one million generations, starting at a state with just two genotypes present (a genotype A_0B_0 at frequency 0.99, and A_0B_1 at frequency 0.01) and using the same parameters values as those in the first two columns. The mutation rate was $\mu = 10^{-5}$. The recombination rate (which does not seem to be an important parameter here) was 0.5; that is, the loci were unlinked. Although there are some similarities between the first two columns and the third, the dynamics with high initial variation is a poor indicator of the dynamics driven by mutation.

First, any changes take much longer. Excluding the cases of very weak preferences (as illustrated by the first rows), it takes hundreds of thousands of generations for the genotype distribution to approach its equilibrium shape. Even after one million generations, the distribution continues to change at a slow, steady rate. This is what is expected because the characteristic timescale of changes due to mutation has order of $1/\mu$ generations.

Second, the final distributions can have a qualitatively different shape. This is illustrated by the second row in both figures, where the male allele distribution is bimodal if initial variation is high, but is unimodal if initial variation is low. The explanation for this difference seems to be the indirect stabilizing selection acting on female alleles. This selection is directed against extreme females, because such females are more likely to mate with deviant males, so their male offspring will inherit the male genes and will have problems finding mates. Although indirect selection is a rather weak force, it can still prevent direct force of mutation (which is weak under realistic conditions) from making the variance of the female allele distribution sufficiently large to lead to the diversification of males.

Third, with mutation the clusters are either much less distinct and have different weights (with a Gaussian preference function) or separate clusters are simply not formed (with a threshold preference function).

Effects of stochasticity. The above conclusions are based on the deterministic analysis. Adding effects of random genetic drift can result in new effects. In Wu's (1985) numerical simulations, random genetic drift occasionally eliminates intermediate genotypes, resulting in the emergence of distinct reproductively isolated groups of genotypes, that is, in sympatric speciation. What seems to be happening in Wu's simulations is that, first, the population diversifies genetically and spreads along the fitness ridge, forming a genetic cluster where genotypes close to the end points are reproductively isolated (an analog of a ring species; see page 99). Then intermediate genotypes get lost because of random drift. After extinction of the intermediates, the remaining distinct clusters are reproductively isolated. Therefore, sympatric speciation in Wu's model requires a large population size and/or high mutation rates (which are necessary for the generation of genetic variation) and random genetic drift (which is necessary for the extinction of intermediates). Because genetic variation was supplied by mutation, the waiting time until speciation was rather long, decreasing with increasing mutation rate and population size. Very large population sizes were necessary for sympatric speciation. These dynamic features of Wu's model are similar to those of the Higgs-Derrida model.

The Nei-Maruyama-Wu model predicts complex genetic clustering within populations, with female alleles typically having higher variation than male alleles. I expect these properties to hold in a multilocus generalization of the model. Some empirical support exists for these predictions (e.g., Palumbi 1999; Seehausen et al. 1999; Sota et al. 2000; Swanson et al. 2001; Vacquier and Moy 1997). For example, the cichlid fish *Neochromis omnicaerules*, which is widely distributed around rocky islands in southern Lake Victoria, exhibits a complicated color polymorphism, with color being the essential determinant of mating preferences (Seehausen et al. 1999). Under natural conditions, there are three common female phenotypes (at frequencies 0.44, 0.25, and 0.31) but only one common male phenotype (at frequency 0.98). I note that more data about the levels of genetic variability in compatibility genes (traits) in natural populations are needed.

Selection for mating success in females. Introducing selection for mating success in females, even a very weak one, dramatically alters the evolutionary dynamics. The conclusions below are based on numerical studies using a linear function of female mating success

$$w_{f,i} = f(\overline{\psi}_i) = 1 + \varepsilon \overline{\psi}_i, \qquad (9.49)$$

where ε is a (small) positive parameter (a default value set at $\varepsilon = 0.005$).

In the case of no mutation and Gaussian preference function, the ultimate outcome of the dynamics is the complete loss of genetic variation in both male and female alleles. A typical sequence of events is (i) a rapid reduction of variation in females (taking a few hundred generations), followed by (ii) a loss of genetic variation in males (taking a few hundred to a thousand generations), followed by (iii) slow loss of genetic variation in females. The loss of genetic variation takes longer with stronger preferences and tighter linkage. With no mutation and threshold preferences, the dynamics are similar, except that in certain cases genetic variation in females can be maintained in a neutral fashion. This happens because all females within mutational distance K from the single remaining male allele have exactly the same preference for this male allele and exactly the same mating success. Therefore, all these females can be maintained in many neutrally stable equilibrium states.

If mating is selective in both sexes, the monomorphic equilibria with matching male and female alleles (that is, monomorphic equilibria with $i = j$) are locally stable. With mutation, the population simply stays at a state of mutation-selection balance close to the monomorphic equilibrium it starts at. Neither evolution of RI nor cluster formation can happen deterministically.

If the population size is finite, random genetic drift can occasionally overcome selection against new alleles, resulting in their fixation (e.g., Nei et al. 1983). If two or more allopatric populations are identical initially, they will slowly diverge genetically, accumulating different substitutions and RI. The dynamics of RI in the Nei-Maruyama-Wu model by a sequence of stochastic peak shifts can be studied using the methods discussed in Chapter 3.

9.3.2 Two polygenic characters

Modeling mating preferences based on two quantitative characters – one of which, say x, is expressed in females only, and another, say y, is expressed in males only – starts with the papers by Sved (1981a,b) and Lande (1981). Let $\psi(y, x)$ be a function measuring preference of females x for male phenotype y. The three most popular choices of $\psi(y, x)$ used in the literature can be written as

open-ended preference: $\qquad\qquad\qquad \psi(y, x) = \exp[\frac{xy}{V_\psi}]$ (9.50a)

absolute preference: $\qquad\qquad \psi(y, x) = \exp[-\frac{(y - x)^2}{2V_\psi}]$ (9.50b)

relative preference: $\qquad \psi(y, x) = \exp\{-\frac{[y - (\bar{y} + x)]^2}{2V_\psi}\}$ (9.50c)

where a positive coefficient V_ψ characterizes the strength of female preferences (Lande 1981). If only positive trait values are allowed, the open-ended preference function implies that males with larger values of the trait are preferred by all females. If both positive and negative trait values are allowed, the open-ended preference function describes disruptive sexual selection, where females with positive x prefer males with large values of y, whereas female with negative x prefer males with large values of $-y$. The absolute preference function describes assortative mating where females prefer males with matching phenotypes.

The relative preference function implies that female x prefers males that have a trait value that is larger (if $x > 0$) or smaller (if $x < 0$) than the average male trait by the amount $|x|$.

As was already mentioned at the beginning of this chapter, if both male and female traits have continuous distributions, then the sums in the equations defining various characteristics of NM are substituted by appropriate integrals. For example, the average preference of female x for males in the population is

$$\overline{\psi}(x) = \int \psi(y, x)\phi_m(y)dy, \tag{9.51}$$

where $\phi_m(y)$ is the distribution of y in the population, and the mating success of males with phenotype y is

$$w_m(y) = \int \frac{\psi(y, x)}{\overline{\psi}(x)}\phi_f(x)dx, \tag{9.52}$$

where $\phi_f(x)$ is the distribution of x in the population. In this formulation it is assumed that all females have equal mating success.

Normal approximation   Let $\overline{x}$ and $\overline{y}$ be the mean trait values and V_x, V_y, and C the additive genetic variances and covariance. The latter can be due to pleiotropy and/or linkage disequilibrium between the loci controlling the traits. A common assumption in modeling quantitative characters is that genetic variances and covariance do not change as a result of NM and selection, but stay constant at a certain level (e.g., controlled by a balance between selection and mutation; Lande 1981). Under this approximation, the population state is completely characterized by two dynamic variables: the mean values $\overline{x}$ and $\overline{y}$. Genetic variances and covariance can be treated as parameters. The changes in $\overline{x}, \overline{y}$ in one generation are described by a multivariate version of the Lande equation

$$\begin{pmatrix} \Delta\overline{x} \\ \Delta\overline{y} \end{pmatrix} = \frac{1}{2} \begin{pmatrix} V_x & C \\ C & V_y \end{pmatrix} \begin{pmatrix} \frac{\partial \ln w_f(x)}{\partial x} \\ \frac{\partial \ln w_m(y)}{\partial y} \end{pmatrix}, \tag{9.53}$$

where the derivatives are evaluated at $x = \overline{x}$ and $y = \overline{y}$ (see Section 2.6 on page 42). The factor $1/2$ accounts for the sex-limited expression of both traits. The assumption that females have equal mating success implies that $\frac{\partial \ln w_f(x)}{\partial x} = 0$. Following Lande (1981), let us assume that

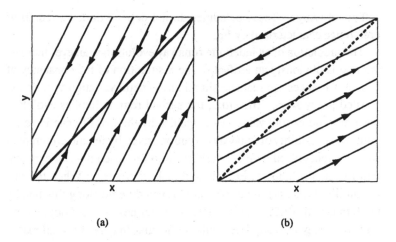

(a)   (b)

FIGURE 9.8. The dynamics of the mean trait values in the Lande model. The lines with arrows show the trajectories on the (x, y)-plane. (a) The line of equilibria (solid line) is stable. (b) The line of equilibria (dashed line) is unstable.

the distributions of x and y in the population are normal. Then the integrals in equations (9.51,9.52) can be easily found. Now equation (9.53) can be rewritten as

$$\Delta \bar{x} = \frac{1}{2} \frac{C}{V_\psi} (\bar{x} - \varepsilon \bar{y}), \qquad (9.54a)$$

$$\Delta \bar{y} = \frac{1}{2} \frac{V_y}{V_\psi} (\bar{x} - \varepsilon \bar{y}), \qquad (9.54b)$$

where $\varepsilon = 0$ if preference is open-ended or relative, and $\varepsilon = 1$ if preference is absolute.

The dynamics of this system are simple. First, all trajectories on the phase-plane $(\bar{x}, \bar{y})$ are straight lines with a slope V_y/C. Second, there is a line of equilibria defined by equation $\bar{x} = \varepsilon \bar{y}$. If $C < \varepsilon V_y$, this line of equilibria is stable and all trajectories are attracted to it (see Figure 9.8(a)). (However, perturbations away from a certain point on the line will not generally return to the same point.) This regime takes place if covariance C is negative for all three models and if C is positive but small for the absolute preference. If $C > \varepsilon V_y$, then the line of equilibria is unstable and all trajectories run towards infinite trait values with an exponentially increasing speed (see Figure 9.8(b)). This regime always takes place if preference is open-ended or relative. It also occurs for the absolute preference if covariance C is positive and sufficiently large. These outcomes do not depend on the strength of preference characterized by V_ψ.

The prediction of an indefinite runaway evolution at an ever increasing rate is obviously an artifact of the assumption of the constancy of genetic variances. In realistic situations, genetic variation will be exhausted pretty soon. Therefore, the second regime implies evolution towards extreme trait values accompanied by the loss of genetic variation. The possibility of runaway evolution driven by female preferences was first recognized by Fisher (1930), who proposed it as an explanation for the evolution of exaggerated male characters. When the line of equilibria is stable, finite populations can diverge along this line by random genetic drift. That is, under both regimes, isolated populations may diverge genetically, accumulating RI that may result in allopatric speciation (Lande 1981).

Kiester et al. (1984) developed a modification of the above model for describing coevolutionary interactions between plants and their specialized pollinators such as orchids and orchid bees, figs and fig wasps, and yuccas and yucca moths (e.g., Thompson 1994; Cook and Rasplus 2003). Their results show that divergence of isolated pairs of mutualistic populations can occur by random genetic drift along the corresponding lines of neutrally stable equilibria. This divergence can be significantly accelerated by sexual selection in a pollinator species.

Note that the normal approximation postulates from the very beginning that trait distributions are unimodal. Therefore, genetic cluster formation and the evolution of within-population RI and sympatric speciation cannot be described by this framework.

No assumptions about distributions   Models with no a priori assumptions about the distributions of male and female characters have been studied using numerical simulations.

For example, Kondrashov and Shpak (1998) used a modification of their models discussed on page 307 with two separate traits x and y. In their interval model, the disappearance of intermediate phenotypes (that is, sympatric speciation) was observed only for the case $C = 0$ and no more than eight loci. In their threshold model, speciation was not observed. Recall that in both models mating is nonselective.

Higashi et al. (1999) numerically studied a model of a diploid species where female and male traits x and y were controlled in an additive way by different sets of five triallelic loci (with effects $-1, 0$, and 1). They used the open-ended preference function (9.50a), with parameter α ranging from 0.02 to 0.2 and assuming that all females mate (that is, mating was nonselective in females). Higashi et al. assumed that intially the mean values $\bar{x}$ and $\bar{y}$ were close to zero, whereas genetic variances in both traits were substantial. The population size was a parameter of the model, ranging from 100 to 1,000. Higashi et al. observed that larger populations with stronger female preferences tend to split into two groups. Within each group, the male trait has extreme value (10 or -10), whereas females remain polymorphic and have the trait value of the same sign as the males. This implies that females have almost zero preference for the males from the other group. Thus, this outcome corresponds to sympatric speciation. In contrast, with weaker preference and smaller population size, random genetic drift overcomes disruptive sexual selection, resulting in males becoming monomorphic for an extreme value of the trait. These results remain robust to introducing weak direct viability selection acting on both traits.

9.3.3 One locus, one character

Turner and Burrows (1995) These authors introduced a model of a diploid population where a male display trait ("shade") is controlled by four additive loci with equal effects (0 and 1). The trait ranges from "black" (all zeros) to "white" (all ones). Female preference is controlled by a single diallelic locus with complete dominance. Females with genotype **pp** choose males on the basis of the rule "court with n males and mate with the palest." Females with genotypes **PP** and **Pp** follow the rule "court with n males and mate with the darkest." This mating scheme implies that all females mate and that there is no direct selection on the female locus. All loci were unlinked. It was also assumed that males are subject to stabilizing selection directed mostly against the two extreme phenotypes. (In the numerical examples used by Turner and Burrow, the mortalities of the "white" and "black" phenotypes were approximately 20 times higher than those of the intermediate male phenotypes, which were approximately equal.) Strong selection against these phenotypes promoted the maintenance of genetic variation in males.

Turner and Burrows used stochastic individual-based simulations of finite populations (with size on the order of 200 to 1,000 individuals). Initially, the population was fixed for the recessive allele **p**. Then a

TABLE 9.9. Relative mating preferences ψ_{ij} used by Takimoto et al. (2000).

Female	Male trait		
genotype	0	1	2
A_0	1	$1 - \alpha$	$1 - \beta_1$
A_1	$1 - \alpha$	$1 - \alpha$	$1 - \alpha$
A_2	$1 - \beta_2$	$1 - \alpha$	1

small number of **P** alleles were introduced (at the frequencies on the order of 0.003 to 0.02). Their results showed that at high values of n (=32 and 64), occasionally the following outcome took place. Heterozygous females **Pp** were eliminated while homozygous females **pp** and **PP** persisted in large numbers. Intermediate males rapidly became scarce. Linkage disequilibrium rapidly developed, so that dark and pale male genotypes became associated with alternative female alleles. This outcome, which is interpreted as sympatric speciation, occurred at frequencies on the order of 1% to 4% at the population sizes used. With higher population sizes, speciation occurred less often. With smaller values of n, speciation was not observed. These results parallel those observed by Gavrilets and Boake (1998) in a simpler model discussed starting on page 299.

Takimoto et al. (2000)   These authors studied a haploid model of nonselective mating in which all females mate. Female mating preference is controlled by a locus with three alleles A_0, A_1, and A_2. Male additive quantitative character is controlled by two diallelic loci with effects 0 and 1, which results in three possible phenotypic values: 0, 1, and 2. Mating preferences ψ_{ij} used by Takimoto et al. are given in Table 9.9, with $0 \leq \beta_1, \beta_2 \leq \alpha \leq 1$. These preferences describe fitness landscape with two peaks. Numerical results of Takimoto et al. (2000) suggest that if $\beta_1 = \beta_2 = 1$, then the population can split into two reproductively isolated groups, in one of which female allele A_0 is associated with male trait 0, whereas in the other female, allele A_2 is associated with male trait 2. This outcome, interpreted as sympatric speciation, is promoted if there is is substantial initial variation in female locus with both extreme alleles present at high frequencies. If mating between these two groups is not completely prohibited (i.e., if $\beta_1 < 1$ or $\beta_2 = 1$), the population becomes monomorphic for a female allele.

TABLE 9.10. Vertical transmission of a two-state cultural trait.

| Mating type | | Probability of | |
mother	father	H child	h child
H	H	b_3	$1 - b_3$
H	h	b_2	$1 - b_2$
h	H	b_1	$1 - b_1$
h	h	b_0	$1 - b_0$

9.3.4 General conclusions on matching-based nonrandom mating

Matching-based NM is usually selective in males, which makes it more difficult to maintain genetic variation. If matching-based NM is also selective in females and females pay costs for being choosy, genetic variation is not maintained and formation of stable genetic clusters within population does not occur. In contrast, if mating is nonselective in females, matching-based NM can lead to genetic clustering within the population especially if initial genetic variation and/or mutation rates are high. However, under realistic biological conditions, RI due to matching-based NM is much more likely to evolve as a result of random genetic drift or runaway sexual selection in allopatric populations than as a result of genetic cluster formation within a single population.

9.4 NONRANDOM MATING CONTROLLED BY A CULTURALLY TRANSMITTED TRAIT

In some birds, female mating preferences are acquired by learning rather than controlled by genotype (e.g., Grant and Grant 1996; Freeberg 1998). For example, in Darwin's finches, females prefer to mate with males that sing the species-specific song they learned early in life and most males sing songs similar to their father's (Grant and Grant 1996, 1997, 1998). Therefore, in these species, RI is largely controlled by culturally inherited factors.

Following Cavalli-Sforza and Feldman (1981, Chapter 2), assume that there are two states (*memes*) H and h that a cultural trait can take. Let p_t and q_t be the frequencies of these memes in the population at generation t. Let the probabilities of an H and an h child resulting from a mating be as given in Table 9.10. This is a model of *vertical transmission* (Cavalli-Sforza and Feldman 1981). For example, if an offspring always inherits the cultural trait of his/her father, $b_0 = 0, b_1 = 1, b_2 = 0$, and $b_3 = 0$. Further, assume that a fraction α of matings are obliged to be with the same cultural type, with $1 - \alpha$ matings being random. Then the dynamics of p_t are described by equation

$$p_{t+1} = p_t^2 B + p_t C + b_0 + \alpha p_t q_t B, \qquad (9.55)$$

where $B = b_3 - b_2 - b_1 + b_0$ and $C = b_2 + b_1 - 2b_0$. With a paternally inherited trait, $B = 0, C = 1$, and $b_0=0$, so that $p_{t+1} = p_t$ independently of α. That is, the frequencies of the memes do not change and both memes are maintained at their initial frequencies, which implies a stable level of within-population prezygotic RI.

This model of assortative mating based on a culturally inherited trait is similar to the O'Donald model discussed at the beginning of this chapter. It will be used in the next chapter as a component of a model of sympatric speciation. More complex models of culturally inherited traits have been built. These include models in which both the male trait and female trait are culturally transmitted, models that combine genetically and culturally transmitted traits, and models that account for horizontal transmission of mate choice (e.g., Richerson and Boyd 1989; Laland 1994; Kirkpatrick and Dugatkin 1994; Nakajima and Aoki 2002; Ihara et al. 2003; Ihara and Feldman 2003). I do not discuss these models here because they have not been used in speciation research.

9.5 SUMMARY

• Successful reproduction requires correspondence of males and female traits and/or loci controlling reproduction. NM is essential in ensuring such correspondence.

• There are two different types of NM. Mating is nonselective if all males have equal mating success and all females have equal mating success. Nonselective mating implies that genotypes pay no costs for being choosy, even if their preferred mates are very rare. Nonselective mating does not change allele frequencies. Mating is selective if genotypes (males or females or both) vary in mating success. This can happen if genotypes pay certain costs for being choosy (for example, if they have reduced mating success in situations where their preferred mates are rare). Selective mating results in a change in allele frequencies.

• Within the context of speciation, there are two general scenarios leading to NM. First, there can be genetically controlled differences in the probabilities of encounter of different individuals (e.g., if individuals have preferences for habitats in which they subsequently mate, if they differ in the timing or location of mating, or if they differ in pollinators). Second, there can be genetically controlled differences in the probabilities of actual mating given an encouter has occurred (e.g., if mating between certain genotypes is difficult because of the absence of genetic, ecological, or behavioral compatibility).

• There are two approaches for modeling the genetics of NM. First, mating patterns can be controlled by a single trait (or a single set of loci) expressed in both sexes (similarity-based NM). Second, mating patterns can be controlled by two different sex-linked traits (or sets of loci) (matching-based NM).

• The dynamics of NM are expected to depend on the initial conditions. The most appropriate choice of initial conditions for studying the *origin* of new species is a very low level of genetic variation. Highly polymorphic initial conditions are most appropriate for studying whether the *fusion* of genetically diverged groups will occur after their secondary contact.

• There are two general (and antagonistic) effects of NM: If mating is nonselective and initial genetic variation is high or if rare phenotypes have an advantage, then NM can lead to the splitting of the population into genetic clusters that are reproductively isolated to a certain degree. Characteristic features of the cluster formation are (i) reduced frequencies of heterozygotes (or hybrids), (ii) positive linkage disequilibrium between loci, (iii) increased variance of additive phenotypic characters, and (iv) the formation of gaps in the distribution of phenotypic characters. In certain situations, NM results in the formation of genetic clusters that are completely reproductively isolated with no intermediate genotypes present. If mating is selective and organisms pay costs for being choosy, NM leads to the (quick) loss of genetic variation.

• The dynamical systems describing nonselective NM are characterized by the existence of lines (and planes) of equilibria and are structurally unstable. These sets of equilibria correspond to ridges of high-fitness values in fitness landscapes for mating pairs. Different weak and/or indirect forces (e.g., mutation or linkage disequilibria) can play a profound role in controlling the dynamics of such systems, which in general are characterized by very long transients.

• Allopatric populations can diverge by random genetic drift along these equilibria accumulating prezygotic RI. The rate of divergence will decrease with increasing the population size. Allopatric divergence and speciation can also be driven by runaway sexual selection away from unstable equilibria. In this case, the rate of divergence increases with the population size.

• Among existing models, sympatric speciation by NM was possible only if mating was nonselective or if rare phenotypes had a reproductive advantage. Under selective NM resulting in costs of choosiness, sympatric speciation appears to be very unlikely.

• If mating is selective, modifiers of mating behavior that make it less selective are expected to invade and spread in the population. This process has not yet been studied in detail.

• Under nonselective mating, increasing the number of loci allows for higher equilibrium levels of within-population RI.

• NM mediated via, for example, habitat choice, timing of mating, or pollinator preference is more likely to result in no (or low) costs of choosiness than NM mediated via genetic preferences or compatibilities.

• Counterintuitively, the scenario that most easily leads to sympatric origin of RI also implies the absence of intrinsic genetic incompatibilities with regard to mating preferences. Therefore, the groups exhibiting clear separation with regard to, say, habitat or timing of mating under natural conditions are not expected to show RI under laboratory conditions.

• Under realistic conditions (i.e., low initial genetic variation, presense of costs of being choosy, not extremely strong sexual selection), sympatric speciation by sexual selection alone appears to be unlikely.

9.6 CONCLUSIONS

Nonrandom mating can be treated within the framework of (frequency-dependent) fitness landscape for mating pairs and groups. Mathematical models have identified important factors controlling the plausibility and patterns of allopatric, parapatric, and sympatric speciation driven by nonrandom mating. In the parapatric and sympatric cases, speciation is most plausible when nonrandom mating does not result in selection for mating success and there is plenty of genetic variation. How common such situations are in natural populations is an open question which must be answered empirically.

CHAPTER TEN

Interaction of disruptive selection and nonrandom mating

In this chapter, I consider the joint effects of disruptive selection (DS) and nonrandom mating (NM) on the possibility of speciation using explicit genetic models operating in terms of genotype frequencies. As discussed in Chapter 8, DS can arise for a number of reasons including spatially heterogeneous selection and competition. Under certain conditions, DS creates situations in which intermediate genotypes are constantly re-created by segregation and recombination in spite of being constantly removed by selection. In these situations, any mechanism that would reduce the production of low-fitness intermediates and, thus, increase the mean fitness of the population will (at least potentially) be favored by selection. Evolution of reproductive isolation (RI) between the organisms residing at alternative optima is one such mechanism. Chapter 9 considered several scenarios generating NM, including preferential joining of mating pools and preferential mating. Here, I will bring the models of DS and the models of NM into a single framework.

Although the results I present below are relevant for all three geographic modes of speciation, my main focus will be on sympatric and parapatric scenarios. Whether conditions for sympatric and parapatric speciation are broad or narrow is a very controversial question even after more than 40 years of theoretical research. The persistence of this argument reflects the limitations of the methods used in theoretical speciation research (discussed in Chapter 1), the most significant of which is a reliance on numerical simulations. To really understand how broad is "broad" and how narrow is "narrow" one needs to know both the details of the model's formulation and the details of its analysis. These two will be the focus of this chapter. As I already mention above, to assure generality of theoretical conclusions, it is necessary to demonstrate that similar conclusions emerge from similar models.

Here I will review a large number of previous models and supplement this review with my own new results. I will show that significant analytical progress is indeed possible. In particular, I provide new analytical results on four classical models first formulated by Maynard Smith (1966), Udovic (1980), Felsenstein (1981), and Diehl and Bush (1989); a model studied recently by Kisdi and Geritz (1999) and Geritz and Kisdi (2000); a model of sympatric speciation by sexual conflict (Gavrilets and Waxman 2002); four new models of sympatric speciation driven by selection on a "magic trait"; and a new model of sympatric speciation with NM controlled by a culturally transmitted trait.

In presenting these results, I will follow the logic of the previous chapter by classifying the models according to the mechanism underlying NM. My justification for such an approach is that establishing NM is the most crucial step in speciation. First, I consider the models in which NM is similarity-based. Then, I discuss the models in which NM is matching-based. Within each of these two classes of models the genes (or traits) underlying DS and NM will be separate. Next, I discuss a class of models in which the same set of loci (or traits) is responsible for both DS and NM. Then, I consider a class of models of the modifiers of mating. Within each of the four classes I will progress from the simplest to more complex models, where complexity is measured in terms of the genetics of NM.

Below I will refer to the loci (and alleles) underlying the traits subject to disruptive selection as the DS loci (and alleles). I will refer to the loci (and alleles) underlying the traits subject to nonrandom mating as the NM loci (and alleles). I will use a number of different criteria for (the progress towards) speciation, such as the heterozygote deficiency index I (see equation (9.8) on page 287), the normalized linkage disequilibrium D (see equation (8.31) on page 266), the difference Δ_x in the frequency of an allele x between two niches, and the hybrid deficiency index I (see equation (9.33) on page 302).

10.1   DISRUPTIVE SELECTION AND SIMILARITY-BASED NONRANDOM MATING

Here, I will assume that NM is based on a set of loci (or traits) that are equally expressed in both sexes. Another set of genes will be subject to disruptive natural selection. For example, the first and second sets of genes can control the flowering time and adaptation to contaminated soil in edaphic endemics (e.g., McNeily and Antonovics 1968; Macnair and Gardner 1998) or the preference for a specific host species and adaptation to this host in the apple maggot fly (Bush 1969, 1975; Feder et al. 1994; Feder 1998; Filchak et al. 2000) or the pea aphids (Via 2001; Hawthorn and Via 2001).

10.1.1 Single locus

Throughout this subsection, I will assume that there is a single NM locus and one or two DS loci. Natural systems with very simple genetics underlying both NM and DS are known. For example, the two sibling insect species of *Chrysopa carnea* and *C. downesi* occur in the northeastern United States. The species differ in microhabitats and in seasonal periods of reproduction. Adaptive cryptic coloration in the species' microhabitat is controlled by one dominant diallelic locus, whereas reproduction periods are controlled by two unlinked dominant loci (Tauber and Tauber 1977). I note that RI between these two species is augmented by the differences in courtship songs (Henry 1985).

I. The Maynard Smith Model The paper entitled "Sympatric speciation," which was published by John Maynard Smith in 1966, has been extremely influential. Its first part analyzes the conditions for the maintenance of genetic variation in a two-deme version of the Levene model (see Section 8.1.1 starting on page 235) that incorporates a possibility of habitat choice by females. The second part is devoted to the following question: given that a stable polymorphism exists under disruptive selection, can RI evolve between the two "extreme" morphs, so that no low-fitness intermediates are produced? Maynard Smith listed four possible mechanisms that can lead to the evolution of RI:

- "Habitat selection," by which he meant the tendency of organisms to mate in the same niche where they were born.
- "Pleiotropism," by which he meant the situations in which the alleles that adapt the individuals to local conditions also control mating behavior.
- "Modifier genes," by which he meant the establishment of selectively neutral genes that cause pleiotropism in the genes underlying local adaptation.

- "Assortative mating genes," by which he meant genes that cause assortative mating regardless of the genotype at the selected loci.

Maynard Smith commented that habitat selection should "perhaps be regarded as a form of allopatric speciation in which isolation is behavioral rather than geographic" (p. 643). He also mentioned that pleiotropism is unlikely from a biological consideration and that the main difficulty of the modifier genes mechanism is to "imagine how a gene ... could influence mating in this way..." (p. 644). Maynard Smith commented that the assortative mating genes mechanism is "more plausible" (p. 644) than the pleiotropism and modifier genes mechanisms. Note that three sections in this chapter closely correspond to the last three mechanisms of Maynard Smith. An additional mechanism that he did not mention is matching-based NM (i.e., sexual selection), to be discussed in the next section.

Maynard Smith proceeded to formulate the following model of speciation via the mechanism of assortative mating genes. Consider a diploid population with discrete nonoverlapping generations inhabiting two distinct niches. Assume that there are two diallelic, possibly linked loci, with r being the recombination rate. The DS locus with alleles A and a controls fitness (viability) in a local environment according to the following scheme:

$$
\begin{array}{cccc}
 & \mathbf{AA} & \mathbf{Aa} & \mathbf{aa} \\
\text{in niche 1} & 1 & 1 & 1 - s_1 \\
\text{in niche 2} & 1 - s_2 & 1 - s_2 & 1.
\end{array}
\qquad (10.1)
$$

Here the coefficients s_1 and s_2 measure the strength of viability selection within each niche ($0 \le s_1, s_2 \le 1$). The dominant allele A is advantageous in niche 1, whereas the recessive allele a is advantageous in niche 2. Maynard Smith assumed that surviving adults form a single randomly mating population. After mating, each female lays a fraction $(1 + H)/2$ of her eggs in the niche where she was raised and lays the remaining fraction $(1 - H)/2$ in the other niche ($0 \le H \le 1$). The coefficient H measures the degree of habitat selection by females. $H = 0$ implies no habitat selection, as in the original Levene model (see page 235). $H = 1$ implies complete habitat selection. Note that, alternatively, one can think of mating as taking place *within* the niches, with males migrating randomly between the niches prior to mating and females either staying in the niche where they were born or going to the other niche with probabilities $(1 + H)/2$ and $(1 - H)/2$, respectively.

The NM locus with alleles **B** and **b** controls assortative mating according to the O'Donald model with dominance (see page 287). That is, individuals **BB** and **Bb** have the same phenotype which is recognizably different from the phenotype of individuals **bb**. With probability α, an individual mates with another individual that has the same phenotype. With probability $1 - \alpha$, the individual is engaged in random mating.

For the special case of $H = 1$ (i.e., complete habitat selection by females), $\alpha = 1$ (i.e., no mating between genotypes **BB**/**Bb** and **bb** is allowed), and $r = 1/2$ (i.e., the loci are unlinked), Maynard Smith presented a numerical example illustrating that an initial difference in the frequencies of NM alleles between the niches gets progressively amplified. One NM allele gets associated with the locally advantageous allele **A** in niche 1, and the other NM allele gets associated with the locally advantageous allele **a** in niche 2. As a result of this, the individuals raised in a niche will tend to mate together. The eventual outcome of the evolutionary dynamics is the increase in frequencies of two reproductively isolated genotypes **AA**/**BB** and **aa**/**bb** (or **AA**/**bb** and **aa**/**BB**) and the complete disappearance of all intermediate genotypes. This outcome is naturally interpreted as sympatric speciation.[1] Although Maynard Smith did acknowledge that the parameter combination he used was very favorable for speciation but unrealistic, he thought that more realistic conditions would merely slow down the speciation process but not prevent it completely. According to his 1966 paper, the crucial process in sympatric speciation is the establishment of a stable polymorphism under DS rather than the subsequent evolution of RI, which he believed was likely (p. 649).

Haploid version of the Maynard Smith model   To get a better understanding of the interactions between disruptive selection and nonrandom mating, let us consider a haploid version of the Maynard Smith model. Let the DS locus control viability according to the following scheme:

$$
\begin{array}{lcc}
 & \mathbf{A} & \mathbf{a} \\
\text{viability in niche 1} & 1 & 1 - s \\
\text{viability in niche 2} & 1 - s & 1,
\end{array} \qquad (10.2)
$$

---

[1] Although in the Maynard Smith model females always lay eggs in the niche where they were born, speciation is sympatric according to our definition (see page 12) because males move randomly between the niches.

where coefficient s measures the strength of selection within each niche ($0 \leq s \leq 1$). Let the NM locus control assortative mating according to the O'Donald model. That is, with probability α, an individual mates with another individual that has exactly the same allele (**B** or **b**) at the NM locus. With probability $1 - \alpha$, the individual mates irrespective of genotype. Assume that males migrate between the niches with probability m per generation, whereas each female always lays the eggs in the niche where she was born. In this model, there are four different genotypes (**AB, Ab, aB,** and **ab**) and two different niches. Thus, the population state has to be characterized by $2 \times (4 - 1) = 6$ independent variables. In spite of this complexity, significant analytical progress is possible by utilizing the symmetry of the model.

No linkage, random male migration. Assume first that the loci are unlinked ($r = 1/2$) and males migrate randomly ($m = 1/2$). In this model, as expected, the population always evolves to an equilibrium state where the locally advantageous allele has a higher frequency than the locally deleterious allele. With regard to other features, there are two possible regimes. In the first regime, the frequencies of the alleles at the NM locus are equal between the niches. This implies that no association between the DS alleles and the NM alleles develops and, thus, speciation does not occur. The equilibrium frequencies at the NM locus depend on initial conditions. In mathematical terms, this regime implies the existence of a line of neutrally stable equilibria for the frequencies at the NM locus. The difference in the frequencies of an allele in the DS locus between the niches is

$$\Delta_{DS} = \frac{s - 2 + \sqrt{(3s)^2 + 4(1 - s)}}{4s}. \tag{10.3}$$

This value grows from 0 to $1/2$ as s increases from 0 to 1. The allele frequencies in the DS locus are $1/2 + \Delta_{DS}/2$ for a locally advantageous allele, and $1/2 - \Delta_{DS}/2$ for a locally deleterious allele.

In the second regime, the population evolves to one of two states (*speciation equilibria*) with strong differentiation in both loci between the niches. At each of these states, the differences in the frequencies of the alleles between the niches are

$$\Delta_{DS} = \frac{\alpha + s - 1}{s} \tag{10.4a}$$

for alleles at the DS locus, and

$$\Delta_{NM} = \sqrt{\frac{(1 + \alpha)(2\alpha^2 + 3s\alpha - 2\alpha - s)}{4\alpha s}} \tag{10.4b}$$

for alleles at the NM locus. The two equilibria differ only in which of
the NM alleles is associated with the locally advantageous DS allele.
These two equilibria exist and are stable if

$$\alpha > \alpha_c = \frac{1}{2} + \frac{\sqrt{(3s)^2 + 4(1 - s)} - 3s}{4}. \tag{10.5}$$

The critical value α_c decreases from 1 to $1/2$ as s grows from 0 to 1.
That is, inequality (10.5) requires sufficiently strong assortative mating.
The population switches from the first regime to the second because this
results in increasing its average fitness. In this model, the gain in the
average fitness can be written as $(\alpha - \alpha_c)/8$.

Linkage and limited male migration. Similar results are readily found
for an arbitrary recombination rate r and an arbitrary male migration
rate m. Although the resulting equations are rather cumbersome, I
still present them below for two reasons. First, I wish to illustrate the
mathematical difficulties involved in finding conditions for speciation
that are present even in this simple model. Second, this model is the only
explicit population genetic model in which the effects of migration rate
on the conditions for parapatric speciation have been found analytically.

I will concentrate on the speciation equilibria. These are characterized by a high degree of symmetry. Specifically, in both niches the
frequencies of locally advantageous alleles are the same, the frequencies
of locally common alleles at the NM locus are the same, and the linkage
disequilibria are the same. The difference Δ_{DS} in the frequencies of an
allele in the DS locus between the niches is given by a solution of the
quadratic

$$s^2(1 - m)\Delta_{DS}^2 + s(2 - s)m(1 + r - m)(1 - \alpha)\Delta_{DS}$$
$$+ rm(1 - \alpha)[s^2 + 2(1 - \alpha)(1 - s)]$$
$$- s^2(1 - m)[1 - m(1 - \alpha)] = 0 \tag{10.6}$$

satisfying $0 \leq \Delta_{DS} \leq 1$. Given Δ_{DS}, the absolute value of the
difference Δ_{NM} in the frequencies of an allele in the NM locus between
the niches can be found from

$$\Delta_{NM}^2 = s[1 - m(1 - \alpha)]^2 \frac{s\Delta_{DS}^2 + m(2 - s)\Delta_{DS} - s(1 - m)}{C_2\Delta_{DS}^2 + C_1\Delta_{DS} - C_0},$$

where

$$C_2 = s^2(1 - 2m)^2,$$
$$C_1 = s(2 - s)m[1 - 4m(1 - m)(1 - \alpha)],$$
$$C_0 = s^2(1 - m)\{1 - 4m[1 - m(1 - \alpha)]\} - 4m^2\alpha(1 - \alpha)(1 - s).$$

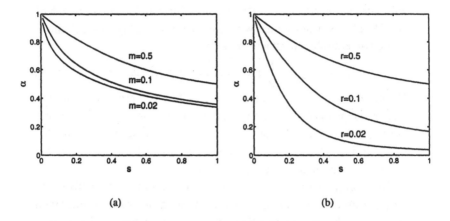

(a)   (b)

FIGURE 10.1. Conditions for sympatric speciation in the haploid version of the Maynard Smith model. (a) $r = 1/2$ and 3 different values of male migration rate. (b) $m = 1/2$ and 3 different values of recombination rate. Speciation occurs for parameter values above the corresponding curve.

Finally, the normalized linkage disequilibrium within each niche is

$$D = \frac{(1 - \alpha)m}{s[1 - m(1 - \alpha)]} \Delta_{DS}(2 - s + s\Delta_{NM}).$$

The speciation equilibria exist whenever the solutions of the above equations are biologically meaningful (i.e., $0 \leq \Delta_{DS}, |\Delta_{NM}| \leq 1$). Figure 10.1 illustrates these conditions. Although reducing both male migration and recombination rates facilitates speciation, recombination has a larger effect. Figure 10.2 illustrates the level of between-niche divergence Δ_{NM} in the NM locus achieved at equilibrium. One can see that once the conditions for sympatric speciation are satisfied, the equilibrium value of Δ_{NM} approaches 1 very rapidly with increasing s and α.

Finally, I note that the assumption that females do not move is very important. If one allows for equal migration of both sexes, then no speciation occurs, even when migration and recombination rates are small and selection and assortative mating are very strong. Therefore, although speciation in this model is sympatric, spatial subdivision and restrictions on migration are crucial for its success.

Dickinson and Antonovics (1973) Thorough analysis of the diploid version of the Maynard Smith model was undertaken by Dickinson and

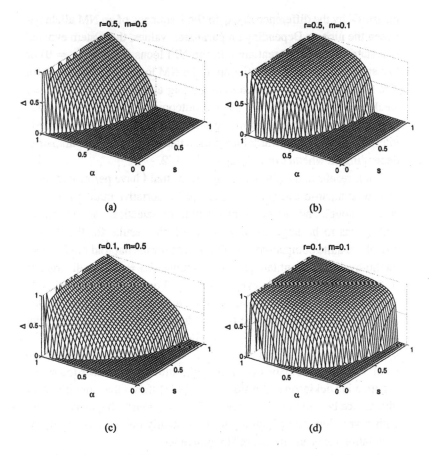

FIGURE 10.2. The difference Δ in allele frequencies in the NM locus between the niches in the haploid version of the Maynard Smith model. Migration and recombination rates are shown on the graphs. Speciation occurs for parameter values such that $\Delta > 0$.

Antonovics (1973) as a part of their study of a number of different models. They assumed that the alleles at the DS locus act additively (see equations (8.4) on page 237), and allowed for a limited migration of males between the niches. The females did not migrate. Dickinson and Antonovics numerically studied the progress towards speciation during the first 400 generations as a function of the strength of selection for local adaptation s, the strength of assortative mating α, and the male migration rate m (see their Figure 5). The progress toward speciation was measured by the difference Δ_{NM} in the frequency of an NM allele between the niches. Depending on parameter values, the system evolved to a state with no differentiation in the NM locus (i.e., $\Delta_{NM} = 0$) or to a state with some differentiation in the NM locus (i.e., $|\Delta_{NM}| > 0$). The first outcome is promoted by large migration, weak selection, and weak assortative mating. The second outcome is promoted by low migration, strong selection, and strong assortative mating. This outcome is interpreted as (a step towards) speciation that is sympatric or parapatric, depending on whether $m = 1/2$ or $m < 1/2$.

Their results and additional simulations that I have performed show that for sympatric divergence to occur, (i) assortative mating has to be strong enough and (ii) the joint strength of selection and assortative mating has to be large enough. Overall, the results for the diploid model are quite compatible with the results for the haploid model, both qualitatively and quantitatively. For example, if $\alpha = 0.4$, no divergence in the NM locus evolves, even if $s = 1$. For $\alpha = 0.5, 0.6, 0.7, 0.8$, and 0.9, sympatric divergence requires s to exceed $0.9, 0.6, 0.4, 0.2$, and 0.1, respectively. For the parapatric case, the range of parameter values leading to divergence in the NM locus becomes broader. For example, if $\alpha = 0.4$ and $m = 0.05$, then $s = 0.6$ is sufficient to result in divergence. However, if the migration rate becomes too small, Figure 5 in Dickinson and Antonovics (1973) shows that conditions for divergence become more stringent. This apparently happens because with a very low rate of immigration of locally deleterious genes, the population lacks an "incentive" to evolve RI.

Overall, Dickinson and Antonovics confirmed Maynard Smith's conclusion that sympatric speciation is theoretically possible. However, they also showed that the maintenance of genetic variation under DS does not necessarily lead to the evolution of RI. Some additional conditions have to be satisfied as well.

Caisse and Antonovics (1978)   These authors extended the two-deme results of Dickinson and Antonovics (1973) to the case of 10 demes arranged along a line. In one version of their model, the rate of male migration between demes decreases exponentially with the distance between them. The other version assumes that males migrated only between the neighboring demes (i.e., as in the stepping-stone model). The strength of selection acting on the DS locus increased linearly from deme 1 to deme 10 or undertook a step change between demes 5 and 6 (see Section 8.1.4 for an illustration of the effects of this type of selection on allele frequencies). Most of the results given are for unlinked loci and $\alpha = 0.9$, that is, very strong assortative mating. Slow convergence towards equilibria and the relatively short duration of the simulations (typically 400 generations) make interpretation of their results difficult because there is no guarantee that they correspond to equilibrium rather than transient states. Overall, however, their results appear to be quite compatible with those of Dickinson and Antonovics. In particular, strong selection ($s > 0.1$) and high levels of assortative mating ($\alpha > 0.4$) are necessary for a divergence at the NM locus.

An interesting question is whether introducing multiple demes makes parapatric divergence in the NM locus easier. A priori, introducing additional demes is expected to result in increased divergence in the DS locus (because the demes away from the boundary will be shielded from migration of locally deleterious alleles). However, migration of deleterious genes is a necessary incentive for the evolution of RI, and the overall effect of reduced migration on the NM locus is not clear. Unfortunately, the results of Caisse and Antonovics do not clarify this issue.

II. The Udovic model   Udovic's paper published in 1980 was the only one in an almost 40-year-long time interval that got very close to analytically finding conditions for sympatric speciation in a nontrivial model. However, both his approach and results were mainly ignored, being overshadowed by other papers describing numerical simulations that biologists apparently found easier to understand. I am positive that theoretical speciation research would have been at a much more advanced stage now had theoreticians paid more attention to Udovic's paper. In this section, I describe and extend Udovic's ground-breaking results.

Consider a very large population with discrete, nonoverlapping generations. There are two diallelic loci. The DS locus with alleles **A** and **a** is subject to symmetric frequency-dependent selection favoring rare genotypes. This implies that if mating is random, the population evolves to a stable equilibrium where the allele frequencies are equal to $1/2$ and the heterozygotes experience a relative fitness loss which we will denote as s ($0 \leq s \leq 1$; see Section 8.4.1). The NM locus with alleles **B** and **b** controls assortative mating according to the symmetric version of the O'Donald model (see page 287). That is, each individual mates with probability α with another individual that has the same genotype at the NM locus. With probability $1 - \alpha$, the individual is engaged in random mating. Let r be the recombination rate.

With two diallelic loci, there are 10 possible diploid genotypes. Because mating is not random and there is selection, in general one cannot expect the population to be at Hardy-Weinberg proportions and at linkage equilibrium. Therefore, the population state has to be characterized by nine independent dynamic variables. In spite of this complexity, the symmetry of the model allows for significant analytical progress. In my derivations, I used a clever transformation to the natural coordinate system proposed by Feldman and Liberman (1984) and Liberman and Feldman (1985).

The first thing to notice about the Udovic model is that there always exists a line of equilibria where the allele frequency at the NM locus can be arbitrary, the deficiency of heterozygotes at the NM locus is given by equation (9.10), the allele frequencies at the DS locus are at 1/2, the genotype frequencies at the DS locus are in Hardy-Weinberg proportions ($I_{DS} = 0$), and the linkage disequilibrium between the loci is absent ($D = 0$). In other words, on this line of equilibria, the loci behave as completely independent. Generically, this line of equilibria is locally stable if DS and assortative mating are weak and linkage is loose.

Under certain conditions, the line of equilibria becomes unstable. If this happens, the population evolves to one of two alternative speciation equilibria, at which the frequencies of all four alleles are equal to 1/2, heterozygotes are in deficiency, and there is a statistical association between the alleles in different loci. The two equilibria differ in the sign of linkage disequilibrium, and which one is approached by the population depends on initial conditions. At these equilibria, the population splits into two genetic clusters. The loci mutually reinforce their effects, leading to stronger RI between the clusters. Evolution towards such an equilibrium represents (a step toward) sympatric speciation.

No linkage. If the loci are unlinked ($r = 1/2$), sympatric speciation occurs if and only if

$$\alpha + s > 1. \tag{10.7}$$

This result is rather remarkable in its simplicity and the clarity of biological interpretation. Considering each locus in isolation, conditions $s = 1$ and $\alpha = 1$ imply complete postzygotic RI and complete prezygotic RI, respectively. Thus, inequality (10.7) tells us that sympatric

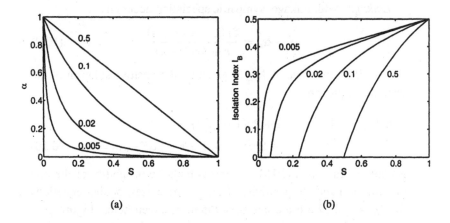

FIGURE 10.3. The Udovic model. (a) Conditions for sympatric speciation. Speciation occurs for parameter values above the corresponding curve.
(b) Isolation index I_{DS} for $\alpha = 0.50$. The recombination rates are given
next to the corresponding curves. (From Gavrilets (2003b), fig.8.)

speciation occurs only if the cumulative strength of viability selection
against hybrids and of assortative mating, $s + \alpha$, is larger than the
threshold strength for each of them to cause complete RI when acting
in isolation.

If condition (10.7) is satisfied, then at equilibrium the heterozygote
deficiency index in the DS locus is

$$I_{DS} = \frac{s + \alpha - 1}{S}. \tag{10.8a}$$

Notice that the condition for sympatric speciation (10.7) coincides with
the condition for I_{DS} to be positive. Index I_{DS} increases with S from
$I_{DS} = 0$ at $s = 1 - \alpha$ to $I_{DS} = \alpha$ at $s = 1$. The heterozygote
deficiency index at the NM locus is

$$I_{NM} = \frac{(s + 2\alpha)\alpha^2}{\alpha^2 s + \alpha^2 + \alpha s - s + 1}. \tag{10.8b}$$

Index I_{NM} increases with S from $I_{NM} = \alpha/(2 - \alpha)$ at $s = 1 - \alpha$ to
$I_{NM} = \alpha$ at $s = 1$. The normalized linkage disequilibrium is

$$D' = \mp \frac{\sqrt{I_{NM}I_{DS}}}{\alpha}. \tag{10.8c}$$

The absolute value of D increases with s from $|D| = 0$ at $s = 1 - \alpha$ to
$|D| = \alpha$ at $s = 1$.

Linkage. With linkage, sympatric speciation occurs if

$$\alpha > \alpha_c = \frac{2r(1-s)(2-s)}{s+4r(1-s)}. \qquad (10.9)$$

The equilibrium value of I_{DS} is given by the positive solution of the quadratic equation

$$s^2 I_{DS}^2 + s\left[2 - s - \alpha + 2r(1-s)\right] I_{DS}$$
$$+ 2r(1-s)(2 - s - 2\alpha) - \alpha s = 0. \qquad (10.10)$$

As before, condition (10.9) guarantees both the instability of the line of equilibria and the positivity of I_{DS}. The corresponding equations for I_{NM} and D are too cumbersome to be given here. Figure 10.3 illustrates the effects of parameters on the possibility of sympatric speciation and the degree of resulting genetic differentiation. This figure and inequality (10.9) show that close linkage (i.e., small r) and strong selection (i.e., large s) both make sympatric speciation easier and result in stronger genetic differentiation. As in the Maynard Smith model, the population switches to a speciation equilibrium because this increases its average fitness. In the Udovic model, the gain in the average fitness can be written as $(\alpha - \alpha_c)/2$.

I note that the equations (10.8a) and (10.10) are given in the appendix of Udovic's original paper. The fact that positivity of I_{DS} is required for the stability of the equilibrium is also apparent from the numerical simulations reported by him. I will illustrate the above results using two simple models.

Example 1. Let us consider the case of unlinked loci and quadratic frequency-dependent selection with fitnesses

$$W_{BB} = 1 + \tilde{s}p^2,$$
$$W_{Bb} = 1,$$
$$W_{bb} = 1 + \tilde{s}q^2,$$

where p and q are the frequencies of the corresponding alleles at the DS locus and $\tilde{s} > 0$. This is one of the simplest cases of frequency-dependent selection (e.g., Gavrilets and Hastings 1995b, 1998) in which rare alleles have a fitness advantage over common alleles. For example, genotypes using an underexploited resource can get a competitive advantage over other genotypes (see Section 8.4.2). In this model, an equilibrium with $p = q = 1/2$ exists and is stable for any $\tilde{s} > 0$. At this

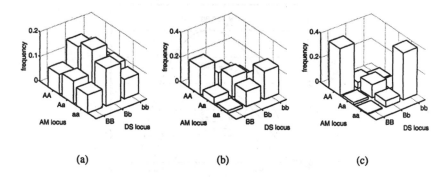

(a)   (b)   (c)

FIGURE 10.4. The distributions of genotype frequencies after 200 generations starting with a random distribution. $\tilde{s} = 4, s = 0.5$. The frequencies of double heterozygotes **AB/ab** and **Ab/aB** are pulled together. (a) $\alpha = 0.4$, so that condition (10.7) is not satisfied and no association of **A** and **B** evolves ($I_{NM} = 0.25, I_{DS} = 0, D = 0$). (b) $\alpha = 0.6$, so that condition (10.7) is satisfied but the association of **A** and **B** is weak ($I_{NM} = 0.46, I_{DS} = 0.20, D = 0.50$). (c) $\alpha = 0.8$, so that condition (10.7) is satisfied and the association of **A** and **B** is strong ($I_{NM} = 0.72, I_{DS} = 0.60, D = 0.82$).

equilibrium, the relative fitness loss of heterozygotes is $s = \tilde{s}/(4 + \tilde{s})$, and condition (10.7) for sympatric speciation can be rewritten as

$$\alpha > \frac{4}{4 + \tilde{s}}.$$

For example, if $\tilde{s} = 1$ (that is, selection is weak), then α has to be larger than 0.8 (that is, very strong assortativeness in mating is required), whereas if $\tilde{s} = 16$ (that is, selection is very strong), then α has to be larger than 0.2 (that is, weak assortativeness in mating is sufficient). Alternatively, if $\alpha = 0.2$ (that is, weak assortativeness in mating), then $\tilde{s}$ has to be larger than 16 (that is, very strong selection is required) and if $\alpha = 0.8$ (strong assortativeness in mating), then $\tilde{s}$ has to be larger than 1 (that is, weak selection is sufficient). Figure 10.4 illustrates the distributions of genotype frequencies for three different combinations of parameters obtained by numerical iterations of the dynamic equations.

Example 2. Let us consider a symmetric two-niche model analyzed by Geritz and Kisdi (2000) (see page 271). Recall that in this model there is a diallelic locus controlling an additive trait z that experiences Gaussian stabilizing selection with optimum values $-\theta$ and θ. Recall also that if $\theta^2 > V_s$, where V_s is a parameter characterizing the strength of stabilizing selection within the niche (see equation (8.11)), then $z = 0$ is a branching point and genetic variation can be maintained.

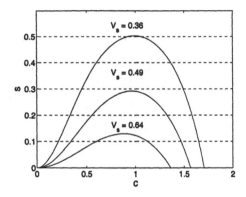

FIGURE 10.5. The normalized strength of selection against heterozygotes, s, in the symmetric version of the Kisdi and Geritz (1999) model as a function of the trait value in homozygotes, c, for $\theta = 1$ and three values of V_s.

To apply our results, which are based on a symmetry assumption, let us assume that genetic variation in the DS locus is maintained by selection and that the allelic effects are $c/2$ and $-c/2$. In this symmetric case, the homozygotes, which have trait values c and $-c$, have equal fitness. Whether heterozygotes, which have trait value 0, have lower fitness than the homozygotes depends on θ, V_s and c. Figure 10.5 shows the relative fitness deficiency of heterozygotes s as a function of c for three different values of V_s assuming that $\theta = 1$. To interpret this figure, let us assume that $\alpha = 0.8$. Figure 10.5 shows that the curve corresponding to $V_s = 0.64$ lies below $s = 0.2$. Therefore, in this case, $s + \alpha < 1$ and the fitness loss of heterozygotes is not big enough to result in sympatric speciation. In contrast, the curve corresponding to $V_s = 0.49$ lies above $s = 0.2$ for $0.55 < c < 1.35$. Therefore, if c is within this range, then $s + \alpha > 1$ and sympatric speciation will occur. However, if $\alpha = 0.7$, sympatric speciation will not occur because s is always smaller than 0.3, meaning that $s + \alpha < 1$. If $V_s = 0.36$, there is a narrow range of c values for which sympatric speciation will occur if $\alpha = 0.5$.

Geritz and Kisdi (2000) analyzed their model using numerical simulations and concluded that "the evolution of assortative mating does not require very strong selection against heterozygotes" (p. 1676). DS indeed does not have to be very strong, but only if α is large enough. In Geritz and Kisdi simulations, α was $\approx 0.8 - 0.9$, which explains why weak DS managed to drive the population to a speciation equilibrium.

III. The Felsenstein Model The paper by Felsenstein published in 1981 was very influential in identifying recombination as a major factor opposing sympatric and parapatric speciation. Felsenstein numerically studied a system of two subpopulations experiencing different selection regimes and exchanging migrants at a constant rate m. Most of his analysis was done for the case of haploid populations, which I will consider here. According to Felsenstein, the diploid version of his model behaves similarly to the haploid one, which is comparable to the properties of the Maynard Smith model.

There are three diallelic loci. Viability is controlled by two diallelic loci A and B (the DS loci) according to the two-locus, two-allele haploid version of the Levene model considered starting on page 238. Recall that in this model parameter s measures the fitness reduction of hybrid genotypes Ab and aB, whereas parameter σ measures the difference in fitness between the extreme genotypes AB and ab within each niche. A NM locus with alleles C and c controls mating according to the haploid version of the O'Donald model. That is, a fraction α of individuals mate with their own type, and a fraction $1 - \alpha$ mates at random. Let the loci be arranged in order ABC, and let r_1 and r_2 be the rates of recombination between A and B and B and C, respectively. I will also assume that recombination events at different positions take place independently. The following is a summary of my results on the sympatric case ($m = 1/2$). I obtained these results analytically by utilizing the symmetry of the model and a clever transformation to "natural" coordinates proposed by Feldman et al. (1974).

With three diallelic loci, there are eight different haploid genotypes. Therefore, in the sympatric case the state of the population is specified by seven independent variables. Let p_A, p_B, p_C be the frequencies of alleles A, B, C; D_{AB}, D_{AC}, D_{BC} the pairwise linkage disequilibria between the corresponding loci; and D_{ABC} the third-order linkage disequilibrium. The third-order linkage disequilibrium is defined as $D_{ABC} = x_{ABC} - p_A D_{BC} - p_B D_{AC} - p_C D_{AB} - p_A p_B p_C$, where x_{ABC} is the frequency of gamete ABC (e.g., Bennett 1954; Slatkin 1972). The first thing to notice about the Felsenstein model is that there always exists a line of equilibria at which the two types of loci behave as completely independent. That is, p_C is arbitrary, the allele frequencies at the DS loci are equal: $p_A = p_B = 1/2$, and the normalized linkage disequilibrium between the selected loci is given by the positive solution of equation (8.8a). In contrast, there is no statistical association between the NM alleles and the DS alleles ($D_{AC} = D_{BC} = D_{ABC} = 0$). Generically, this line of equilibria is locally stable if selection and assortative mating are weak and σ is sufficiently small (see condition (8.8b)).

As s or α increases, the line of equilibria can become unstable. If this happens, the population evolves to one of the two speciation equilibria with equal allele frequencies at all three loci ($p_A = p_B = p_C = 1/2$) and certain pairwise linkage disequilibria between the NM locus and the DS loci. The two equilibria differ in the sign of these linkage disequilibria. The third-order disequilibrium D_{ABC} is zero.

No linkage ($r_1 = r_2 = 1/2$). The two speciation equilibria exist if

$$\alpha > \alpha_c = \frac{1}{2} + \frac{\sqrt{(3s)^2 + 4(1 - s)} - 3s}{4}. \tag{10.12}$$

Note that $\alpha_c > 0.5$ always. Figure 1 in Felsenstein (1981) illustrates this relationship, which he obtained numerically. The normalized pairwise linkage disequilibria are

$$D'_{AB} = \frac{\alpha + s - 1}{s}, \tag{10.13a}$$

for the DS loci, and

$$D'_{AC} = D'_{BC} = \pm\sqrt{\frac{(1 + \alpha)[2\alpha^2 + (3s - 2)\alpha - s]}{4\alpha s}}, \tag{10.13b}$$

for the disequilibria between the NM locus and each DS locus. Surprisingly, condition (10.12) is identical to the condition for speciation in the haploid version of the Maynard Smith model (see inequality (10.5)), and the expressions for disequilibria (10.13) are identical to the expressions (10.4) for the differences in allele frequencies in that model. This indicates some deeper relationships between the two models which, however, I have not explored.

Linkage between the DS loci, no linkage between the NM locus and the DS loci ($r_1 = r \leq 1/2, r_2 = 1/2$). In this case, the two speciation equilibria exist if

$$\alpha > \alpha_c = (1-s)(1-r) - \frac{rs}{2} + \frac{1}{2}\sqrt{s^2(2 - r)^2 + 4r^2(1 - s)}. \tag{10.14}$$

Note that $\alpha_c > 1 - r$ always. That is, as already noticed by Felsenstein, linkage between the DS loci makes the evolution of associations between the NM locus and the DS loci more difficult. The normalized pairwise linkage disequilibrium between the DS loci is still given by

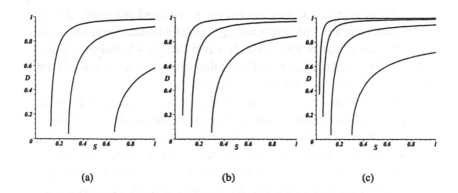

FIGURE 10.6. Normalized linkage disequilibrium D_{AB} as a function of s and recombination rate r_2. $r_1 = 1/2$ in all figures. (a) $\alpha = 0.25$. (b) $\alpha = 0.50$. (c) $\alpha = 0.75$. Different lines correspond to $r_2 = 0.004$ (the left-most line), $0.02, 0.1$, and 0.5. In the case of unlinked loci, the speciation equilibrium does not exist if $r_2 = 1/2$ and $\alpha = 0.25$ and $\alpha = 0.5$. Therefore, only three lines corresponding to $r_2 = 0.004, 0.02$, and 0.1 are shown.

equation (10.13a). The normalized disequilibria between the NM locus and each DS locus are

$$D_{AC} = D_{BC} =$$
$$\pm \sqrt{\frac{(1+\alpha)[2\alpha^2 + (3s-2)\alpha - s] - \frac{1-2r}{2}[3s - 2 + \alpha(2-s)]}{4\alpha sr}}.$$

$$(10.15)$$

Arbitrary linkage. Although the algebraic equations determining conditions for the existence of speciation equilibria and the corresponding linkage disequilibria for arbitrary r_1 and r_2 are readily found, they are too cumbersome to be given here. Instead, I illustrate them graphically in Figure 10.6. As expected, linkage between the NM locus and the DS loci promotes sympatric speciation. Figure 10.6 shows that once the threshold value of s is exceeded, the absolute value of D_{AB} increases very rapidly and can achieve high values, even if linkage between the loci A and B is not extremely tight. Interestingly, as far as the above patterns are concerned, the coefficient σ is not important as long as it is large enough to guarantee the instability of the monomorphic equilibria with extreme genotypes fixed (see condition (8.6)).

Felsenstein also studied the effects of restricted migration. As expected, parapatric speciation occurred under a broader range of conditions than sympatric speciation. It would be very interesting to generalize the above analytical results for the parapatric case to evaluate the effects of restricted migration in more detail.

In reviewing models of sympatric speciation in existence at that time, Felsenstein introduced a distinction between two kinds of speciation. One involves speciation by the substitution of the same allele in the two nascent species ("one-allele model"). The models describing this mechanism will be discussed in Section 10.4. The other involves speciation by the substitution of different alleles in the two nascent species ("two-allele model") as in the three models considered so far in this chapter. Felsenstein identified recombination as a major force acting against speciation in the two-allele models by randomizing the association between the NM locus and the DS loci. According to Felsenstein, "...speciation would be nearly impossible unless it were based..." on a one-allele mechanism (p. 135). His paper was apparently interpreted as a strong theoretical objection to the possibility of sympatric speciation.

IV. The Diehl and Bush model The life cycle in the Felsenstein model consisted of (i) natural selection within the niches, followed by (ii) random migration between the niches and (iii) assortative mating within each niche. Diehl and Bush (1989) modified the haploid version of the Felsenstein model by assuming that (i) natural selection within the niches is followed by (ii) genetically based habitat preference and (iii) random mating within each niche.

In their model, the two DS loci control viability in the same way as in the Felsenstein model. The NM locus controls the probability of entering the first or second niche as defined by Table 9.2 on page 294. Effectively, this modification establishes strong correlation between habitat preference (which controls mating) and selection experienced. Indeed, offspring surviving in, say, niche 1 will tend to have both the genes providing local adaptation (i.e., alleles A, B) and the gene for the preference of this niche (i.e., allele C), which brought its parents to this niche. In contrast, in the Maynard Smith, Udovic, and Felsenstein models, such a correlation emerges only if linkage disequilibrium between the NM and DS loci evolves.

In the Maynard Smith and Felsenstein models, the contribution of each niche to offspring was proportional to the fraction of the population entering the niche. In contrast, Diehl and Bush assumed that both niches

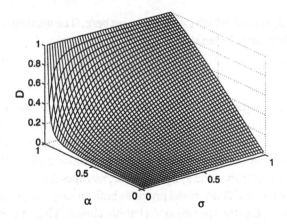

FIGURE 10.7. Normalized linkage disequilibrium between the NM locus and a DS locus in the additive version of the Diehl-Bush model with unlinked loci as function of σ and α.

contribute equally to the overall offspring. As I already mentioned in Chapter 8, this assumption effectively introduced frequency-dependent selection in favor of a rare allele at the NM locus, which resulted in a quick approach of a state with allele frequencies at $1/2$.

As a result of these modifications, there are no threshold values for the strength of selection and/or habitat preference that have to be exceeded for linkage disequilibrium between the NM locus and the DS loci to evolve. Diehl and Bush used numerical simulations to study their model. Significant analytical progress can be achieved as well. Assume that the loci are unlinked and that selection is additive (i.e., $s = 0$). Note that in this case none of the monomorphic equilibria are stable. Then the population always evolves to a state where all allele frequencies are at $1/2$ and there is some association between the NM locus and the DS loci. The normalized linkage disequilibrium between the NM locus and each DS locus is given by the positive real root of a cubic

$$6\sigma^2\alpha^2 D^3 + \sigma\alpha(7-5\alpha^2)D^2 + (2-2\alpha^2-6\sigma^2\alpha^2)D - 2\sigma\alpha = 0. \quad (10.16)$$

Figure 10.7 illustrates the dependence of D on α and σ. This figure shows that although linkage disequilibrium is always present, large values of D require, as before, strong habitat selection (i.e., large α) and strong viability selection (i.e., large σ). With arbitrary linkage, the equilibrium values of linkage disequilibria are given by roots of a fifth-order polynomial, which I do not present here. The normalized linkage disequilibrium between the DS loci is

$$D'_{BC} = \frac{\alpha^2 D + 2\sigma\alpha D}{1 + 2\sigma\alpha D},$$ (10.17)

with D as in equation (10.16). Diehl and Bush also show that assuming multiplicative rather than additive fitnesses does not result in qualitative differences. They also studied the case of parapatric speciation (with $m = 0.1$) and observed that the resulting linkage disequilibrium is (much) larger than in the sympatric case, as expected.

The Diehl and Bush model provides both an easy and general mechanism for sympatric speciation via habitat choice. Their model belongs to Felsenstein's class of two-allele models, yet the recombination in the Diehl and Bush model is largely irrelevant. Further results on different modifications of this model are described in Kawecki (1996, 1997).

10.1.2 Single quantitative character

Here, I review several models in which NM is controlled by an additive character.

Crosby (1970)   Crosby was the first to use individual-based simulations for studying speciation. His model was designed for plants. Each diploid individual was characterized by three additive characters controlled by different sets of unlinked diallelic loci with equal contributions (equal to 0 and 1). A DS character z was controlled by eight "G-loci." Correspondingly, the range of possible trait values was from 0 to 16. Fertility (i.e, the total number of flowers produced by a plant) reduced approximately linearly from a maximum value at the extreme values of z (at 0 and 16) to approximately 25% of the maximum value in the middle of the trait range (at $z = 8$). A character "time of commencement of flowering" was controlled by three "D-loci." This character ranged from week 1 to week 6. A character "number of consecutive weeks during which a plant is flowering" was controlled by two "W-loci." This character ranged from 2 to 6 weeks. A plant's fertility was independent of the number of weeks during which it flowered, meaning that this character was neutral. Initially, the population consisted of two groups. In one group all individuals had $z = 0$ and in the other all individuals had $z = 16$. Alleles in the NM loci (i.e., the D-loci and W-loci) were assigned randomly with probability 0.5. The expectation was that the population will avoid the production of individuals with intermediate values of z (which have low fitness) by a change in flowering period which can be achieved by shifting the time of commencement of flowering or by shortening the period of flowering.

Crosby considered two scenarios: a single population with no spatial structure, and a population where individuals occupy vertices of a two-dimensional lattice. In the first scenario, the overall number of individuals was 512. After a few generations, the genetic differentiation in the G-loci was lost as a result of hybridization, as expected. However, within approximately 80 generations, differentiation in the G-loci was re-created and differentiation in the flowering time (controlled by the D-loci) was established. Fixation of alleles 0 in the W-loci had occurred in one group. Overall, the population had split into two groups, one of which had become late-flowering and the other early-flowering. Both groups shortened the length of their flowering periods. There was very little overlap in flowering and, consequently, there was very little hybridization. This outcome is naturally interpreted as sympatric speciation. In the second scenario, with spatial structure, initial conditions corresponded to two separated genetic groups meeting in the middle of a rectangular area. Crosby assumed different migration patterns for pollen and seeds dispersal. The overall number of individuals was about 2,560. His results were basically the same as in the first case: almost complete separation in the D-loci across the whole population range, which took a few hundred generations. This outcome is naturally interpreted as reinforcement.

Crosby's pioneering paper provided the first theoretical example of reinforcement and sympatric speciation driven by selection against "gamete wastage." His simulations demonstrated that selection against low fertility caused by hybridization could bring about almost complete separation in flowering time, both by shortening the flowering period and by changing the date of flowering, and could allow the two subspecies to intermingle freely with little gene flow between them. He also made two important observations, which were confirmed numerous times in the following years. The first is that hybridization often leads to the quick loss of genetic variability rather than to the evolution of prezygotic RI. To prevent this from happening, Crosby occasionally replaced when necessary 16 randomly chosen organisms with organisms having all alleles 0 or all alleles 1 in the G-loci. The second is that because selection involved in the avoidance of hybridization is indirect, its effects are rather weak and, thus, can be easily overcome by a variety of factors producing direct selection, for example, for mating success. In Crosby's simulations, giving a 4% fertility advantage to plants flowering in excess of 2 weeks resulted in fixation of allele 1 in the W-loci, which prevented both speciation and reinforcement.

Stam (1983)   RI in the model studied by Crosby (1970) evolves as a response to DS. Stam (1983) has shown that divergence in flowering time can occur even without DS. The idea is as follows. Assume that the only difference between two adjacent habitats 1 and 2 is that habitat 1 induces a shift toward earliness in flowering time. Then, even if pollen is transported randomly between the habitats, gene flow will be nonrandom. Genes for late flowering times will migrate preferentially from habitat 1 to habitat 2, whereas genes for early flowering times will migrate preferentially from habitat 2 to habitat 1. Stam used individual-based simulations to demonstrate the possibility of significant divergence in flowering time via this mechanism of preferential gene flow. In a version of his model, a selective difference between the habitats was included. Specifically, Stam introduced another independent set of loci controlling an additive trait subject to spatially varying selection. Here, one can think of a gradient of metal concentration in soil and a set of genes determining metal tolerance. His results show that the joint action of preferential gene flow and spatially heterogeneous selection can result in rapid divergence in both flowering time and metal tolerance, that is, in parapatric speciation.

Van Dijk and Bijlsma (1994)   Crosby (1970) completed only a few numerical experiments. Van Dijk and Bijlsma performed a much more detailed study of a similar model. These authors constructed an individual-based model where flowering time was controlled by 14 additive unlinked loci: 3 loci with large effects, 4 loci with intermediate effects, and 7 loci with small effects. The hybrids were inviable, implying much stronger selection against hybridization than in Crosby (1970). Van Dijk and Bijlsma also accounted for the fact that flowering time itself is likely to be under direct selection against displacement, because less frequent genotypes are less likely to find mates in the population. They did this by using the ratio of ovule/pollen (equal to 0.01, 0.40, and 1.00 in their simulations) as an additional parameter of the model and by prohibiting self-fertilization. This created extra selection against rare genotypes in pollen-limited situations. Overall results of van Dijk and Bijlsma are compatible with those expected from Crosby's results and interpretations: reinforcement and evolution of RI are possible in the pollen-saturated situations (when mating is essentially nonselective), but do not happen in the pollen-limited situations (when there is selection for mating success, so that mating is selective). In the latter situations, the direct effects of selection for mating success, which acts against rare genotypes, overcome the indirect effects of selection against hybridization. This was true for both sympatric and parapatric cases.

Johnson et al. (1996) These authors have significantly extended the results of Diehl and Bush (1989). They considered a diploid population inhabiting a two-habitat system. There are several diallelic loci: one or two loci controlling habitat-specific survival, two loci controlling habitat preferences, and two loci controlling assortative mating (non-habitat specific). It was assumed that individuals mate and place their offspring in the selected habitat. Assortative mating was described using a version of the O'Donald model. All individuals succeeded in mating (meaning that there was no selection for mating success). Initial conditions were set in such a way that at each locus one allele was common (at frequency 0.99) and the other one was rare (at frequency 0.01). As in Diehl and Bush (1989), the population was readily splitting into two separate groups adapted to different habitats, with low gene flow between them. Johnson et al. presented numerical results illustrating the dependence of the waiting time to speciation (that is, the time of achievement of a stable association between the alleles at different loci) on model parameters. In general, speciation is rapid (on the time scale of a few thousand generations) and occurs under a range of conditions. These results clearly demonstrate the power of habitat choice in eliminating gene flow between the diverging races and in causing speciation when mating is nonselective.

Fry (2003) Fry modified the Diehl-Bush model by increasing the number of loci controlling habitat choice and local adaptation (up to 16 loci for each trait). In his model, the probability of entering a niche was equal to the proportion of alleles in the (haploid) genotype giving the preference of the niche. Viability in a niche was given by a product of the contributions of individual loci underlying local adaptation. Fry's choice of preference function allowed for the evolution of complete RI (observed when only two "antagonistic" genotypes, each of which has complete preference and maximum adaptation for one of the two niches, are observed in the population). Achieving such a state, naturally interpreted as sympatric speciation, required DS to be sufficiently strong (coefficient of selection per locus > 0.1 to 0.5). Increasing the number of NM loci makes sympatric speciation more difficult as recombination produces intermediate genotypes which make formation of separate genetic clusters more difficult.

In summary, the evolution of associations between the DS loci and NM loci in sympatric and parapatric context requires either (i) the joint strength of DS and NM to exceed a threshold, or (ii) the presence of a habitat choice mechanism. Costs of choosiness can easily prevent speciation.

10.1.3 Sympatric speciation with culturally transmitted mating preferences

Assume that a DS locus with alleles A and a is subject to frequency-dependent selection favoring rare genotypes. Let s be the relative fitness loss of heterozygotes Aa at the stable polymorphic equilibrium with equal allele frequencies. Let there also be a cultural trait with two states (memes) H and h. Assume that with probability $1 - \mu$, offspring acquires the meme of its father, and with probability μ, it acquires the meme of a randomly chosen male. That is, μ is the probability of *oblique* transmission (e.g., Cavalli-Sforza and Feldman 1981). Finally, assume that mating is based on the culturally transmitted trait as specified in the model described in Section 9.4 on page 327. That is, mating is assortative with respect to the cultural trait or random with probabilities α and $1 - \alpha$, respectively.

This model is similar to the Udovic model, with the difference that mating preferences are transmitted culturally rather than genetically. In the Udovic model, establishing linkage disequilibrium between genes underlying viability and mating increases the mean fitness of the population. In a similar way, formation of associations between genes and memes can decrease the production of unfit heterozygotes. Let $\phi_{A|H}$ and $\phi_{a|H}$ be the frequencies of alleles A and a among offspring with meme H, and $\phi_{A|h}$ and $\phi_{a|h}$ the frequencies of alleles A and a among offspring with meme h. Let p_A, p_a, p_H, and p_h be the frequencies of

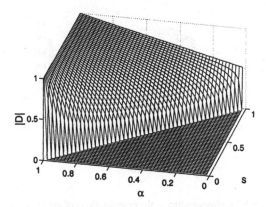

FIGURE 10.8. Association between genes and memes as a function of σ and α with perfect transmission of memes from fathers to offspring ($\mu = 0$).

the corresponding genes and memes in the population. The degree of association between genes and memes can be characterized by index

$$D = \phi_{A|H}\phi_{a|h} - \phi_{a|H}\phi_{A|h}, \qquad (10.18)$$

which is analogous to the normalized linkage disequilibrium between two loci defined by equation (8.7) on page 239. The absolute value of D changes from 0 (no association of genes and memes) to 1 (complete association).

To analyze this model, one needs to work with a system of five (equal to six combinations of genes and memes minus one) difference equations. Luckily, because of the symmetry of the model, its equilibria can be found relatively easily. One can show that sympatric speciation occurs (i.e. a stable association of genes and memes develops) if

$$s + \alpha > 1 + \mu(1 + \alpha). \qquad (10.19)$$

As expected, weaker selection against hybrids and weaker assortative mating (i.e., smaller s and α) and higher probability of error in copying the cultural trait (i.e., larger μ) decrease the plausibility of sympatric speciation. Note that if the memes are transmitted perfectly from father to offspring (i.e., if $\mu = 0$), condition (10.19) becomes the same as in the Udovic model. Condition (10.19) also guarantees that the fusion of two populations that have diverged in genes and memes will not occur once they start hybridizing.

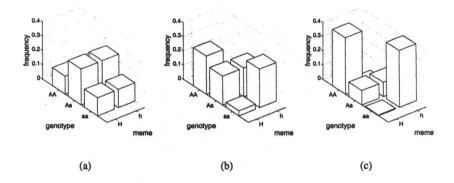

(a)   (b)   (c)

FIGURE 10.9. The frequencies of the combinations of genes and memes after 200 generations starting with a random distribution. $\bar{s} = 4, s = 0.5, \mu = 0$. (a) $\alpha = 0.4$, so that condition (10.19) is not satisfied and no association of **A** and **B** evolves ($D = 0$). (b) $\alpha = 0.6$, so that condition (10.19) is satisfied but the association of **A** and **B** is weak ($D = 0.46$). (c) $\alpha = 0.8$, so that condition (10.19) is satisfied and the association of **A** and **B** is strong ($D = 0.78$).

At the speciation equilibrium, the frequencies of all alleles and memes are at $1/2$ and the absolute value of D is given by

$$|D| = \frac{(1+\alpha)(1-\mu)}{2} \sqrt{\frac{s+\alpha-1-\mu(1+\alpha)}{s\alpha}}. \tag{10.20}$$

If inequality (10.19) is not satisfied, no association between genes and memes develops, so that $D = 0$. Figure 10.8 illustrates the dependence of $|D|$ on s and α, assuming that $\mu = 0$. This figure shows that the equilibrium value of $|D|$ grows quickly with both s and α once condition (10.19) is satisfied. If the probability of oblique transmission μ is small, the value of D at equilibrium will be affected only slightly. Figure 10.9 illustrates the distributions of the frequencies of six different combinations of genes and memes assuming that $\mu = 0$ and using the same frequency-dependent fitness function as in *Example 1* for the Udovic model (see page 344).

In a study of Darwin's finches, in a sample of 482 females of *Geospiza fortis* and *G. scandens*, all but two mated with males that sang their fathers' song (Grant and Grant 1998). In terms of our model, this corresponds to α being very close to one. In *G. fortis*, 70% to 80% of sons sang their fathers' song (Grant and Grant 1996). This corresponds to $\mu \approx 0.2$ to 0.3. Then the model predicts strong association between genes and memes if $s > 0.4 - 0.6$. Selection against hybrids that strong has not been observed in Darwin's finches. Note that if $\alpha \approx 1$, the inequality (10.19) can be rewritten as $s > 2\mu$. That is, the selection coefficient has to be at least twice as large as the probability of oblique transmission. This shows that imprecision in copying a culturally inherited trait can easily prevent sympatric speciation. I conclude that more quantitative data on the strength of assortative mating, the probability of oblique transmission, and the strength of selection against hybrids are necessary to see if cultural inheritance can lead to sympatric speciation or prevent fusion of hybridizing populations under natural conditions.

10.2 DISRUPTIVE SELECTION AND MATCHING-BASED NONRANDOM MATING

In this section, I consider models in which NM is controlled by two sets of loci (or traits), one of which is expressed in males only and another is expressed in females only. Sometimes these traits are referred to as the male's *display trait* and the female's *preference* for the male trait. In addition, there will be loci controlling a trait subject to DS, for example, due to spatially heterogeneous selection. For example, one can think of the *Drosophila* populations inhabiting the opposing slopes in an "Evolution Canyon" which diverge both in mating preferences and in traits controlling local adaptation (Nevo 1997; Korol et al. 2000; Iliadi et al. 2001). Other examples are the two morphs of the marine snail *Littorina saxatilis*, one on the upper-shore barnacle belt and the other in the lower-shore mussel belt, which undergo differential adaptation and simultaneously diverge in mating preferences (Johannesson et al. 1995). DS can also be due to competition for a resource as hypothesized by Rosenzweig (1978) and Pimm (1979).

10.2.1 Two loci

The Waxman and Gavrilets model   The Nei-Maruyama-Wu model discussed starting on page 313 describes prezygotic RI in terms of incompatibilities between alleles at a male locus and a female locus. David Waxman and I have considered the following modification of this model. There is a three-locus, multi-allele diploid population. A DS locus controls an additive trait z that determines an individual's viability. We assume that individuals experience stabilizing selection and are also engaged in competitive interactions that reduce their fitness (viability). An individual's viability is specified by fitness function (8.23a) on page 259, with both the competition function $\xi(z' - z'')$ and the carrying capacity $K(z)$ given by Gaussian functions (see equations (8.24a) and (8.24b)). Surviving organisms produce gametes that encounter each other randomly. The two NM loci control male and female traits involved in gamete compatibility as specified by the Nei-Maruyama-Wu model with Gaussian preference function (9.46) on page 315. From the results presented in Figures 9.6 and 9.7 on pages 316 and 317, respectively, we already know that in the Nei-Maruyama-Wu model prezygotic RI can cause clustering in male alleles. Would adding the third locus subject to DS result in any new dynamics? In particularly, is sympatric speciation possible?

To answer these questions, David Waxman and I have performed deterministic simulations. Because now there are three rather than two multi-allele loci, the simulations proceeded much more slowly than those for the Nei-Maruyama-Wu model. This resulted in only a small number of runs being feasible. Throughout the runs to be discussed, the parameters of the fitness function were chosen in such a way that the maintenance of genetic variation in the DS locus was guaranteed. The following is a summary of our observations and interpretations.

• If all three loci are unlinked, the addition of the DS locus makes little difference in the patterns of diversification in the NM loci. Particularly, sympatric speciation was never observed. The first row in Figure 10.10 describes one of the patterns observed, which directly corresponds to Figure 9.6 with $V_\psi = 2$. There is splitting in the male locus accompanied by the maintenance of variation in the DS locus, but no associations between the NM alleles and DS alleles are established.

• Sympatric speciation was observed only if linkage between the loci was very strong. The second row in Figure 10.10 describes one of the patterns corresponding to sympatric speciation. This row shows splitting of the population into two groups.

• Adding very small costs of assortative mating (i.e, selection for mating success) as described on page 320 results in a quick loss of genetic variation in the NM loci (see the third row in Figure 10.10). Some variation in the DS locus is maintained by DS resulting from

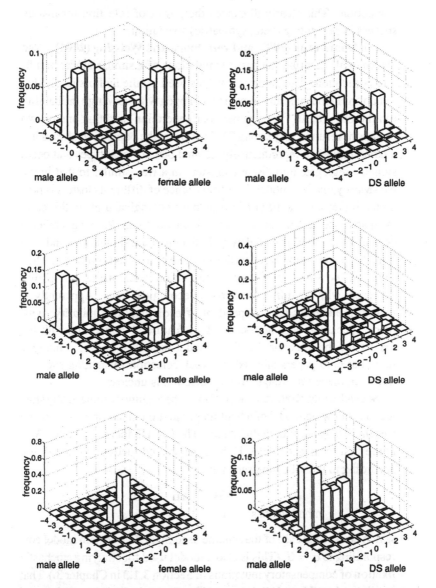

FIGURE 10.10. The distributions of genotype frequencies in the three-locus, multi-allele model assuming $V_\psi = 2$, $V_C = 4$, and $V_K = 100$. First row: unlinked loci, no selection for mating success in females. Second row: closely linked loci ($r_1 = r_2 = 0.01$), no selection for mating success in females. Third row: closely linked loci ($r_1 = r_2 = 0.01$), weak selection for mating success in females ($\varepsilon = 0.02$). Initial distribution is slightly perturbed uniform; no mutation.

competion. This clearly illustrates the power of selection for mating success in preventing (here, sympatric) speciation.

van Doorn et al. (2001) and van Doorn and Weissing (2001)   Van Doorn et al. describe stochastic simulations of a similar model of a finite population. The parameter values used by the authors imply that mutations in the male and female loci happen with very high probability (on the order of 10^{-2} per gamete per generation) and that one or two mutations result in very strong RI. Therefore, the population always has a large number of reproductively isolated pairs of genotypes that occasionally segregate into discrete clusters in a way similar to that in the Nei-Maruyama-Wu model. The coexistence of different clusters is promoted if they also happen to diverge in the ecological trait. In this case, sympatric speciation was observed. Van Doorn and Weissing studied a diploid version of this model using both numerical simulations and analytical approximation based on the adaptive dynamics framework. They report the possibility of sympatric speciation when mating probability decreases with the difference between ecological traits. Apparently, if mating preferences are independent of ecological dissimilarity, sympatric speciation does not occur. This feature is compatible with the results for the Waxman-Gavrilets model described above. Finally, I note that both papers allowed for weak selection for mating success; however, the exact strength of this selection is unclear.

Servedio and Kirkpatrick (1997)   These authors numerically studied the following model of a haploid population inhabiting a two-niche system. There are four diallelic loci. The two DS loci with alleles A, a at the first locus and alleles B, b at the second locus control viability according to the following scheme:

	AB	Ab	aB	ab
in both niches	1	$1-s$	$1-s$	1,

which is a special case of the scheme used in the Felsenstein model corresponding to $\sigma = 0$. (This is also a model used for studying stochastic fixation of compensatory mutations in Section 3.1.3 in Chapter 3.) That is, alleles A and b and alleles a and B are genetically incompatible and their simultaneous presence in the genotype results in reduced viability. A male trait locus with alleles Y and y and a female preference locus with alleles X and x control mating behavior according to the haploid version of the Nei-Kirkpatrick model (see page 311). That is, females X are more likely to mate with males Y, and females x are more likely to mate with males y. Recall that in the Nei-Kirkpatrick model, all females mate. Servedio and Kirkpatrick also assumed that males are subject to spatially heterogeneous selection, so that the corresponding fitness components are

	Y	y
in niche 1	1	$1 - s_m$
in niche 2	$1 - s_m$	1

with $0 < s_m < 1$ (which is similar to the selection scheme used in the haploid version of the Maynard Smith model). The fitness components due to epistatic selection and due to selection on the male trait combine multiplicatively to determine the fitness of a given male.

Servedio and Kirkpatrick assumed that initially the population in niche 1 has genotype $\mathbf{ABXY}$ and the population in niche 2 has genotype $\mathbf{abXy}$. That is, the populations have diverged in the DS loci and male trait, but in both populations females have the same ancestral preference allele $\mathbf{X}$. If selection against hybrid genotypes $\mathbf{Ab}$ and $\mathbf{aB}$ (measured by coefficient s) and selection for the locally advantageous male trait (measured by coefficient s_m) are sufficiently strong relative to migration, the system will remain polymorphic in all three loci (see Section 8.2.1 for an analysis of a related model). Servedio and Kirkpatrick studied whether a new female preference allele $\mathbf{x}$ can get established in the system if introduced at a low frequency in niche 2. If this happens, females in niche 2 will discriminate against males $\mathbf{Y}$. The establishment of allele $\mathbf{x}$ was interpreted as the initiation of RI and the success of reinforcement.

The major result of Servedio and Kirkpatrick is that allele $\mathbf{x}$ gets established relatively easily with a weak female preference if migration between the niches is symmetric (as may be the case in a two-island system). If migration occurs only from niche 1 to niche 2 (as may be the case in a continent-island system), establishment of $\mathbf{x}$ requires very strong selection against hybrid genotypes $\mathbf{Ab}$ and $\mathbf{aB}$ and/or very strong female preference. The difference between the outcomes of symmetric and one-way migration is explained by the effects of migration on the frequency of allele $\mathbf{x}$. In the symmetric case, migration does not change the overall frequency of this allele in the system, as the allele moves from one population to another. In contrast, in the one-way migration case the direct effects of migration push the frequency of $\mathbf{x}$ to zero, preventing it from responding to the indirect effects of selection for reinforcement. Unfortunately, Servedio and Kirkpatrick report neither the level of resulting linkage disequilibrium nor other indexes of RI.

10.2.2   Two polygenic characters

Sved (1981a,b)   Sved (1981a,b) was the first to consider the reinforcement scenario assuming NM based on quantitative characters expressed separately in males and females and controlled by two separate sets of genes. He used normal approximations for the distributions of female and males traits, x and y, respectively, and a Gaussian preference function (9.50b) on page 321 defining the probability of mating given female x and male y have met. Sved studied the consequences of matings within and between two different strains A and B, assuming that hybrids (that is, offspring resulting from between-strain matings) are lethal or sterile. His results show that the outcome of hybridization significantly depends on whether or not there is selection for mating success in females. If all females eventually mate, which happens if the potential number of encounters is effectively infinite, the change in the average female trait for a strain is directly proportional to the difference in the average male traits between the strains. This implies continuous divergence between the females, given an initial difference between the males from the two strains. In contrast, if females are allowed only one encounter, the change in females is proportional to the difference in the average trait values between the males and females within the strain. In this case, females stop evolving once the male and female traits match. Sved argued that in large populations the divergence will be very slow, whereas in small populations, genetic variation will be quickly lost, so that final divergence will not be great.

Spencer et al. (1982)   These authors extended Sved's approach in several directions. In particular, they allowed for more than one quantitative character controlling the probability of mating, explicitly modeled the dynamics of the strain sizes allowing for extinction of a strain, and studied the effects of the number of males encountered by a female on the evolutionary dynamics. They concluded that for reinforcement to work, selection against hybrids has to be very strong, initial genetic variation and initial divergence between the strains has to be high, some mechanism (for instance, frequency-dependent selection) must allow for the coexistence of the strains throughout the period in which selection is acting, and costs for being choosy must be absent or very small. Spencer et al. emphasized that because the costs of being choosy play such a profound role, they cannot be ignored in any realistic analysis. They also noticed that as divergence progresses, the selective pressure for further divergence becomes weaker and weaker, which makes it unlikely that selection alone could complete the process.

Liou and Price (1994)   These authors used individual-based simulations to model the dynamics of reinforcement in a system of two sympatric populations whose densities are regulated independently. They assumed that there are two different sets of five diploid loci with three alleles each (with effects -1,0,1) that determine a male secondary trait y and a female preference x. The relative preference of female x for male trait y was given by Gaussian function (9.50b) on page 321. It was assumed that each female mated once; that is, there were no costs for females being choosy. Initially, all allele frequencies were set to the same value within each population. This value, however, was different between the populations. As a result, the mean values of male and female traits were the same within populations but differed between the populations. In addition to the 10 NM loci, there were 2 diallelic DS loci. It was assumed that "parental" genotypes 00/00 and 11/11 had fitness 1 while all other genotypes (hybrids) had reduced (and equal) fitness. In the DS loci, initially genotype 00/00 was at frequency 1 in one population and genotype 11/11 was at frequency 1 in the other population.

If hybrids have zero fitness, the two possible outcomes are genetic divergence in the NM loci (interpreted as success of reinforcement) or extinction of one of them. In the case of nonzero fitness of hybrids, there is an additional possible outcome – the coexistence of the populations with hybridization. Liou and Price's results show that reinforcement is possible if selection against hybrids is strong and initial divergence between the populations is large. The latter implies substantial divergence in allopatry prior to the secondary contact. High female discrimination, high levels of initial genetic variation, and tight linkage between the DS loci and NM loci promote reinforcement. Liou and Price also considered a case in which the populations initially overlap in only a small part of their range. The results were similar to those under sympatry.

Kondrashov and Kondrashov (1999)   This paper considers two models of sympatric speciation. In the first model, similarity-based NM was controlled by a quantitative trait (color) equally expressed in both sexes. In the second model, matching-based NM was controlled by two different sex-limited traits (male color and female preference for male color). In both models, an independent quantitative character interpreted as size was subject to disruptive frequency-dependent selection. Specifically, equal proportions of phenotypes with the largest and smallest traits have fitness 1, whereas all other phenotypes have smaller fitness. This *rectangular fitness valley* assumption effectively implies that the genotypes with the largest and the smallest trait values are ecologically independent, which may be the case if they use very different resources. In both models, each trait was controlled up to 16 diallelic loci. The main conclusions of Kondrashov and Kondrashov is that if the fitness valley is sufficiently deep, the number of loci controlling size is sufficiently large, whereas if the number of loci controlling mating preferences is sufficiently small, sympatric speciation is very likely. Speciation was more likely with similarity-based than with matching-based NM. Rather than doing individual-based simulations or iterating exact dynamic equations, the authors used an approximate method (a so-called hypergeometric model) that assumes that all possible genotypes within each phenotypic class are always at equal frequencies (Shpak and Kondrashov 1999; Barton and Shpak 2000). Although solutions satisfying these conditions can exist, their stability is not guaranteed a priori. Because Kondrashov and Kondrashov did not present any evidence for the validity of the approximation they used, their numerical analysis of the model remains tentative. There are also two components of the model itself that are questionable from biological considerations. First, selection for mating success was (practically) absent. Second, the frequency-dependent fitness scheme they used was not only unrealistic, but also resulted in extremely strong disruptive selection and explicitly promoted the maintenance of genetic variation. It appears that under more realistic assumptions, the level of genetic variation in the Kondrashov-Kondrashov model will be low and sympatric speciation will be very difficult to achieve.

Kirkpatrick and Servedio (1999); Kirkpatrick (2000, 2001) Kirkpatrick and Servedio developed an analytical approach predicting the degree of divergence in mating preferences of island and continent females in the case of one-way migration. This approach assumes that a certain divergence in male trait between the populations is maintained by some unspecified factors. Kirkpatrick and Servedio characterized NM in terms of a phenotypic correlation between the female preference and the male trait among mating pairs and considered the case when hybrids have reduced fitness because of epistatic interactions between different alleles. Kirkpatrick (2000) generalized the approach for the case of NM based on similarity at a trait equally expressed in both sexes. Kirkpatrick (2001) further extended the approach for the case when hybrids have reduced fitness because of ecological rather than genetic factors. The overall conclusions are that reinforcement (i.e., divergence in female preference) is promoted by stronger selection against hybrids and larger phenotypic differences between the populations, which is quite compatible with both previous work and biological intuition. The overall effects are, however, rather small. For example, Figure 1 in Kirkpatrick and Servedio (1999) reports a relative divergence in female preference at equilibrium of about 6% of that for the male display trait. The divergence in the frequency of an allele controlling female preference reported in Figure 2 in the same paper is less than 3%. The analytical approximations used by Kirkpatrick and Servedio (1999) and Kirkpatrick (2000, 2001) are based on a method developed by Barton and Turelli (1991), which heavily relies on the assumption that selection is weak. Because selection is assumed to be weak, significant change cannot be studied by this method.

Sadedin and Littlejohn (2003)   These authors used individual-based simulations to study the possibility of preventing the fusion of two populations. It was assumed that the populations have diverged in an ecological trait prior to their contact. The ecological trait as well as a male trait and a female preference were treated as additive characters, each controlled by 10 diallelic loci. The population inhibited a two-dimensional habitat. Initially, genotypes at the male and female loci were assigned randomly. All females mated with their most preferred male, and offspring with intermediate values of the ecological trait had reduced viability or fertility. The population was prevented from becoming homogeneous with respect to one of the two optimum values of the ecological trait by culling individuals with the most common ecological trait when necessary. Thus, the only two possible outcomes of model dynamics were the formation of a hybrid swarm or the maintenance of two clusters of genotypes reproductively isolated to some extent. Sadedin and Littlejohn (2003) show that preventing hybrid swarm formation requires sufficiently strong selection against hybrids, and that if hybrid swarm formation is prevented, the two populations diverge in female preferences (that is, reinforcement occurs). They also show that hybrid swarm formation was prevented more easily if hybrids were infertile rather than inviable and if the two populations had initially parapatric distributions rather than sympatric.

In summary, the evolution of associations between the DS loci and MN loci in sympatric and parapatric contexts is much more difficult under matching-based NM than under similarity-based NM. Typically, such association could develop only if DS is very strong, initial genetic variation is very high, and females pay no cost for being choosy.

10.3 "MAGIC TRAIT" MODELS

In the models considered so far in this chapter, speciation requires the establishment of correlations between the DS alleles and the NM alleles. This requirement disappears if the same set of loci pleiotropically control both the trait subject to DS and the trait used in mating, or if the same trait underlies both processes. For example, in the threespine stickleback fish, the limnetic and benthic forms have diverged in size, apparently as a result of adaptation to foraging in different ecological niches. Laboratory experiments have revealed that in this fish mating is assortative and based on size (e.g., McKinnon and Rundle 2002). In seahorses, mating is assortative and based on size, which also may be subject to disruptive natural selection (Jones et al. 2003). In Darwin's finches, the adaptive diversification of beak morphology and body size has apparently resulted in the correlated divergence of songs that are the main factor of prezygotic RI (Podos 2001). In the pea aphids, the quantitative traits loci that control performance on different hosts are closely linked to the loci that produce assortative mating via host choice (Hawthorn and Via 2001). In the *Heliconius* butterflies, hybrids of the two sympatric species *H. cydno* and *H. melpomene* have reduced mating success and, simultaneously, are less protected from predators because of reduced resemblance to the parental species, which are both unpalatable and warningly colored (Naisbit et al. 2001). In this section, I consider several models, which I call *magic trait models*, describing such situations. These are the models of pleiotropism in the terminology of Maynard Smith (1966). I note that several magic trait models were already studied in Chapters 5 and 6.

10.3.1 Single locus

Moore (1981)   Moore was the first to introduce a model of speciation where a single diallelic locus is both under disruptive viability selection and controls NM. He considered a diploid population inhabiting a one-dimensional stepping-stone system with nine demes. The fitness of heterozygotes was normalized to be 1 in all demes. The fitnesses of homozygotes aa increased linearly from 0.2 at deme 1 to 1.8 at deme 9. The fitnesses of homozygotes AA decreased linearly from 1.8 at deme 1 to 0.2 at deme 9. NM was described using the Moore model discussed on page 295. That is, mating probabilities were equal to $1, 1 - \alpha, 1 - \beta$, depending on whether the two organisms share 2, 1, or 0 alleles at the locus. Moore used numerical simulations, starting with a population monomorphic for genotype AA and allowing for mutation at a low rate ($\mu = 10^{-5}$). He assumed that mating discrimination was very strong by choosing $\beta - 1$ (which implies that different homozygotes do not mate) and $\alpha = 1/2$. His numerical results show that homozygotes aa rapidly increase in frequency in the right-most deme (where they are most advantageous) and then rapidly get close to fixation in the three adjacent demes. The system becomes subdivided into two areas with four demes each dominated by different homozygotes. These results are easily understood. Because in the right-most deme viability of Aa is five times larger than that of resident homozygote AA, natural selection for allele a overcomes selection for mating success acting against this allele when it is rare (see equation (9.23) on page 296) and the swamping effect of gene flow from the deme on the left. Subsequently, the same happens in the three other demes where allele a is advantageous.

   Slatkin (1982)   Slatkin analyzed the possibility of parapatric speciation in a haploid population. He considered a population with discrete, nonoverlapping generations that is subject to immigration at rate m of individuals with ancestral allele a. The new allele A is advantageous in the local environment. Viabilities of A and a are $w_A = 1$ and $w_a = 1 - s$, respectively ($s > 0$). Slatkin assumed that the surviving adults mate randomly but that the A × a pairs produce a fraction $1 - f$ offsprings relative to those of the A × A and a × a pairs. This assumption implies fertility selection against mixed matings (which can be due to pre- or postzygotic factors). Slatkin's analytical results show that if there is a large reduction in the fertility of mixed matings (i.e., if f is large enough), then the locally advantageous allele A cannot deterministically increase from a low frequency. If the population were completely isolated long enough to let A increase to a high frequency before secondary contact, then the gene flow does not necessarily eliminate allele A. If the population has never been isolated, an allele adapting the population to local conditions could increase from a low frequency only if the reduction in fertility f is small relative to the strength of viability selection s. Reduction in fertility is a measure of RI. Thus, in this model strong RI cannot evolve in the parapatric case.

10.3.2   Two loci: speciation by sexual conflict

In the models considered so far in this chapter, sympatric speciation resulted as an outcome of disruptive natural selection emerging from ecological interactions between individuals. Here, I discuss a model where sympatric speciation is a consequence of sexual conflict.

Sexual conflict occurs when characteristics that enhance the fitness components of one sex reduce the fitness of the other sex. Numerous examples of sexual conflict resulting from the costs of mating, polyspermy, and sensory exploitation are known (e.g., Parker 1979; Arnqvist and Rowe 1995; Chapman and Partridge 1996; Rice 1996; Stockley 1997; Parker and Partridge 1998; Holland and Rice 1998; Howard and Berlocher 1998; Stutt and Siva-Jothy 2001). For example, peptides contained in the seminal fluids of *Drosophila melanogaster* males increase female death rate (Chapman and Partridge 1996), mating in bed bugs results in severe physical harm to females (Stutt and Siva-Jothy 2001), and if more than one sperm fertilizes an egg, the egg usually dies (Howard et al. 1998). These detrimental effects of mating on female (or egg) fitness components can be reduced by females evolving "resistance" to male (or sperm) pre- and postmating manipulations (Holland and Rice 1998). Sexual conflict over the mating rate can result in continuous between-sexes coevolution in which males try to increase their mating rates, whereas females try to keep the mating rate at an optimum value. This coevolution was hypothesized to be an important engine of continuous divergence of isolated populations which could lead to rapid allopatric speciation (Parker and Partridge 1998; Holland and Rice 1998). Here, I show that sexual conflict can be important in causing both allopatric and sympatric speciation.

Let us return to the Nei-Maruyama-Wu model (see page 313). That is, we consider a large sexual haploid population with distinct nonoverlapping generations. We concentrate on two possibly linked multiallelic loci, assuming stepwise mutation. The alleles A_i at the first locus are only expressed in females (or eggs), and the alleles B_j at the second locus are only expressed in males (or sperm). Assume that the probability ψ_{ij} of compatibility between a female (or egg) carrying an A_i allele and a male (or sperm) carrying a B_j allele is given by the Gaussian function (9.46) on page 315.

Let $P_i = \sum \psi_{ij} \Phi_{m,j}$ be the proportion of males in the population that are compatible with females carrying allele A_i. As before, $\Phi_{m,j}$ is the frequency of the male allele j in the population. Note that P_i is mathematically equivalent to the average preference $\overline{\psi}_i$ of female i for the males in the population as defined by equation (9.4) on page 285. I will use variable P_i rather than $\overline{\psi}_i$ to emphasize the difference in interpretation: here P_i will be interpreted as a proxy of the female mating rate. We assume that males can be involved in multiple mating and that they compete for fertilization opportunities.

To formalize the idea of sexual conflict over mating rate, let us assume that the overall probability f_i that an A_i female leaves offspring (see page 284) is a unimodal function of P_i: $f_i = f(P_i)$ that reaches a maximum at a certain value $P_{opt} < 1$ (Gavrilets 2000a; Gavrilets et al. 2001; Gavrilets and Waxman 2002). For example, in sea urchins, egg fitness is maximized at a level of sperm density that is much smaller than levels common under natural conditions (Franke et al. 2002). In our model, the female mating rate is directly proportional to the proportion of compatible males. The assumption $P_{opt} < 1$ formalizes the idea of sexual conflict over mating rate, because for the males it is optimal to have $P_{opt} = 1$, since then all females are susceptible to fertilization by any male. To clarify the implications of the above assumptions, assume that the population is monomorphic for male allele B_j. Then $P_i = \psi_{ij}$ and the females that have the optimum mating rate and the highest overall fitness are those for which $\psi_{ij} = P_{opt}$. Using the definition of ψ_{ij}, it is easy to see that there are two such female alleles: $A_{j+\delta}$ and $A_{j-\delta}$, both at a distance δ from the male allele, where

$$\delta = \sigma \sqrt{\ln\left(P_{opt}^{-2}\right)}. \tag{10.21}$$

This simple model exhibits three general dynamic regimes (Gavrilets and Waxman 2002). The first regime is an *endless coevolutionary chase* between the sexes in which females continuously evolve to decrease the mating rate, while males continuously evolve to increase it (Holland and Rice 1998; Gavrilets 2000a; Gavrilets et al. 2001; Gavrilets and Waxman 2002). In this regime, there is a dynamic compromise between the sexes, and the proportion of compatible pairs is intermediate between P_{opt} and 1. Generically, the coevolutionary chase is observed if the level of genetic variation is not too large. The rate at which the population accumulates mutations in male and female loci is constant and increases with mutation rate and the strength of sexual conflict as measured by P_{opt}. Different isolated populations will diverge at different rates or in different directions, accumulation RI as a byproduct. A few hundred or thousand generations may be sufficient for substantial RI (Gavrilets and Waxman 2002). Analogous conclusions hold in a model based on quantitative characters (Gavrilets 2000a). Thus, sexual conflict can result in rapid allopatric speciation. Increasing the population density is expected to result in stronger sexual conflict and in more rapid divergence of populations. This prediction was recently verified experimentally using replicate fly populations (Martin and Hosken 2003).

The two other regimes are observed when the population size or mutation rates are sufficiently large. In the *Buridan's Ass regime* (Gavrilets and Waxman 2002), there is very low variation in male alleles maintained by mutation, whereas female alleles split into two clusters, both at the optimum distance δ from the male allele (see Figure 10.11(a)). In this regime, males get trapped between the two female subclusters and have relatively low mating success, a situation resembling the fabled Buridan's Ass.[2] A potential example of the Buridan's Ass regime is provided by a spider species, *Araneae: Pholcidae*, where females show dimorphism in the size of genitalia whereas males have a low degree of variation (Huber and González 2001).

In the *sympatric speciation regime* (Gavrilets and Waxman 2002), males answer the diversification of females by diversifying themselves and splitting into two clusters that start evolving towards the corresponding female clusters. This happens in a way similar to that in the model

---

[2]Since the Middle Ages, this ass, associated with the name of Buridan (John Buridan; c. 1295/1300 to c. 1360) though not referred to in his extant writings, has been invoked in discussions concerning free will and determinism. The hungry animal stood between two haystacks that were indistinguishable in respect of their delectability and accessibility. Unable to decide from which stack to feed, the ass starved to death (Bro 1995).

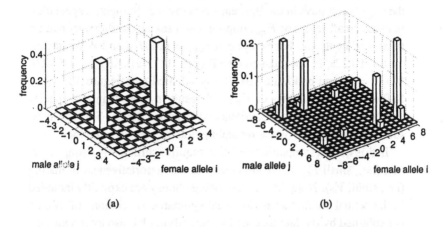

(a)   (b)

FIGURE 10.11. Population genetic states in the sexual conflict model. Parameters: mutation rate $\mu = 10^{-5}$, recombination rate $r = 0.5$, $\sigma^2 = 10$, and female overall fitness $w_f = \exp[-(P - P_{opt})^2]$. (a) The Buridan's Ass regime ($P_{opt} = 0.6$; the average of values P_i is 0.64; there is a single male allele B_0 and two female alleles A_3 and A_{-3}). 5,000 generations of selection. (b) The sympatric speciation regime ($P_{opt} = 0.4$; the average of values P_i is 0.42). 8,000 generations of selection. (From Gavrilets (2003b), fig.10.)

of competition for renewable resources (see page 272). As a result, the initial population splits into different genetic clusters (species) that are reproductively isolated and have emerged sympatrically (see Figure 10.11(b)). The regime of a coevolutionary chase within species ends after increasing genetic variation in female alleles leads to the splitting of female alleles into two subclusters within each species. By contrast, genetic variation in male alleles remains very low within each species. At equilibrium, female P_i values are close to P_{opt}, whereas males get trapped between two female subclusters and have low mating success.

The probability ψ_{ij} can be written as a function $\psi(d)$ of the genetic distance $d = i - j$ between the female and male alleles. In the limit of low mutation rates, sympatric speciation occurs if

$$\psi(\delta - 1) + \psi(\delta + 1) > 2\psi(\delta) \qquad (10.22a)$$

If ψ is a Gaussian function with zero mean and variance V_ψ, this inequality can be rewritten as

$$V_\psi < \delta^2 \qquad (10.22b)$$

(Gavrilets and Waxman 2002). If the above conditions are not satisfied, the population stays in the Buridan's Ass regime. Sympatric speciation requires small values of P_{opt}, implying that sexual conflict over mating rates must be strong. For example, if $P_{opt} = 0.637$, then $\delta = 3$ and condition (10.22a) is not satisfied (see Figure 10.11(a)). If $P_{opt} = 0.450$, then $\delta = 4$ and condition (10.22a) is satisfied (see Figure 10.11(b)). Sufficiently small values of P_{opt} can result in more than two species emerging sympatrically (Gavrilets and Waxman 2002). The above results are not affected by the recombination rate between the loci.

In this model, sympatric speciation requires sufficiently strong selection (i.e., small P_{opt}) and sufficiently strong assortativeness in mating (i.e., small V_ψ). Note that costs of being choosy are explicitly included in the sexual conflict scenario. That sympatric speciation still occurs is explained by the fact that the loci underlying RI also experience direct selection for diversification induced by sexual conflict. Therefore, selection does not have to overcome the homogenizing effect of recombination that otherwise can prevent sympatric speciation.

Theoretically, sexual conflict shows a great potential for driving the evolution of RI and causing speciation. Important quantitative characteristics of sexual conflict are the shape of female fitness function $f(P)$, the optimum proportion of compatible males P_{opt}, and the maximum reduction in female fitness from too many compatible males (Gavrilets 2000a). These characteristics need to be measured in the laboratory or in natural populations if we are to understand whether sexual conflict is as significant in evolution as some believe it is.

10.3.3 Single polygenic character

Here I consider several models in which the "magic trait" is controlled by several loci acting additively. First, I describe four simple two-locus models that can be studied analytically. Then I describe some numerical models with more than two loci.

Sympatric speciation in two-locus, two-allele haploid models Let us consider a sexual haploid population. There are two diallelic loci with effects 0 and 1 controlling an additive trait z. The rate of recombination is r. Let $\phi_1, \phi_2, \phi_3,$ and ϕ_4 be the frequencies of genotypes ab, Ab, aB, and AB. Neglecting the effects of environment on z, the corresponding trait values are $z_1 = 0, z_2 = z_3 = 1,$ and $z_4 = 2$. Assume that trait z

is subject to viability selection and, simultaneously, controls mating. I will consider two models for viability selection and two models of NM.

Viability selection. The first model of viability selection is a version of the Levene model which was used above in analyzing the Felsenstein model (see page 347). That is, there are two niches with different selection regimes. The viabilities of genotypes ab, Ab, aB, and AB in the first and second niche are $1 - \sigma, 1 - s, 1 - s, 1 + \sigma$ and $1 + \sigma, 1 - s, 1 - s, 1 - \sigma$, respectively ($0 < \sigma, s < 1$). Coefficient s measures the strength of selection against hybrid genotypes, whereas coefficient σ measures the difference in fitnesses between the extreme genotypes. Surviving adults disperse randomly between the niches and then mate.

The second model of viability selection is the Bürger model of stabilizing selection and intraspecific competition within a single population (see page 262). Recall that this model is characterized by two coefficients: s, which measures the strength of stabilizing selection ($0 \leq s \leq 1$), and c, which measures the strength of competition ($c \geq 0$).

Nonrandom mating. The first model of NM is a version of the Wright model considered starting on page 301. That is, with probability α, an individual mates with another individual with exactly the same value of the trait, and with probability $1 - \alpha$, it mates randomly. Recall that in this model mating is nonselective.

The second model of NM is a version of the Moore model specified by Table 9.6 on page 303 and equations (9.36). That is, the probability of mating decreases with increasing the difference in the phenotypic traits between the potential mates. Recall that in this model mating is selective.

The resulting four models are simple enough to allow some analytical investigation. The general strategy is to rewrite the dynamics equations in terms of genotype frequencies by using some natural coordinates as proposed for two-locus, two-allele systems by Karlin and Feldman (1970) and Bürger (2000). This allows one both to identify polymorphic equilibria and, in some cases, investigate their stability. Note that in these models, the hybrid deficiency index (defined by equation (9.33) on page 302) at the symmetric polymorphic equilibrium is given by the normalized linkage disequilibrium D.

Model I. Consider first the Levene model of spatially heterogeneous selection coupled with the Wright model of NM. As in the case of random mating, the monomorphic equilibria with hybrid genotypes Ab and aB fixed cannot be stable. The monomorphic equilibria with ex-

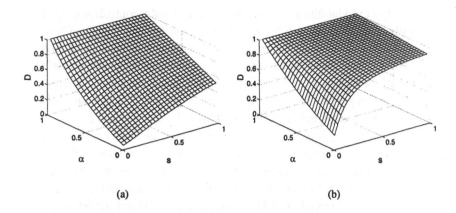

(a)   (b)

FIGURE 10.12. The normalized linkage disequilibrium in the haploid two-locus, two-allele model with spatially heterogeneous selection according to the Levene model and NM according to the Wright model. (a) $r = 0.5$. (b) $r = 0.1$.

treme genotypes **AB** and **ab** fixed can be stable if condition (8.6) on page 238 is satisfied and α is sufficiently small. That is, strong NM (i.e., large α) can destroy the stability of these equilibria. Besides the monomorphic equilibria, there always exists a symmetric polymorphic equilibrium with allele frequencies equal to $1/2$ and the normalized linkage disequilibrium D given by a positive root of quadratic

$$sD^2 + r(2 - S - \alpha)D - s(1 - r) - \alpha r = 0. \tag{10.23}$$

Note that if mating is random ($\alpha = 0$), this equation simplifies to equation (8.8a) on page 239. The stability condition of this equilibrium is still given by equation (8.8b). Intuition and numerical simulations suggest that no other equilibria can be stable. Figure 10.12 illustrates the dependence of D on parameters. In this model, linkage disequilibrium is maintained even if mating is random (see Figure 8.1 on page 239). As expected, stronger assortative mating as well as stronger selection and tighter linkage increase the value of D. Note that with $r = 0.1$, D becomes close to its maximum for relatively small values of s and α. In this model, sympatric speciation occurs relatively easy.

Model II. Next consider the Bürger model of competition coupled with the Wright model of NM. Recall that if mating is random, the population evolves to a monomorphic equilibrium if competition is weak and to a symmetric polymorphic equilibrium with positive D if competitions is strong. For a relatively narrow range of intermediate values of c, the population evolves to a polymorphic state with $D = 0$. Nonrandom mating (i.e., if $\alpha > 0$) does not affect the conditions for stability of the monomorphic equilibria. The two polymorphic equilibria with $D = 0$ exist under slightly broader conditions and their position is shifted (the difference in the frequencies of A and B is increased). A symmetric polymorphic equilibrium with allele frequencies equal to 1/2 exists always. At this equilibrium, the normalized linkage disequilibrium D is given by a positive root of the cubic equation

$$
-2scD^3 + (2rc + 2c - 2s + r\alpha sc - scr - r\alpha c)D^2 + (2s + 2r
$$
$$
+ r\alpha sc - sr + 2sc - 2c + r\alpha s - r\alpha c - \alpha r - scr)D - r = 0.
$$
$$
(10.24)
$$

Numerical simulations show that a somewhat stronger competition is required for the symmetric polymorphic equilibrium with $D > 0$ to be stable under NM. Figure 10.13 illustrates the dependence of D on parameters when the stabilizing selection is absent (i.e., $s = 0$), so that the symmetric polymorphic equilibrium is always globally stable. This figure shows that linkage disequilibrium increases very rapidly with c and α, especially, if there is some linkage. In this model, sympatric speciation occurs relatively easy.

Model III. Consider the Levene model of spatially heterogeneous selection coupled with the Moore model of NM. In the resulting model, there are two pairs of monomorphic equilibria. The equilibria with hybrid genotypes Ab and aB fixed are locally stable if

$$
s < \alpha, \tag{10.25}
$$

that is, if mating preferences are sufficiently strong. The equilibria with extreme genotypes AB and ab fixed are locally stable if

$$
\sigma^2 < \min\left(1 - (1 - \alpha)(1 - s), \frac{1 - (1 - r)(1 - \beta)}{1 + (1 - r)(1 - \beta)}\right), \quad (10.26)
$$

that is, if the fitnesses of homozygotes do not differ too much.

There can be up to three symmetric polymorphic equilibria with $p_A = p_B = 1/2$, and the normalized linkage disequilibrium given by a

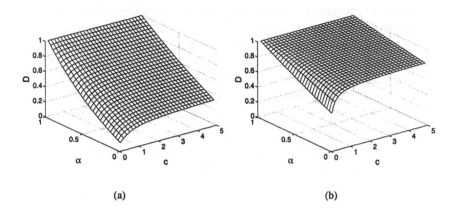

(a)   (b)

FIGURE 10.13. The normalized linkage disequilibrium in the haploid two-locus, two-allele model with competition according to the Bürger model and NM according to the Wright model. (a) $r = 0.5$. (b) $r = 0.1$.

solution of the cubic equation

$$\begin{aligned}
[2s(s - 2\alpha) + 4\alpha - \beta] D^3 & \\
- [2(1 + r)s(s - 2) - \beta(1 + 2r)] D^2 & \\
- \left\{ 2s(s - 2\alpha) + 4\alpha - \beta - 4r[(1 - s)^2 + 1 - \beta] \right\} D & \\
+ [\beta(1 - 2r) - 2(1 - r)(2 - s)s] &= 0. \quad (10.27)
\end{aligned}$$

To analyze this model further, let us assume that the preference function is Gaussian, that is, $1 - \beta = (1 - \alpha)^4$ (see equation (9.36b) on page 304). I did not manage to find exact conditions for stability of these equilibria. However, it appears that only the solution with the largest positive D can be locally stable, and this requires[3] that

$$s > \frac{\alpha(\alpha^2 - 4\alpha + 2)}{2(1 - \alpha)}. \quad (10.28)$$

However, even if this equilibrium is stable, the population does not necessarily end up there.

The bifurcation diagrams in Figure 10.14 illustrate the dependence of the normalized linkage disequilibrium D on parameters observed

---

[3]There is a pair of asymmetric polymorphic equilibria with $\phi_1 = \phi_4$ that exist if $s < \alpha$ and inequality (10.28) is true. These asymmetric equilibria bifurcate from the monomorphic equilibria $\phi_2 = 1$ and $\phi_3 = 1$ at $s = \alpha$ and collide with the symmetric equilibrium when s is equal to the right-hand side of inequality (10.28).

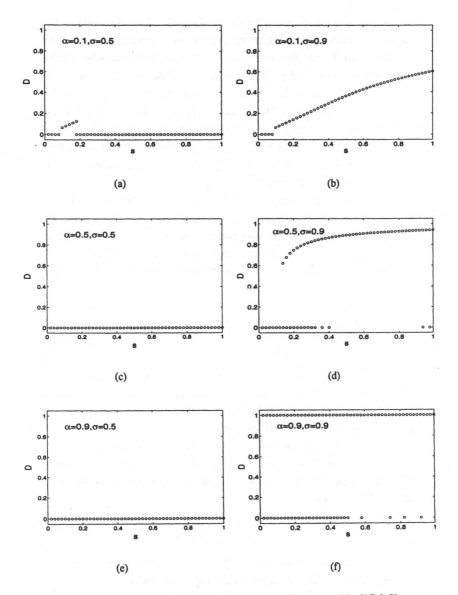

FIGURE 10.14. The Levene model of DS coupled with the Moore model of NM. Shown are the values of D observed after 5,000 generations in 100 runs with random initial conditions and 50 equally spaced values of s. Below, "H" denotes monomorphic equilibria with hybrid genotypes fixed, and "E" denotes monomorphic equilibria with extreme genotypes fixed. (a) H stable for $s < 0.1$, E stable for $s > 0.17$. (b) H stable for $s < 0.1$, E never stable. (c) H stable for $s < 0.5$, E stable for all s. (d) H stable for $s < 0.5$, E stable for all s (but rarely observed). (e) H stable for $s < 0.9$, E stable for all s. (f) H stable for $s < 0.9$, E stable for $s > 0.62$ (but rarely observed). Some locally stable equilibria were not observed or were only rarely observed because of their small domains of attraction.

in numerical simulations. The equilibria with $D = 0$ correspond to monomorphic equilibria. This figure shows that different monomorphic equilibria can be locally stable simultaneously with the polymorphic equilibrium. Generically, for D to be large, both s and α have to be large. However, large values of these two parameters also imply stability of the monomorphic equilibria. Therefore, sympatric speciation will not typically occur if starting with a monomorphic population. However, if two populations with alternative extreme genotypes fixed prior to their contact meet, their genetic fusion will not occur, even if the populations become completely sympatric.

Analysis of speciation models is sometimes performed under the assumption that allele frequencies are maintained at 1/2 by some unspecified mechanisms (e.g., Kirkpatrick and Ravigné 2002). This assumption is attractive because it greatly simplifies the algebra involved. If we make this assumption, then in the case when equation (10.27) has three different equilibria, the model predicts local stability of the two solutions with the largest and the smallest value of D. However, as I already stated above, numerical simulations show that only the one with the largest value of D can be stable. This illustrates the danger of assuming that allele frequencies are somehow maintained at 1/2, without explicitly considering a mechanism responsible for this. This approach results in very unrealistic models. Although in numerical simulations allele frequencies can be easily forced to stay at prespecified values, in natural systems this will not happen.

Model IV. Finally, let us consider the Bürger model of competition coupled with the Moore model of NM. There are two pairs of monomorphic equilibria. The equilibria with a hybrid genotype (Ab or aB) fixed are locally stable if

$$c < \frac{s + \alpha(1 - s)}{(1 - \alpha)(1 - s)}, \qquad (10.29a)$$

that is, if competition is sufficiently weak. The equilibria with an extreme genotype (AB or ab) fixed are locally stable if

$$c < \min\left(\frac{\alpha - s}{1 - \alpha}, \frac{r + \beta(1 - r)}{4(1 - r)(1 - \beta)}\right), \qquad (10.29b)$$

which requires that assortative mating is strong ($\alpha > s$) and competition is sufficiently weak. The right-hand sides of inequalities (10.29) increase with α and β, implying that selective mating increases the domains of stability of monomorphic equilibria, as expected. In principle, there can be up to five symmetric polymorphic equilibria with

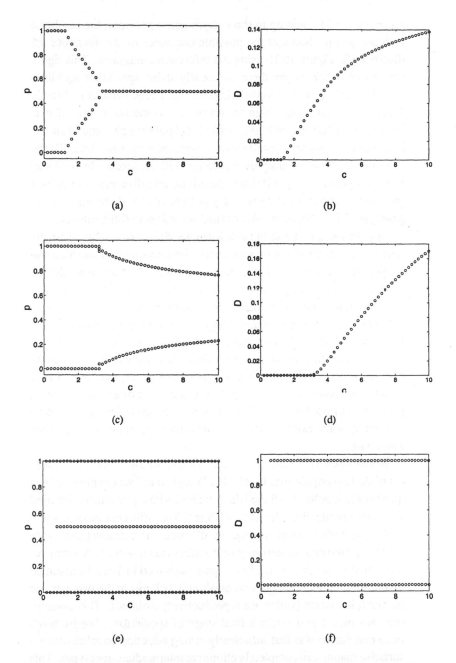

FIGURE 10.15. The Bürger model of competition coupled with the Moore model of NM. Shown are the values of p and D observed after 5,000 generations in 100 runs with random initial conditions and 50 equally spaced values of c, $r = 0.5$, and the strength of stabilizing selection $s = 0.5$. (a) and (b) $\alpha = 0.1$. (c) and (d) $\alpha = 0.5$. (e) and (f) $\alpha = 0.9$.

$p_A = p_B = 1/2$, and up to three asymmetric polymorphic equilibria with $p_A \neq p_B$. Some of the possible outcomes of the dynamics are illustrated in Figure 10.15 using the bifurcation diagrams. This figure shows that in this model there are locally stable speciation equilibria at which linkage disequilibrium is positive. Large values of D (as in Figure 10.15(f)) require high values of α. At the same time, if α is large (see the last row of Figure 10.15), the polymorphic equilibria exist simultaneously with locally stable monomorphic equilibria. Therefore, an initially monomorphic population will not speciate but will stay monomorphic. However, if two populations have diverged prior to their secondary contact, so that one has genotype **ab** fixed and another has genotype **AB** fixed, they will not fuse but will stay differentiated.

Sympatric speciation in multilocus models A number of papers numerically studied the case when both disruptive natural selection and preferential mating were based on a single additive trait controlled by multiple loci. If mating is nonselective and there is a positive correlation between trait values of the mates, the resulting increase in phenotypic variance can be rather substantial, especially if there are many loci (see the discussion of the Wright model starting on page 301). In models of preferential mating in which the probability of mating decreases with the difference in the trait values between potential mates, mating is always selective in males. This typically results in the loss of genetic variation, preventing sympatric speciation. However, the loss of variation can be averted by some external frequency-dependent selection acting on the same trait.

Kondrashov (1983a,b, 1986) numerically studied a series of models of diploid populations in which a "magic trait" was controlled by a number (2, 4, or 8) of unlinked diallelic loci with equal effects. He used five different functions describing DS and five different functions specifying the probability of mating. In all cases, the extreme phenotypes at the opposite ends of the range of possible values were completely reproductively isolated. In the two papers published in 1983, he assumed that initially most of the population already belongs to the two extreme phenotypic classes (which are reproductively isolated). This assumption was meant to describe a final stage of speciation. Kondrashov's main conclusion was that sufficiently strong selection coupled with assortative mating can completely eliminate intermediate genotypes. This conclusion was confirmed in the 1986 paper in which he studied a more general model that explicitly incorporated frequency-dependent DS and allowed for changes in the frequencies of extreme phenotypes.

Similar results were reported by Rice (1984), who numerically studied a four-locus model. Rice used three different functions describing the probability of mating and a single functional form for fitness function that was assumed to be V-shaped with a small slope $|s| = 0.05$. To avoid the loss of genetic variation, one-half of mated females were constrained to have $z \leq 4$, and the other half were constrained to have $z \geq 4$. Rice reported a case in which weak DS coupled with strong assortative mating produced very strong RI, as measured by a very low proportion of matings between organisms with $z \geq 4$ and $z \leq 4$.

Kondrashov et al. (1998) studied a number of models in which only the two most extreme phenotypes were completely reproductively isolated, whereas any premating RI between all other mating pairs was absent. They noticed that to counterbalance the homogenizing effects of random mating, DS must double the phenotypic variance in the population. An effective choice of DS for accomplishing this is a "rectangular fitness valley" (see page 366). Kondrashov et al. showed that if the valley has an appropriate width and is sufficiently deep, all genotypes but the two most extreme (which are reproductively isolated) can be eliminated. The rectangular fitness valley assumption helps delay the loss of variation expected because of the selectivity of mating. In addition, to prevent the population from becoming monomorphic, Kondrashov et al. adjusted fitnesses by slightly reducing the fitness of the most abundant phenotypic class periodically.

Christensen et al. (2002) studied a model in which mating pairs comprised individuals different in less than K genes (as in the Higgs-Derrida model; see equation (9.37) on page 305). The same genes that underlay mating compatibilities also controlled the strength of between-individual competitive interactions. Numerical simulations showed that the population sporadically splits into reproductively isolated clusters that coexist for some time until the next period of hectic reorganization of the population structure. This behavior is similar to that of the Higgs-Derrida model with the difference that competition somewhat stabilizes the system prolonging the periods of coexistence of reproductively isolated genetic clusters.

Jones et al. (2003) used the rectangular fitness valley assumption together with the Gaussian preference function (9.50b) on page 321. Rather than using a diallelic model, Jones et al. assumed a continuum of alleles (e.g., Bürger 2000) at 50 diploid loci. Their results show that the fitness reduction of the valley phenotypes is more than 15% to 20%, sympatric speciation is a very likely outcome, which is promoted by increasing the population size and heritability of the trait.

A typical timescale in all these simulations was on the order of a few hundred generations. These simulations convincingly show that if a lot of genetic variation is present initially, its loss is prevented, and the antagonism between selection and recombination is removed, sympatric speciation is plausible. If costs of choosiness are present and initial genetic variation is low, sympatric speciation is very difficult.

10.3.4 Two polygenic characters: speciation by sexual selection

Here, I return to the Lande model considered in Section 9.3.2. Recall that in this model the distribution of the female trait x and the male trait y in the population is bivariate normal, with fixed variances G_x, G_y and covariance C. The probability of mating between female x and male y is controlled by a preference function $\psi(x, y)$, with parameter V_ψ characterizing mating tolerance (see equations (9.50) on page 321).

In Section 9.3.2, we did not allow for any natural selection in this model. Direct viability selection on male trait can be incorporated in a straightforward way (Lande 1981). Let $w_s(y)$ be the corresponding fitness components for males, and let the overall male fitness be the product of the viability selection component as specificed by $w_s(y)$ and the mating success component as specified by $w_m(y)$ (as defined by equation (9.52) on page 322). In spite of the addition of viability selection in males, all trajectories on the phase plain $(\overline{x}, \overline{y})$ are still straight lines with the slope G_y/C. These trajectories run to or from a curve of equilibria given by equality

$$\overline{x} = \epsilon\overline{y} - a\,\frac{\partial \ln \overline{w}_s(\overline{y})}{\partial \overline{y}}, \tag{10.30}$$

where $\overline{w}_s(\overline{y})$ is the average viability of males, and $a = 1$ for open-ended preference and $a = V_\psi$ for relative and absolute preferences (Lande 1981; Lande and Kirkpatrick 1988). Two papers built on this approach to study the dynamics of speciation.

Lande (1982)   Lande considered a population inhabiting an infinite one-dimensional habitat. Let variable r specify the spatial position and let $\bar{x}(t, r)$ and $\bar{y}(t, r)$ be the average trait values for females and males at time t at spatial position r. Assume that the male trait y is under natural stabilizing with the optimum trait value $\theta(r)$ varying in space:

$$w_s(y) = \exp\left[-\frac{(y - \theta(r))^2}{2V_s}\right],$$

where, as before, parameter V_s characterizes the strength of stabilizing selection. Using a diffusion approximation for describing migration of individuals similar to the one used in the Bazykin model (see equation (8.15) on page 252), Lande showed that the deviations of equilibrium distributions of the average female, $\bar{x}(r)$, and male, $\bar{y}(r)$, trait values from the line of equilibria given by equation (10.30) are $f(r)$ and $C/G_y f(r)$, respectively, where $f(r)$ satisfies to equation

$$0 = \frac{1}{2}\frac{G_y}{V_s}\left[\theta(r) - \frac{\bar{y}}{A}\right] + \frac{\sigma^2}{2}\frac{\partial^2 \bar{y}}{\partial r^2}. \tag{10.31}$$

Here σ^2 is the variance in distance between parent and offspring, and

$$A^{-1} = 1 + \frac{V_s}{V_\psi}\left(\varepsilon - \frac{C}{G_y}\right).$$

Function $f(r)$ characterizes the pattern of geographic variation in the population. If mating is random (i.e., if $V_\psi \to \infty$), then $A = 1$. Note that the condition that guarantees the stability of the line of equilibria in the corresponding spatially homogeneous model is equivalent to the condition $A > 0$. If A is positive and $\theta(r)$ is bounded, then the solution of equation (10.31) can be written as

$$f(r) = \frac{A}{2l}\int_{-\infty}^{\infty}\theta(\eta)\exp(-\frac{|r - \eta|}{l})d\eta, \tag{10.32}$$

where the *characteristic length* $l = \sigma\sqrt{AV_s/G_y}$ has the meaning of the minimum distance over which the equilibrium cline responds to spatial variation in selective forces. If $A < 0$, the trait values evolve in a runaway fashion to extreme trait values across the whole population.

In the absence of female preferences (i.e., if $V_\psi \to \infty$ and, thus, $A = 1$), the equilibrium cline in the male character is simply a weighted spatial average of the optimum phenotype under selection (Slatkin 1978). Female preferences alter both the magnitude of geographical variation and the characteristic length l via the amplification factor A. This is illustrated in Figure 10.16 for several choices of $\theta(r)$ for which the corresponding integral in equation (10.32) can be evaluated exactly. The

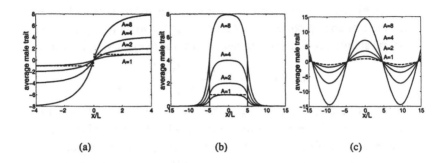

(a)   (b)   (c)

FIGURE 10.16. The equilibrium clines in the Lande model for different values of A. The optimum trait values are given by dashed lines. (a) A step change in the optimum: $\theta(r) = \text{sign}(r)\theta$. (b) The optimum within a spatial pocket of size $2D$ differs from that outside the pocket: $\theta(r) = \hat{\theta}$ for $|r| \leq D$, and zero elsewhere. (c) A periodically varying optimum: $\theta(r) = \hat{\theta}\cos(r/D)$. Reproduced after Figure 1 in Lande (1982). $\theta = 1, D = 5, \hat{\theta} = 1$.

covariance between male trait y and female preference x is expected to be positive. Therefore, $A > 1$ always if preference is open-ended or relative (i.e., $\varepsilon = 0$). If preference is absolute (i.e., $\varepsilon = 1$), then $A > 1$ if $C > G_y$. Larger covariance C, weaker stabilizing selection (i.e., larger V_s), and stronger preference (i.e., smaller V_ψ) will all increase A and, thus, will amplify geographic variation. Significant divergence in the male trait will also imply significant reproductive isolation between spatially adjacent populations. Thus, the mechanism considered here can potentially lead to parapatric speciation.

Lande and Kirkpatrick (1988)   These authors considered a single population concentrating on two homologous characters in females and males, z and y, such as body size and a female mating preference x based on male character. All three characters are genetically correlated (e.g., because of pleiotropy or linkage disequilibrium) and the corresponding variances and covariances are maintained at constant levels by some unspecified factors. Lande and Kirkpatrick assumed that the fitness functions describing viability selection in each sex have two local peaks separated by an fitness valley as in Figure 3.3 on page 60. This assumption was intended to describe the presence of two different ecological niches potentially available for the population. Their analytical and numerical results show that female preference based on ecologically important characters can initiate a sudden shift to a new niche by males alone or by both sexes. In this model of allopatric speciation, prezygotic and postzygotic RI can evolve rapidly. The fact that RI is accompanied by ecological separation suggests that the populations occupying different niches might be able to coexist in the same area after secondary contact.

I note that the analyses of these and similar models (e.g., Iwasa and Pomiankowski 1995; Gavrilets et al. 2001) utilize certain rather strong approximations (e.g., the constancy of genetic variances and covariances and the Gaussian form of the corresponding distributions) as well as certain technical assumptions limiting the applicability of these models to the evolution of strong RI (e.g., weak selection assumption implicit in all derivations). Whether the models' conclusions remain valid under more realistic conditions remains unproven but definitely very plausible.

10.4   DISRUPTIVE SELECTION AND MODIFIERS OF MATING

In this section, I review models incorporating neutral alleles that modify the effects of other alleles controlling mating patterns. I will refer to such alleles (and loci) as the M alleles (and loci). These models belong to the class of modifier genes mechanism as specified by Maynard Smith (1966). They are also one-allele models in the terminology of Felsenstein (1981). Because M alleles are neutral, they are subject to very weak indirect selection. This implies, first, that any changes will be very slow and, second, that indirect selection can be easily overcome by very weak direct selection. Most models considered only the invasion in a randomly mating population. Whether such an invasion will be followed by other invasions making assortative mating stronger and stronger is usually not considered but implied. Some other models assumed very large initial genetic variation in the M loci and studied the sorting of this variation. Note that if there is no selection against hybrid genotypes, modifiers *reducing* assortativeness are expected to invade (Karlin and McGregor 1974).

Balkau and Feldman (1973)   Balkau and Feldman (1973) considered two diploid populations inhabiting niches I and II and exchanging migrants. A DS locus with alleles A and a controls local adaptation according to the fitness scheme (8.4) on page 237. That is, each DS allele is advantageous in one niche and disadvantageous in the other niche. As shown above, in this model the population will quickly reach a state of migration-selection balance where locally deleterious alleles are maintained by immigration. There is also an M locus that controls migration rate. Initially, all individuals have genotype MM and migrate with probability m per generation. Then mutation introduces an allele m which reduces the migration rate. For example, this may occur when the deletion of flagella reduces the mobility of a protozoan. Similarly, the strength with which hydra may be attached to substrata (and, subsequently, their motility) may be under genetic control. Balkau and Feldman showed that any modifier allele m that decreases the migration rate when in a heterozygous state will successfully invade the system, resulting in decreasing migration. Karlin and McGregor (1974) proved that if the heterozygote has the smallest migration rate, a stable polymorphism in the modifier gene can be established. Otherwise, successful invasion of m will result in eventual disappearance of M. Balkau and Feldman note that as migration rate m becomes smaller, the rate of increase of the modifying allele is expected to decrease. Therefore it becomes increasingly more difficult to stop further migration by this mechanism. Because selection for reduced migration is equivalent to selection for geographic isolation, this model can also be interpreted as a model of parapatric speciation (Karlin and McGregor 1974; Felsenstein 1981). If the process of migration reduction were to continue, the outcome would be two allopatric reproductively isolated populations. In this model, there is neither a threshold amount of selection necessary to allow divergence nor any effect of linkage between the loci.

Endler (1977) Endler (1977, Chap. 5) explicitly considered the invasion of alleles for assortative mating in a randomly mating population. Following Endler, let us assume that the presence of a dominant mutant allele M causes assortative mating with respect to a diallelic DS locus according to the O'Donald model. Individuals with resident genotype mm at the M locus mate randomly. Endler stated that allele M will spread to fixation if fitness of the heterozygotes at the DS locus is smaller than the average of the homozygote fitnesses. He also discussed more complicated models with more loci subject to selection and with spatial variation in selection. His overall conclusion was that modifier alleles reducing the production of unfit hybrids invade easily.

Sawyer and Hartl (1981) Neither Balkau and Feldman (1973) nor Endler (1977) explicitly studied the rate of increase in the frequency of modifiers. This was done by Sawyer and Hartl, who developed a stochastic model of two diploid populations (strains) with finite sizes, assuming that hybrids are sterile. Let organisms of the first strain that carry resident genotype mm in the locus under consideration mate indiscriminately. Following Sawyer and Hartl, let organisms of the first strain that carry mutant genotypes mM and MM reject an organism from the other strain given an encounter, with probabilities α_1 and α_2, respectively. After rejection, both organisms return to the mating pool and the process continues until a prespecified number of mating pairs is formed. Sawyer and Hartl showed that for small α_1, α_2, the change in the frequency p of the allele M inducing mating discrimination per generation can be approximated as

$$\Delta p = p(1-p)(1-R)\left[\alpha_1(1-2p) + \alpha_2 p\right], \qquad (10.33)$$

where R is the relative frequency of organisms of the first strain among potential mates. With small p, the right-hand side of the above equation can be approximated as $p(1-R)\alpha_1$. This implies that the modifier allele will increase in frequency slowly (because both p and α_1 are small) and that assortative mating is more likely to evolve first in the strain that is rarer (that is, the one for which $1 - R$ is largest).

Sanderson (1989) Choosiness of mating is likely to have some costs (Moore 1979; see Chapter 9). These costs were completely absent from the models developed by Endler (1977) and Sawyer and Hartl (1981). A model incorporating costs of choosiness was constructed and studied by Sanderson (1989), who considered a two-locus, two-allele population with discrete generations. The life cycle has both haploid and diploid stages. All individuals start as haploids, and these haploids then pair to produce diploids that go through viability selection. Recombination gives rise to the next generation of haploids. Viability selection acts against heterozygotes, with fitnesses being $w_{AA} = w_{aa} = 1, w_{Aa} = 1 - s$ for the three diploid genotypes at the DS locus. Organisms with allele m at the NM locus pair with other haploids randomly. Organisms with allele M form pairs assortatively according to the genotype in the DS locus. Specifically, for mating pairs that have one allele M, the frequency of $B \times b$ matings is reduced to $1 - \alpha/2$ of that expected under random mating. For mating pairs with two alleles M, the frequency of $B \times b$ matings is reduced to $1 - \alpha$ of that expected under random mating. It is assumed that allele M causes a small reduction β in fitness. Using a weak selection approximation (i.e., assuming $s, \beta \ll 1$) and assuming that the alleles at the DS locus are maintained at equal frequencies, Sanderson showed that the change in the frequency of the modifier allele M per generation is approximately

$$\Delta p = p(s\alpha/4 - \beta). \qquad (10.34)$$

Thus, this allele will increase when rare only if the fitness cost of the modifier allele is sufficiently small ($\beta < s\alpha/4$). Sanderson also used numerical simulations to study the dynamics of the modifier allele in a one-dimensional stepping-stone system.

His results indicate that the modifier increases in frequency only in the center of the hybrid zone, that reinforcement is very likely to be prevented by the swamping effects of gene flow from outside of the hybrid zone, and that even if the modifier allele does get established, resulting prezygotic RI is not strong. Sanderson also showed that alleles that increase fitness of heterozygotes can invade as well. This will decrease the gamete wastage by decreasing postzygotic RI rather than by increasing prezygotic RI as implied in the reinforcement scenario.

Kelly and Noor (1996) Endler (1977); Sawyer and Hartl (1981), and Sanderson (1989) considered similarity-based NM where assortative mating was based on a single locus equally expressed in both sexes. A model of reinforcement in which assortative mating was based on the degree of matching between genes that are differentially expressed in the sexes was studied by Kelly and Noor (1996). These authors modeled two distinct diploid populations with sizes regulated independently, assuming that a certain proportion of males in each population migrates to the other population and there attempts to mate with native females.

The model has five diallelic loci. Two DS loci control relative fertilities of hybrid males and females. Four different fitness schemes were considered, each describing strong selection against hybrids, with many hybrid genotypes having zero fitness. Two NM loci control mating behavior: one locus controlling the male trait y and another locus controlling female preference x. Both x and y take values 0, 1, or 2, depending on the number of copies of allele "+" that they contain. Females always accept a preferred male (i.e., a male with $y = x$), and accept an unpreferred male (i.e., a male with $y \neq x$) with probability $\rho < 1$. Females mate repeatedly. There was no direct selection on female mating success; that is, all females mated. The fifth M locus modifies ρ: females with 0, 1, and 2 alleles "+" at this locus are characterized by values ρ_0, ρ_1, ρ_2, respectively. There were no deleterious effects of modifier alleles on female viability. A criterion for reinforcement was the replacement of the resident alleles in the M locus by an allele decreasing ρ. The results of Kelly and Noor (1996) show that the stable coexistence of two populations and subsequent reinforcement required both a relatively high initial mating discrimination (i.e., small ρ) and a relatively small migration rate. Although these authors seemed to interpret their model as showing that reinforcement can happen more easily than in alternative models, their overall results are quite compatible with those of other authors.

Cain et al. (1999)   Cain et al. (1999) used numerical simulations to study a variant of the one-dimensional stepping-stone model of reinforcement first introduced by Sanderson (1989). The major difference between Sanderson (1989) and Cain et al. (1999) was in specifying viability selection. In the former paper there was no spatial variation in viability selection and the homozygotes always had identical fitnesses. In contrast, in the latter paper one homozygote was favored in half of the demes and the other homozygote had a larger fitness in the other half of the demes. The intention of Cain et al. was to study the plausibility of reinforcement in mosaic hybrid zones, which are characterized by abrupt reversals of gene frequencies at diagnostic markers along linear transects through the hybrid zone (Harrison 1990). Such hybrid zones have a broader region in which hybrids could be formed, potentially leading to stronger selection for assortative mating. The results of Cain et al. show that in their model, reinforcement is possible with stronger selection against the modifier allele than in Sanderson (1989). However, unless the coefficient α (which measures the strength of prezygotic RI induced by the modifier; see page 389) is close to 1, relatively weak selection (on the order of 1%) against the modifier allele is still able to prevent its spread (Cain et al. 1999, Figures 7 to 9).

Servedio (2000)   Servedio studied two modifications of the model introduced by Servedio and Kirkpatrick (1997; see page 362). In both modifications, females with allele X mate randomly. In the first, females with allele x mate preferentially with males y. In the second, allele x causes a female to mate preferentially with a male with whom she shares an identical allele at the Y locus. In our terminology, in the first version, NM is matching-based, whereas in the second version, the modifier makes mating nonrandom similarity-based. Servedio shows that overall reinforcement is most likely if migration is symmetric and NM is similarity-based. In the case of symmetric migration, reinforcement is most likely if NM is similarity-based. Reinforcement is more difficult with matching-based NM because female allele x experiences indirect selection against it, induced via its genetic association with male allele y, which is preferred by females but also is deleterious in niche 1. In the case of one-way migration, reinforcement is most likely if NM is matching-based. Reinforcement is more difficult with similarity-based NM because female alleles x are more easily associated with a deleterious male allele Y, to which they do not have preference.

Matessi et al. (2001) The papers by Endler (1977); Sawyer and Hartl (1981); Sanderson (1989); Kelly and Noor (1996), and Cain et al. (1999) studied the invasion of an allele introducing assortative mating into a randomly mating population experiencing DS. The general expectation was that if some level of assortative mating can get established, then further mutations that make assortative mating stronger and stronger would proceed. To check if this intuition is justified Matessi et al. (2001) considered a deterministic two-locus diploid model. A "magic" locus was subject to disruptive frequency-dependent selection, resulting in the maintenance of two alleles at equal frequencies, with the heterozygote having lower fitness than the two homozygotes. The same locus also controlled NM according to the Gavrilets-Boake model discussed in Chapter 9. That is, similar genotypes had a higher probability of mating than different genotypes. Each female had a limited number of attempts to mate, which induced selection for mating success in females. The degree of assortment was controlled by an M locus, where new mutations of small effects were repeatedly introduced. Matessi et al. showed analytically that assortment can always invade in a randomly mating population. However, their numerical results show that the successful invasion of an allele that makes assortment stronger does not necessarily result in the extinction of the resident allele. In fact, the population can reach a polymorphic state with different alleles for partial assortment that is stable to further invasion. Continuous increase in the strength of assortment leading to sympatric speciation occurs only if DS is sufficiently strong. These features were observed even with no costs of assortment. Introducing the latter will easily prevent the establishment of strong assortative mating in the population.

In summary, the spread of modifiers of mating can result in sympatric or parapatric speciation. However, even weak costs of assortative mating can prevent the invasion of modifier alleles. This effect may have contributed to the failure of the riparian ecotype of the desert spider *Agelenopsis aperta* to evolve mating discrimination against the arid-land ecotype in spite of strong natural selection against hybrids (Riechert 1993; Riechert et al. 2001). Mating in this system is random with regard to ecotype, with about 9% of riparian spiders mating with arid-land partners.

Finally, I consider three models that assumed very large initial genetic variation in the M loci and studied the sorting of this variation under the joint action of selection and random drift.

Dieckmann and Doebeli (1999)   The main conclusion of this paper, that "sympatric speciation is a likely outcome of competition for resources," was enthusiastically embraced by many biologists. The paper describes the results of stochastic individual-based simulations of two different models. In the first model, there is an ecological trait controlled by five diallelic loci with equal effects (0 or 1) that is subject to DS mediated via competition similar to that in the Christiansen and Loeschcke model (see Section 8.4.2). This trait is also used in preferential mating with the Gaussian preference function (9.50b) given on page 321. That is, in our terminology, this is a "magic trait" model with similarity-based NM. In the second model, preferential mating (which is still similarity-based) is based on a *marker trait* controlled by five other diallelic loci with equal effects (0 or 1). In this case, speciation requires establishing correlations (i.e., linkage disequilibria) between the two sets of loci. In both models, there is an additional *modifier of mating trait*, also controlled by five diallelic loci with equal effects (0 or 1), the value of which controls the variance V_ψ of the preference function. All loci were unlinked. In numerical simulations, Dieckmann and Doebeli (1999) enforced equal mating success for all organisms, which implies that mating was nonselective. The major conclusion of the authors is that if competition is stronger than stabilizing selection (so that genetic variation in the ecological trait is maintained), then sympatric speciation happens almost always if mating is based on the ecological trait. If mating is based on the marker trait, speciation happens if competition is sufficiently (and not necessarily extremely) strong. In both models, splitting in the ecological (and marker) trait is accompanied by the evolution of the marker trait towards the strongest assortativeness possible.

The conclusions of Dieckmann and Doebeli (1999) are based on a numerical analysis that was unavoidably superficial because of the large number of parameters (more than ten). Whether their observations hold if numerical values of parameters are modified is presently unknown. However, there are three crucial assumptions of the model itself that are questionable from a biological point of view. The first is the enforcement of equal mating success for all organisms. The second is the numerical value of the mutation rate used which was 10^{-3} per allele per generation (which is at least two orders of magnitude higher than biologistically realistic values). The third is the initial conditions used which set the frequencies of two alternative alleles at each locus to $1/2$. Each of these assumptions acts to increase dramatically the level of genetic variation. Generalizing from what is known about speciation models as described in this book, I conclude that introducing even small costs of being choosy, reducing the mutation rate to biologically realistic values, and starting the population with a low level of genetic variation (say, at a mutation-selection balance) will result in very low genetic variation in the population that will not be sufficient to initiate sympatric speciation. Whether this intuition is correct remains to be seen. In any case, as it stands now, the Dieckmann and Doebeli (1999) paper should be interpreted as demonstrating the possibility of preventing the fusion of two previously diverged populations after their secondary contact under certain hardly realistic conditions rather than sympatric speciation per se.

Kawata (2002) This paper describes stochastic individual-based simulations performed within a spatially explicit context. Organisms occupy sites in a two-dimensional lattice with randomly distributed food of different sizes. The food size distribution is bimodal with intermediate food sizes less abundant than small and large sizes. Initially organisms were adapted to small food sizes. Each diploid individual is characterzed by four additive quantitative characters, each controlled by 10 diallelic loci with equal effects (0 and 1). All loci are unlinked. One character controls the optimum size of food. Two other characters are interpreted as a male trait and a female preference (i.e., mating is matching-based). A female x can mate with male y in her spatial mating area if $|x - y| < K$, where K is the value of the fourth character interpreted as a female's mating tolerance. This is a modifier character analogous to that in the Dieckmann and Doebeli model. Kawata's study was the very thorough. He varied initial conditions, the food size distribution, dispersal distances, the size of the mating area, and mating strategies used if no suitable males existed within the mating area. The mutation rate used (10^{-4}) was more realistic than in Dieckmann and Doebeli (1999). Kawata observed four possible dynamic outcomes: speciation (i.e., the appearance of a subpopulation utilizing large food that is reproductively isolated from the initial population utilizing small food), bimodal or unimodal polymorphism with regard to the resource type used with no RI, and monomorphic population. The plausibility of speciation increased with increasing the number of initially polymorphic loci (which was between three and five in the simulations), decreased with the size of the mating area and the dispersal distance, and was affected by the resource type distribution. Speciation was never observed if mating was selective in females or if initial genetic variation was low. Overall, Kawata's results and conclusions are quite compatible with those based on simpler models. In particular, conditions for speciation are restrictive.

Doebeli and Dieckmann (2003)   This paper generalized the models built by Dieckmann and Doebeli (1999) for the case of a spatially distributed population experiencing spatially heterogeneous stabilizing selection with a linearly varying optimum (similar to that one studied in Section 8.1.4). The strength of ecological competition between any two organisms decreased both with their phenotypic divergence and with spatial distance between them. Compared to their 1999 paper, the authors did not enforce equal mating success and allowed for a weak selection for mating success (the exact strength of which is unclear). However, they still used the same mutation rate (10^{-3} per locus per generation) and initial conditions (allele frequencies at one half for all loci) as before. The overall number of parameters has grown at least to 15. First, the authors analyzed the case of an asexual population and demonstrated that when migration rate is intermediate, the population differentiates into two spatially separated and genetically different clusters. This happens even if competition is weak relative to stabilizing selection. Then they considered sexual populations using the same two models as in Dieckmann and Doebeli (1999), again concluding that formation of spatially separated and genetically distinct clusters occurs very easily. The authors interpreted their findings as an illustration of the power of ecological competition to cause branching (see Subsection 8.4.4) in spatial context.

This interpretation appears questionable. In particular, frequency-dependent selection resulting from ecological competition does not seem to be important in the formation of spatially separated clusters. Indeed, in asexual populations such clusters can be formed as a result of boundary effects (see Figure 8.11 on page 264) even without any frequency-dependent selection. In sexual populations, there is another factor at play. Recall that models (see Section 8.1.4) predict that under random mating the interaction of limited migration and spatial gradients in selection will lead to the formation of continuous clines in trait values. If the cline is steep, then the organisms with the intermediate trait values will be present only in the center of clines where they have the highest viability. If mating is similarity-based, nonrandom, and some costs of choosiness are present, then the individuals with intermediate pheno-types will have reduced mating success simply because there are few of them. In this case, selection for mating success can overcome viability selection and remove intermediate phenotypes from the population re-sulting in the formation of separate clusters. Random genetic drift will augment the effects of selection against rare individuals (Endler 1977).

Independently of interpretation, the findings of Doebeli and Dieck-mann (2003) are very intriguing and important. In particular, they re-quire a re-evaluation of results on the structure of clines when mating is nonrandom (such as given in Section 10.3.4). I note that the main implication of Doebeli and Dieckmann's results, that spatial structuring and spatial gradients in selection make speciation more plausible, are well in line with all previous work on parapatric speciation.

Overall, among the models that do not assume very high levels of ini-tial genetic variation, the most conducive conditions for sympatric and parapatric speciation are those that involve habitat choice or a "magic trait."

10.5 SUMMARY

• Both parapatric and sympatric speciation are theoretically possi-ble under certain conditions. In a number of models, conditions for sympatric speciation can be found analytically.

• RI can evolve if it increases the average fitness of the population via reduction of gamete wastage, that is, the production of low-fitness intermediates.

• One mechanism for parapatric and sympatric speciation is the dif-ferentiation in both the DS loci and the NM loci accompanied by the development of associations (i.e., linkage disequilibria) between the DS loci and the NM loci. These associations are promoted by reduced re-combination between the DS loci and the NM loci, increased strength of selection on the DS loci, increased strength of NM, the absence of costs of being choosy, high levels of genetic variation (realized via large initial variation or high mutation rates), and low migration (at least in some models). Evolution of these associations can lead to the splitting of the population into reproductively isolated clusters of genotypes.

• Some mechanisms of parapatric and sympatric speciation produce these associations automatically. They include (i) the models of habitat preference, in which organisms have genetically predetermined preference for entering a specific habitat where they subsequently experience natural selection and mate, and (ii) the "magic trait" models in which the DS loci and the NM loci are the same (or closely linked).

• If speciation (sympatric or parapatric) driven by strong selection (natural or sexual) is to occur, it typically occurs fast (on the order of a few hundred to few thousand generations).

• Sympatric speciation is theoretically possible if mating preferences are controlled by a culturally transmitted trait. This, however, requires that the copying of the parental trait is sufficiently precise.

• Sympatric and parapatric speciation can occur in a number of different models in which traits experiencing DS and underlying NM are controlled by multiple loci.

• Spatial gradients in selection coupled with non-random mating create conditions especially favorable for parapatric speciation.

• Potentially, sexual conflict is an important engine of speciation (allopatric, parapatric and sympatric). More quantitative data on the deleterious effects of high mating rates are necessary to evaluate the evolutionary importance of sexual conflict in natural populations.

• RI can be strengthened by modifier genes that cause other genes to act as a "magic trait." Invasion of modifier genes is facilitated by the fact that differentiation in the modifier genes is not required. The modifiers work on variation that is already present; they simply augment the importance of this variation in RI. However, weak direct selection against the modifiers can easily prevent their spread. Sympatric speciation via sorting of large initial variation in modifier genes is achieved more easily than via the invasion of modifier genes.

• Even moderate costs of being choosy (e.g., arising from reduced mating success if preferred mates are rare) can prevent the differentiation of the population (and, consequently, speciation) if starting with low initial variation. Selection induced by these costs is direct and, thus, is typically much stronger than the indirect selection against gamete wastage.

• In certain situations, DS coupled with NM can prevent the fusion of two populations after their secondary contact given sufficient initial divergence of the populations. Preventing the fusion of already diverged populations requires less stringent conditions than splitting a genetically uniform population into two.

• Some genetic differentiation can be established relatively easy via the mechanism of reinforcement. Evolving strong RI via this mechanism requires more stringent conditions.

10.6   CONCLUSIONS

Mathematical models have identified biologically realistic conditions under which the joint action of DS and NM can result in parapatric or sympatric speciation. Whether the traits underlying DS and NM are controlled by a few or by many loci is not crucial. Typically, conditions for sympatric speciation are much more restrictive than those for parapatric speciation. Often, sympatric speciation requires that the joint strength of DS and NM exceeds a certain threshold. The costs of being choosy are an important factor that can easily prevent sympatric speciation. Evaluating the strength of NM, the levels of genetic variation in NM, and the costs of choosiness in natural populations is an important empirical question.

CHAPTER ELEVEN

General conclusions

It is becoming more and more clear that speciation can occur in different ways.

Dobzhansky et al. (1977)

There are multiple possible answers to every aspect of speciation.

Mayr (1982)

In biology, no pattern or process will typically be considered general and important if it has been demonstrated only in a single group of organisms. In a similar way, theoretical conclusions can be viewed as general only if they emerge from a number of different mathematical models. This means that careful consideration of many models, their connections, modeling details, and general predictions is a must. This is what has been missing from speciation research. This is what I wished to accomplish with this book. Any mathematical model is necessarily based on a number of simplifying and sometimes very questionable approximations. Ideally one would like to know how robust the theoretical conclusions based on a model are under realistic biological settings. Although in speciation modeling, details matter a lot, some generalizations can be made. Various specific conclusions were given at the end of the nine previous chapters. In this final chapter, I summarize some more general observations and discuss some additional issues relevant to theoretical speciation research.

11.1 THE STRUCTURE OF FITNESS LANDSCAPES AND SPECIATION

Speciation can be visualized as the process of formation and subsequent divergence of clusters of organisms in genotype space accompanied by the evolution of RI between the emerging clusters. Fitness landscapes play a prominent role in this picture. Although combining all possible fitness components into a single fitness landscape is straightforward from a mathematical point of view, the resulting construction will be too general to be of any use. It is much more productive to think in terms of different fitness landscapes corresponding to different fitness components. As discussed in Chapter 2, RI can be characterized in terms of a fitness landscape defining viability of individuals and/or a landscape defining the probability of mating between pairs of males and females under a certain set of conditions. In a similar way, selection driving divergence of clusters can also be characterized in terms of other fitness landscapes, such as defining the degree of local adaptation.

In certain situations, details of fitness landscapes are not very important, and predicting major features of the dynamics of speciation can be achieved on the basis of general properties of landscapes. Fitness landscapes describing real biological systems have enormous dimensionality with up to millions of dimensions. An inescapable consequence of this multidimensionality is the existence of nearly neutral contiguous networks of high-fitness genotypes expanding throughout the genotype space. Another general property of fitness landscapes is the increase in RI with genetic divergence. Reproductive isolation between the great majority of pairs of organisms recognized as different species is absolute and does not depend on environment. Although any biologist can easily come up with a list of examples of hybridization between different species, these are but a tiny proportion of pairs that will never be able to produce viable and fertile offspring. Therefore, *typical* fitness landscapes underlying RI can often be treated as constant. These are the approaches used in Part 2 of this book.

In other situations, especially those that concern RI arising at low levels of genetic divergence, details of fitness landscapes matter a lot. In particular, the changes in landscapes of different types (e.g., characterizing viability or mating preferences) with an abiotic and biotic environment, where the latter includes the population itself, can be very important. Understanding the dynamics of speciation in these cases requires detailed consideration of changes in the population genetic structure characterized by genotype frequencies. These are the approaches used in Part 3 of this book.

11.2   ALLOPATRIC SPECIATION

Allopatric speciation is the most straightforward geographic mode of speciation. There are two general mechanisms by which RI leading to allopatric speciation can accumulate. First, RI can emerge as an accidental byproduct of genetic divergence caused by various evolutionary factors including drift, mutation, recombination, and natural and sexual selection. A general framework for modeling this process is provided by the classical Bateson-Dobzhansky-Muller model and its multilocus generalizations considered in Chapters 5, 6, and 7. The resulting RI can be prezygotic, postzygotic, or both pre- and postzygotic. Because *direct* selection does not affect the characters/genes underlying RI, allopatric speciation driven by this *accidental byproduct* mechanism is expected to be generally slow. I speculate that the more the habitats occupied by allopatric populations differ, the more genes will be involved in local adaptation, and the more rapidly genetic divergence will result in RI.

Second, RI can emerge as a direct response to natural or sexual selection on a "magic trait" that also controls mating. For example, if mating is assortative by size, and the optimum size differs significantly between two habitats, then adaptation of two allopatric populations to these habitats will be accompanied by the evolution of prezygotic RI. Moreover, if hybrids are intermediate in size, then their phenotypes will deviate from the optima in either niche, implying postzygotic RI in addition to prezygotic RI. Another example would be divergence driven by sexual conflict when the coevolutionary chase between females and males trying to optimize their mating rates results in the evolution of RI between isolated populations. Yet another example will be divergence driven by runaway sexual selection. Speciation by these mechanisms can be rapid.

11.3   PARAPATRIC SPECIATION

Many species can be viewed as represented by many populations weakly connected by migration rather than by completely isolated populations. The most important consequences of spatial subdivision for parapatric speciation are NM resulting from isolation by distance and spatial variation in selection regimes coming from both abiotic and biotic factors. With explicit spatial structure and gene flow, both modeling and data analysis become much more complicated and various new effects emerge. The two mechanisms of allopatric speciation, that is, the accidental-byproduct mechanism and the selection-on-a-magic-trait mechanism, also work in the parapatric case. However, because of the homogenizing effect of gene flow, these mechanisms are much less effective than in the allopatric case. In particular, parapatric speciation driven by mutation and random genetic drift becomes practically impossible if there are more than a few loci underlying RI and if migration is higher than only a few immigrants per population per generation. However, strong, spatially heterogeneous selection, for example, in the form of a spatial gradient, can lead to speciation. Besides these two mechanisms, an additional mechanism of parapatric (and sympatric) divergence is selection against low-fitness hybrids or intermediate genotypes. In this selection-against-gamete-wastage mechanism, migration plays a complex role. On the one hand, increasing migration increases the resulting gene flow, which opposes genetic divergence and the evolution of RI. On the other hand, migration and gene flow are essential for producing low-fitness hybrids and intermediate genotypes on which selection can act. As a result of these two conflicting effects, the migration rate most favorable for parapatric speciation may be intermediate. (An example supporting these predictions is provided by a study of walking-stick insects *Timema cristinae* in which female mating discrimination against males from populations adapting to alternative hosts is strongest at intermediate rates of migration (Nosil et al. 2003).) Moreover, with migration and resulting population mixing, ecological competition becomes an additional factor promoting divergence. The strength of selection resulting from competition can increase with migration, but genetic divergence becomes more difficult. Therefore, the overall effect of migration on divergence is difficult to evaluate without knowledge of specific details. Still, overall parapatric speciation is much more plausible than sympatric speciation.

I note that low interest in parapatric speciation exhibited so far by both empiricists and theoreticians is puzzling, especially given the ubiquity of spatially structured populations and species in nature. Hopefully this situation will change.

11.4 SYMPATRIC SPECIATION

> The question whether and to what extent sympatric speciation occurs among animals is one of the most controversial subjects in the field of speciation. Unfortunately, most of the discussions of this subject have been largely speculative, and, if one attempts to gather well-substantiated data, one is surprised to find how little concrete knowledge exists (Mayr 1942, p. 190).

> It is rather discouraging to read this perennial controversy because the same old arguments are cited again and again in favor of sympatric speciation, no matter how decisively they have been disproved previously (Mayr 1963, p. 450).

> One would think it should no longer be necessary to devote much time to this topic, but past experience permits one to predict with confidence that the issue will be raised again at regular intervals. Sympatric speciation is like the Lernaean Hydra which grew two new heads whenever one of its old heads was cut off (Mayr 1963, p. 451).

These three quotes illustrate Mayr's influential views on sympatric speciation as well as the lack of precise knowledge of it in spite of the huge interest in this speciation mode that biologists had in the middle of the last century. Sympatric speciation still remains an extremely popular, albeit controversial, topic. I think that it is also fair to say that the level of precise knowledge is still disproportionately small relative to the interest.

Sympatric speciation was defined in Chapter 1 as speciation occurring when mating is random with regard to the birthplace of mating partners. Not many natural systems will fit this definition, as spatial subdivision and resulting restrictions on mating are almost always present. In fact, most putative examples of sympatric speciation in the literature are treated as such based on a different definition in which *sympatric* basically means "nonallopatric" or "occurring in well-defined geographic area." Sympatric speciation in the precise sense (as used here and in most modeling work in general) represents more of a theoretical abstraction, which, however, is very useful for identifying factors promoting and opposing speciation in natural populations and in uncovering general rules and patterns of speciation.

As Chapters 9 and 10 show, sympatric speciation can be analyzed using simple mathematical models that can be treated analytically. Contrary to common claims in recent theoretical papers, conditions for sympatric speciation are not wide and sympatric speciation does not occur easily. However, as the numerous analytical and numerical models discussed in the previous chapters convincingly demonstrate, and contrary to the beliefs of Mayr and others, under certain biologistically realistic conditions, sympatric speciation is indeed theoretically possible. Generally speaking, all three mechanisms that can lead to parapatric speciation (accidental byproduct, selection on a "magic trait," and selection against gamete wastage) can result in sympatric speciation. Mathematical models corresponding to each of these mechanisms exist.

Under any mechanism, speciation is a consequence of genetic divergence. One route to sympatric speciation is the divergence in both the DS loci and the NM loci. In this case, a correlation between the two sets of loci needs to be established and maintained. Recombination is a major mechanism that opposes the establishment of such correlations (Felsenstein 1981). However, certain realistic speciation scenarios (e.g., "magic trait" or habitat choice) establish these correlations automatically. Therefore, recombination is not always that important. Another route to sympatric speciation is the divergence in only the DS loci, with subsequent spread in both diverging populations of modifier genes that transform the DS loci into a "magic trait." Under this scenario, the divergence in only the DS loci is necessary and recombination is again largely irrelevant.

Mayr's (1947) rejection of the possibility of sympatric speciation was based largely on his contention that complex traits, such as habitat preference or mate choice, were under the control of multiple loci, and thus verbal models that assumed simple genetic control were unrealistic. The assumption of simple genetic control may have been unrealistic (which remains to be seen), but it turns out that it does not really matter. The models work even if one assumes polygenic control of traits experiencing DS and underlying NM.

The factor that is always relevant in opposing speciation is selection for mating success in females (e.g., resulting from the costs of being choosy). Most models, in which sympatric speciation was shown to be possible, explicitly or implicitly assume that this selection is absent. In most models where this selection is present, sympatric speciation does not occur. Selection for mating success in females acts against deviating organisms, thus preventing them from increasing in numbers, which is the necessary first step of speciation. Selection for mating success in females appears to be a largely neglected area of evolutionary biology. The search of the database of the Institute for Scientific Information using "female mating success" as the search keywords produces only 2 papers in which these words were used in the title and 13 papers in which they were used in the title, keywords, or abstract. In contrast, the corresponding numbers using "male mating success" as the search keywords are 121 and 492 papers, respectively (as of May of 2003). The extent of this difference is puzzling. I hope that the significance of female mating success in speciation revealed by mathematical models will stimulate more empirical work on this important component of sexual selection. Especially important will be studies of the strength of this selection in natural populations comparable to those existing for natural selection (Endler 1986; Kingsolver et al. 2001).

Although mathematical models do show that sympatric speciation can occur under biologistically realistic conditions, the almost complete absence of unequivocal examples of sympatric speciation in nature (reviewed by Via 2001 and Berlocher and Feder 2002) in spite of enormous interest to find them makes one wonder whether these conditions are often satisfied in natural populations. Sympatric speciation is the area where the mathematical foundations are the most developed. Incorporation of theoretical insights into empirical approaches will be a crucial step towards resolving this controversy.

Which of the geographic modes (i.e., allopatric, parapatric or sympatric) is more conducive for *rapid* speciation depends on a delicate balance of different factors with the most important being the strength of selection and the level of migration. Occasional claims in the literature that allopatric speciation must be slower than parapatric and sympatric speciation (e.g., Bush 1975; Kondrashov and Mina 1986; McCune 1997; Via 2001) are not justified. For example, if RI results from selection on a "magic trait," then reducing the migration rate between two diverging populations to zero is expected to make speciation much easier and/or much faster. Therefore, given everything else the same, allopatric speciation by this mechanism is expected to be much faster and/or much more probable than parapatric or sympatric speciation. In general, in any of the three geographic modes, speciation driven by mutation and random genetic drift will, typically, be slow, whereas speciation driven by selection can be rapid.

11.5 SOME SPECIATION SCENARIOS AND PATTERNS

Mathematical models support the notion that a large variety of scenarios can lead to speciation within each of the three geographic modes.

Founder effect speciation and adaptive radiations Stimulated by empirical patterns of increased diversification and speciation on islands, the theory of founder effect speciation, which focuses on stochastic transitions in small founder populations, has intrigued many biologists for a long period of time. However, mathematical models show that this theory faces major theoretical difficulties. A major argument against founder effect speciation used to be that the depth of the fitness valley the population has to cross in order to speciate is strongly linked to the resulting RI. Because stochastic transitions across deep fitness valleys are improbable (see Chapter 3), the conclusion was that a single founder event is unlikely to result in a stochastic peak shift producing strong postzygotic RI. However, this objection is not really strong. If there are fitness ridges (which should be true in general as argued in Chapter 4), then the depth of the valley actually crossed by the population and the degree of resulting RI are decoupled (Gavrilets and Hastings 1996) and a single stochastic transition can trigger subsequent genetic changes resulting in strong postzygotic RI. Moreover, models exist in which prezygotic RI can evolve by random genetic drift without the need to cross any fitness valleys (Turner and Burrows 1995; Gavrilets and Boake 1998). A stronger objection to the importance of stochastic transitions following founder events lies with the small probability of having enough genetic variation in a founder population for a significant genetic change. Indeed, low genetic variation in populations is usually interpreted as evidence that they have passed through a population bottleneck. Therefore, the probability that a lot of genetic variation is present initially in a founder population is very small. For this reason, a single founder event is very unlikely to cause a significant phenotypic change or significant RI even if no fitness valley has to be crossed. A sequence of founder effects is more likely to produce significant change than a single founder event. In particular, the evolution of RI by stochastic transitions may be more likely in a metapopulation comprising a large number of subpopultions that are subject to continuous extinction and recolonization processes. However, in any case, stochastic transitions during founder event(s) appear to be less likely to lead to strong RI than other factors to be discussed next.

Colonizing a new environment often implies encountering a new selection regime and reduced mortality because of the absence of predators or competitors. A new selection regime can cause significant genetic changes (as modeled in Chapters 8 and 10). Reduced mortality can allow the population to persist longer and, thus, enable it to accumulate more genetic changes (as modeled in Chapter 7). The importance of these two effects is well recognized (e.g., Mayr 1963; Schluter 1998). There is an additional effect. The common textbook-type wisdom is that large populations are more responsive to selection than small populations. However, in natural populations, one usually observes strong positive correlations between the population density, dispersal ability, and the species range size (e.g., Gaston 1996, 1998). This implies that the most abundant species typically face the most heterogeneous environment and have the highest dispersal ability. Ohta (1972) argued that the probability that a mutant allele is advantageous on average across a range of environmental conditions experienced by a population should decrease with population size: "... the greater the population size, the greater is the habitat diversity; the greater the diversity, the smaller is between-mutant variance of selection coefficients; the smaller the variance..., the smaller is the probability that a new mutant will behave as if it were advantageous" (pp. 308–309). Ohta's conclusion was that the rate of advantageous gene substitutions should be higher in small populations, even in spite of the fact that the number of mutants arising in such populations is smaller than in large populations. Eldredge (1995, 2003) uses a similar reasoning to argue that abundant species will exhibit evolutionary stasis, and that speciation driven by selection is expected to occur after isolation of small (peripheral) populations rather than in large, widespread populations. The question of spatial heterogeneity in selection is also relevant in the context of a recent debate on Wright's shifting balance theory and its alternatives (briefly discussed in Chapter 3). Wright's theory does appear to have serious problems. However, the alternative hypothesis advanced by some researchers (e.g., Coyne et al. 1997; Coyne and Orr 1998) – simple mass selection – hinges on the questionable assumption that species experience identical selection regimes across their geographic ranges. Although some mutations are advantageous throughout the range of a species, such mutations are expected to be rare compared with those adapting a local population to the specific ecological conditions it encounters (e.g., Futuyma 1987). In summary, small populations entering specific new habitats can be more responsive to selection, which can promote explosive speciation (i.e., adaptive radiation) following a colonization event. Some support for the conclusion that small populations respond faster to selection than large populations is provided by Gavrilets and Gibson (2002), who studied the fixation probabilities in a spatially heterogeneous environment. Much more theoretical work on this is necessary.

I conclude that founding a new population in a new geographic location is indeed an event favorable for subsequent speciation and adaptive radiation. However, this is so not by virtue of stochastic rearrangements of genetic structure of populations as argued by the adherents of founder effects speciation, but because of the three deterministic factors discussed in the previous paragraph. Strong empirical support for this point of view is provided by studies of deterministic consequences of colonization of novel environments (e.g., Losos et al. 1997, 1998, 2001; Schluter 2000). For example, Losos et al. (1998) examined the evolutionary radiation of *Anolis* lizards on the four islands of the Greater Antilles. Morphometric analysis indicated that the same set of ecomorphs (i.e., habitat specialists) occurs on all four islands, whereas phylogenetic analysis indicates that the ecomorphs originated independently on each island. This shows that adaptive diversification in similar environments can overcome historical contingencies and produce very similar evolutionary outcomes. In a similar way, experimental studies, which typically impose new selection regimes on small populations under controlled conditions preventing extinction, show that selection can produce simultaneous divergence in traits underlying local adaptation and in traits underlying mating patterns (e.g., Rice and Hostert 1993).

Centrifugal speciation and peripatric speciation The question of whether speciation is most probable in central or peripheral parts of a spatially subdivided population has so far received almost no attention as far as concrete modeling is concerned. Individual-based simulations of speciation driven by mutation and drift under the assumption of the threshold function of reproductive compatibility (see Chapter 7) suggest that the answer depends on the relative strength of different factors. In particular, central subpopulations are more likely to produce new species if new genetic variation is the major limiting factor of speciation. If genetic variation is abundant and migration is the major limiting factor of speciation, then peripheral subpopulations will be more likely to split off from the rest of the species. Much more concrete modeling incorporating natural selection is necessary for a better understanding of these processes.

Punctuated equilibrium, turnover pulses, and coordinated stasis
Mathematical models of speciation strongly support the view (e.g., Eldredge 1971; Eldredge and Gould 1972; Gould 2002) that the periods of time during which new species arise are much shorter than the periods of time during which the species persist unchanged. Speciation is a rare event which requires either a simultaneous action of many weak forces (such as mutation and random genetic drift) coincidental with some additional circumstances (i.e., restrictions on migration), or an action of a very strong force (such as natural selection). In either case, the changes in the population genetic structure must be fast while favorable conditions last. Strong and nonlinear dependence of the conditions for speciation and the waiting time to speciation on model parameters whose values, in turn, depend on biotic and abiotic environmental factors suggest that speciation will in general be triggered by changes in the environment. In this case, the patterns of the turnover pulses and coordinated stasis should indeed be common, as already argued by paleontologists (e.g., Vrba 1985; Brett and Baird 1995; Brett et al. 1996; Eldredge 2003). An intriguing theoretical observation is that if speciation is driven by weak stochastic factors, then the absence (or very low abundance) of intermediate forms is expected only in the parapatric case. In the case of allopatric speciation, plenty of intermediate forms are expected to be present, as discussed on page 62.

Chromosomal speciation Speciation by chromosomal rearrangements is only plausible if heterokaryotypes experience small reduction in fitness (i.e., if the corresponding fitness valleys are very shallow). Accumulating strong postzygotic RI isolation by this scenario requires multiple fixations of rearrangements. The most favorable spatial structure for chromosomal speciation is that of many small (semi)isolated populations subject to frequent extinction and recolonization. Genetic substitutions leading to strong RI between chromosomal races can also occur *after* fixation of a weakly underdominant chromosomal rearrangement. This can happen because reduced (or absent) recombination between rearranged chromosomes promotes the accumulation of genes for RI on these chromosomes by establishing a barrier to gene flow as discussed in Chapters 5 and 6.

Reinforcement Reinforcement is the process of divergence driven by selection against gamete wastage. Reinforcement should be viewed as a special case of parapatric or sympatric speciation that corresponds to the initial conditions describing two populations that have already diverged in genes causing reduced fitness of hybrids. In a sense, the reinforcement scenario assumes that the system starts closer to a speciation equilibrium (if it exists) which, therefore, can be approached easier than if the initial level of genetic variation is low. The three general outcomes of the secondary contact are the fusion of the two populations into one, the coexistence of the two forms in a kind of a hybrid zone, and speciation/reinforcement. Fusion is promoted when mating is selective and when spatial heterogeneity in selection is not present during the secondary contact. Coexistence is promoted if migration is low and spatial heterogeneity in selection is present during the secondary contact, or if the sizes of the two hybridizing populations are independently regulated. Reinforcement is promoted by stronger selection against hybrids, by the absence of costs of being choosy, and by higher levels of initial divergence. Some of the argument on whether reinforcement occurs under broad or narrow conditions stems from the differences in a way to measure its success. Reinforcement in the sense of the establishment of some NM or additional genetic divergence between hybridizing populations occurs under much broader conditions than reinforcement in the sense of the evolution of strong RI.

Competitive speciation   In the scenario originally proposed by Rosenzweig (1978) and Pimm (1979), the population is first driven by directional selection to a point in phenotype space corresponding to the maximum resource utilization. Once the population is there, selection becomes disruptive as a result of stronger competition. If selection is very strong, it can split the population into two (or more) distinct clusters utilizing different resources. If hybrids have reduced fitness (e.g., if they suffer from competition with both parental forms), then prezygotic RI can evolve as a response to ecological selection against gamete wastage in a way similar to that in the reinforcement scenario. This inherently sympatric scenario does not require the crossing of fitness valleys, generates multimodal distributions in phenotype space, and provides a mechanism for ecological differentiation. High population densities and strong competition promote competitive speciation. A number of different models support the plausibility of different components of this scenario as well as the whole scenario. Although currently empirical support does not appear to be strong, one should expect competitive speciation to occur in certain situations.

Speciation along environmental gradients This noncontroversial scenario of parapatric speciation has received significant attention from theoreticians, starting with classical works by Endler (1977), Caisse and Antonovics (1978), Moore (1981), and Lande (1982). Spatial heterogeneity in selection is ubiquitous, selection gradients are common, and restrictions on migration are almost always present. All these factors are very powerful in creating conditions for the maintenance of significant genetic variation (even under weak selection) and for subsequent evolution of prezygotic RI by both the selection-on-a-magic trait mechanism and the selection-against-gamete-wastage mechanism.

The hybrid zones (i.e., the areas where hybrids between two diverged groups are formed) represent one of the intermediate steps of this process. Although numerous hybrid zones are well documented, their existence cannot be considered as unambiguous evidence for parapatric speciation because they are also expected to be formed after secondary contact of two groups partially diverged in allopatry (Endler 1977). Still, from theoretical considerations, speciation via this scenario is expected to be very common.

Speciation via host shift This scenario is usually cast in terms of sympatric speciation. However, because both spatial separation of hosts is essential and host fidelity is strong in putative examples (e.g., *Rhagoletis*), speciation via host shift should probably be viewed as parapatric. If both mating and viability selection occur on the preferred host, the need to overcome recombination to establish correlations between the DS loci and the NM loci is removed. This implies that speciation according to this scenario can proceed more easily than in alternative models. Still, the evolution of strong RI requires strong DS and substantial NM from the beginning. As with all other types of speciation, NM that does not involve the costs of being choosy is the most favorable kind of mating for speciation via host shift. In particular, this implies that the pairs of species emerging by this mechanism most likely will not have inherent genetic incompatibilities affecting the probabilities of mating given an encounter, or of successful fertilization given that mating has taken place.

Speciation by runaway sexual selection and by sexual conflict Sexual selection can lead to the divergence of allopatric populations either by random genetic drift and mutation along the corresponding lines (or planes) or equilibria, or by Fisher's runaway process in which the male trait is driven to extreme values by female preferences. In the former case, genetic divergence is expected to be slow, whereas in the latter case it can be rapid. A less likely outcome is transient genetic divergence and convergence of allopatric populations as predicted by some models of sexual selection exhibiting cycling dynamics (Iwasa and Pomiankowski 1995; Gavrilets et al. 2001). In the parapatric case, sexual selection can significantly augment the divergence induced by selection gradients (Lande 1982). In the sympatric case, sexual selection is expected to induce more clustering in male traits (or genes) than in females traits (or genes). However, natural viability selection acting on male traits and female preferences can easily prevent diversification and speciation.

Strong sexual conflict is a very powerful engine of speciation. Under sexual conflict, both natural and sexual selection interact in a way that under certain conditions allows for rapid divergence of allopatric populations or for the sympatric formation of genetically separate and reproductively isolated clusters within the population. Increasing population density is expected to intensify sexual conflict. Thus, the periods when local population densities are high will be the most conducive for sympatric speciation. Therefore, geographic areas of high productivity and, subsequently, of high population densities may be a major source of new species. Subsequent ecological differentiation or some spatial segregation are required for stable coexistence of these species.

11.6   GENERAL RULES OF EVOLUTIONARY DIVERSIFICATION

Both the existing empirical data and theory show that genetic, ecological, and environmental details have profound effects on the dynamics of speciation and diversification and that no universal rules or patterns of speciation exist. At the same time, one should expect that some relatively common trends or tendencies of evolutionary diversification in related groups of organisms can be identified. For example, in birds of the New Guinea mountains the diversification with respect to habitat elevation preceded diversification with respect to other characteristics (Diamond 1986; see Schluter 2000 for discussion of similar data). In some fishes, birds, and lizards a common sequence of events in evolutionary diversification is divergence, first, in habitat, then in food type, and, finally, in mating signals (Streelman and Danley 2003). What does mathematical theory have to say about such trends and tendencies?

I will differentiate between *macrohabitats* (i.e., general environmental conditions existing in a geographic location) and *microhabitats* (i.e., different ecological niches). I start with three observations. First, selective forces acting throughout the microhabitats existing at a given geographic location should typically be more important than those acting only within specific microhabitats. Second, selective forces acting earlier in the life cycle should typically be more important than those acting later on. Third, before adults can engage in competition for food or for mates, they have to be sufficiently adapted to their macrohabitats. Therefore, given that everything else is the same, adaptation to the macrohabitat is expected to precede adaptation to other factors. In terms of speciation, this suggests that divergence with respect to macrohabitat will precede divergence with respect to microhabitat, food type, foraging method, or traits involved in mating.

Sympatric coexistence of closely related species requires both NM and some ecological divergence (e.g., Mayr 1963). Independently of the geographic mode of speciation, the easiest way to achieve both of these is via (i) NM that does not involve costs of being choosy and (ii) a mechanism that simplifies the establishment and maintenance of associations between traits underlying NM and ecological separation. Although both the habitat choice mechanism and the magic trait mechanism are effective with regard to the latter requirement, the costs of being choosy can be avoided more easily with the habitat choice mechanism. Therefore, given that everything else is the same, divergence in traits underlying microhabitat choice will precede divergence with respect to magic traits and other traits controlling survival and reproduction. Note that the evolution of habitat choice is easier for organisms of relatively limited vagility, whereas divergence of organisms of larger vagility should be more dependent on "magic traits." The latter can be involved either in adaptation to microhabitat or in competition for a limited resource.

Diversification with regard to traits involved exclusively in mating and fertilization (and not affected by other selective forces) would require the generation and maintenance of large levels of corresponding genetic variation. This can be most easily achieved in large, relatively stable, spatially distributed systems of populations between which only a limited amount of gene flow is occurring.

Overall, the argument given above suggests the following sequence of diversification events: (i) divergence with respect to macrohabitat, (ii) evolution of microhabitat choice and divergence with respect to microhabitat, (iii) divergence with respect to "magic traits," and (iv) divergence with respect to other traits controlling survival and reproduction. Of course, each of these stages is possible only if both the corresponding selective forces and the necessary genetic variation are present. Different exceptions are expected to exist as well. For example, if competition is very important at early stages of the life cycle (e.g., at the larval stage), then diversification with respect to traits mediating competition is expected to precede divergence with regard to microhabitat.

One common pattern of diversification in flowering plants was first identified by Grant (1949) and then reformulated by Schluter (2000) in the following way: Differentiation in floral traits precedes vegetative divergence in lineages pollinated by specialized animals, but reverse sequence holds in lineages pollinated by unspecialized animals or by abiotic agents (wind, water). This pattern can be explained as follows. If divergence and speciation are to occur in parapatric or sympatric spatial setup, the joint strength of DS and NM have to be sufficiently large. If NM mating is strong, then weak DS might be sufficient for speciation, whereas if NM is weak, then strong DS selection is necessary. Specialized animal pollinators induce stronger NM. As a consequence, weaker DS can be sufficient for speciation and less vegetative change will be caused by weak selection. In contrast, nonspecialized pollinating agents produce only weak NM, and thus stronger DS is necessary. Such selection will cause larger vegetative change.

11.7 WHY SPECIES?

Once one has a good understanding of *how* new species can be formed, one can attempt to answer the question of *why* species are formed. In other words, why are living beings clustered into distinct groups with typically no intermediates present rather than being spread in genotype space more or less continuously? This theoretically important question has received relatively little attention (e.g., see discussions by Dobzhansky et al. 1977; Maynard Smith and Szathmáry 1995; Coyne and Orr 1998; Avise 2000; Turelli et al. 2001; Barraclough et al. 2003).

Assume first that any spatial structure and heterogeneity in selection is completely absent. Obviously, not all possible combinations of genes can code for organisms that can successfully develop. If successful combinations of genes (i.e., developmental niches) are distinct and very different from each other, species occupying different developmental niches will be discrete as well. However, the major point of Chapter 4 is that high-fitness combinations of genes are expected to form continuous ridges extending throughout genotype space. If so, why are these ridges not occupied by organisms more or less continuously? Although there are not enough organisms to represent all possible genotypes, theoretically it is possible to have a swarm of genotypes in a certain area of genotype space with no cluster structure.

In asexual species, clusters are expected to arise naturally because of the common ancestry, even if evolution occurs on a flat fitness landscape. Indeed, different organisms that can be traced to a common ancestor will necessarily cluster together in genotype space (see Box 2.3 on page 52). If evolution occurs on a holey fitness landscape, the cluster structure is maintained but with less fragmentation (see Section 4.7.2 on page 112).

Next let us consider a sexual species evolving on a holey fitness landscape. Even if any restrictions on possible matings are absent, offspring of different high-fitness genotypes may have low fitness. In this case, organisms are expected to cluster into groups minimizing the production of low-fitness offspring. However, restrictions on possible matings are always present. Therefore, one should expect organisms to form groups whose members have high fitness, are mutually compatible as far as mating is concerned, and whose offspring have high fitness as well. In other words, discrete cluster formation is a way to increase fitness components such as individual fitness, fitness of mating pairs, or offspring fitness. In this sense, clustering (i.e., the formation of discrete species) is "adaptive". In principle, one can also imagine a situation where different reproductively isolated genotypes are connected by a chain of mutually compatible genotypes, as is observed in a number of known ring species (see discussion on page 99). Mathematical models however indicate that such a sexual continuum is very unstable to fluctuations in the population genetic structure (e.g., due to random genetic drift) and will tend to break into separate clusters (Noest 1997; Gavrilets et al. 1998). Therefore, clustering is also the *most likely* adaptive outcome of evolutionary dynamics.

The factors discussed so far work even in a highly idealized situation when any spatial structure or heterogeneity in selection is absent. The existence of discrete ecological niches and the spatial heterogeneity in selection will further promote cluster formation. Moreover, geographic and/or sexual isolation will result in different clusters (i.e., species) becoming evolutionarily independent and able to explore fitness landscapes independently. At this stage, the complexity of fitness landscapes will imply almost definite divergence of emerging clusters along fitness landscapes, which will further increase fragmentation of the overall pool of genotypes. I conclude that discrete species formation is an unavoidable consequence of a multitude of biotic and abiotic factors.

11.8   SOME OPEN THEORETICAL QUESTIONS

Forty years of theoretical research have laid a solid foundation for a comprehensive theory of speciation. Still, a number of important areas have not received the attention they deserve; Hopefully these oversights will be corrected in the near future.

Analytical models of speciation discussed in this book necessarily concentrate only on a small number of factors affecting the process of speciation. Bringing many different factors into a single analytical model will not be possible because of the mathematical difficulties involved. However, detailed models can be studied using numerical simulations. Unfortunately, there have been only a few attempts to build realistic models of speciation, and those in existence have not been thoroughly examined. In evolutionary biology, comprehensive studies of a few *model organisms* have been very successful in identifying and understanding general evolutionary mechanisms and principles. In a similar way, comprehensive numerical studies of a few *model models* of speciation will greatly benefit theoretical research. Ideally, these would be models adapted to describe specific biological organisms for which there is a wealth of detailed information, for example, on genetics of RI, selection regimes acting, and spatial structure.

Spatial subdivision of biological populations is ubiquitous and can have profound effects on different aspects of the evolutionary process, including speciation. Explicitly incorporating spatial structure in mathematical models results in the need to overcome a great deal of mathematical difficulties. This is an area of speciation research that would greatly benefit from both analytical and numerical studies.

Another area that has been almost completely neglected by theoreticians is the modeling of speciation and diversification at larger temporal and spatial scales than those covered by the models considered in this book. For example, adaptive radiation is a process of great interest for biologists, and the empirical data on repeatable patterns in adaptive radiation keep growing. Developing models of adaptive radiation that explicitly account for the existence of multiple niches, spatially heterogeneous selection, and extinction and recolonization dynamics will provide a wealth of useful information.

Building models of diversification at higher taxonomic levels studied by paleontologists, and linking microevolutionary processes with macroevolutionary patterns is another important direction.

Finally, developing statistical methods for testing different hypotheses about speciation on the basis of empirical data will be crucial. Some very promising steps in this directions have been made recently using genomic data (e.g., Kliman et al. 2000; Machado et al. 2002, Osada and Wu, personal communication).

I hope this book will stimulate some serious theoretical work in these directions.

11.9   FINAL THOUGHTS

In the 150 years after the publication of Darwin's most important book, speciation research has seen a steady growth of empirical facts and observations, ideas, and verbal theories. Both bringing genetics into speciation studies by Dobzhansky (1937) and the thorough analysis of existing data and logic underlying possible scenarios of speciation by Mayr (1942) have started fruitful traditions in speciation research which by now have resulted in an impressive body of knowledge on different aspects of speciation. Inevitably, as in any other emerging science, this growth of knowledge has been accompanied by the occasional spread of confusion and numerous controversies, for example, about the plausibility and generality of sympatric speciation and reinforcement, the importance of central versus peripheral populations, the role of random genetic drift in speciation, and the effects of genetic architecture on the plausibility of speciation.

The development of an adequate quantitative theory based on mathematical models is a necessary step in the growth of most scientific disciplines. A simple reason for this is that intuition, common sense, and verbal reasoning are not working well and are unreliable when one deals with complex nonlinear processes influenced by myriads of interacting factors. Although verbal reasoning might work in some situations, such as in dealing with the electorate, its utility in science is always questionable. The dynamics of speciation is an extremely complicated process which cannot be understood just by verbal models and generalizations about data. Mathematical models are the best tool to clarify verbal arguments, conclusions, and controversies; classify key factors and components; identify timescales of the corresponding dynamics; and answer many other important questions.

The "Modern Synthesis" of the 1930s and 1940s became possible only after the foundations of theoretical population genetics were built by Fisher, Haldane, and Wright. However, theoretical studies of speciation did not become a part of these foundations. Evolutionary biology is concerned with two major processes: (i) adaptation and changes over time within a given population, and (ii) the origin and multiplication of species. The problems of adaptation and evolutionary changes within populations have received a great deal of attention both during the Modern Synthesis and in subsequent years. The second great evolutionary problem – the multiplication of species, or the origin of biodiversity – was largely left outside of the mathematical framework of population genetics. This contributed to the feeling of some biologists that the synthesis was not finished (e.g., White 1978; Eldredge 1985; Carroll 2000; and Mayr's introduction to the 1999 edition of his *Systematics and the Origin of Species*).

Although mathematical research on speciation still lags behind empirical studies, the past 40 years have seen remarkable advances. In a very influential book published in 1974 Lewontin was right to point that "we have no quantitative theory of speciation at all" (p. 28). As this book aims to show, a theoretical component of speciation research is emerging now that is becoming comparable in its breadth and depth to the mathematical frameworks developed for studying changes within populations. I believe that we really do now have a comprehensive and general mathematical theory of speciation. This is very encouraging and shows that we are making significant progress towards achieving the synthesis started in the first half of the last century.

References

Abramowitz, M. and Stegan, I. A. 1965. *Handbook of mathematical functions*. Dover Publications, New York.

Abrams, P. A. 2001. Modelling the adaptive dynamics of traits involved in inter- and intraspecific interactions: An assessment of three methods. *Ecology Letters* 4:166–175.

Aguade, M., Miyashita, N., and Langley, C. H. 1992. Polymorphism and divergence of the mst 355 male accessory gland region. *Genetics* 132:755–770.

Ajtai, M., Komlós, J., and Szemerédi, E. 1982. Largest random component of a k-cube. *Combinatorica* 2:1–7.

Alipaz, J. A., Wu, C.-I., and Karr, T. L. 2001. Gametic incompatibilities between races of *Drosophila melanogaster*. *Proceedings of the Royal Society London B* 268:789–795.

Allmon, W. 1992. A causal analysis of stages in allopatric speciation. *Oxford Survey in Evolutionary Biology* 8:219–257.

Allmon, W. D., Morris, P. J., and McKinney, M. L. 1998. An intermediate disturbance hypothesis of maximal speciation. In M. L. McKinney and J. A. Drake, editors, *Biodiversity dynamics*, pages 349–376. Columbia University Press, New York.

Altenberg, L. 1991. Chaos from linear frequency-dependent selection. *American Naturalist* 138:51–68.

Amitrano, C., Peliti, L., and Saber, M. 1989. Population dynamics in a spin-glass model of chemical evolution. *Journal of Molecular Evolution* 29:513–525.

Andersson, M. B. 1994. *Sexual selection*. Princeton University Press, Princeton.

Arnold, M. L. 1997. *Natural hybridization and evolution*. Oxford University Press, Oxford.

Arnold, M. L., Hamrick, J. L., and Bennett, B. D. 1993. Interspecific pollen competition and reproductive isolation in *Iris*. *Journal of Heredity* 84:13–16.

Arnold, S. J., Pfrender, M. E., and Jones, A. G. 2001. The adaptive landscape as a conceptual bridge between micro- and macroevolution. *Genetica* 112:9–32.

Arnold, S. J. and Wade, M. J. 1984a. On the measurement of natural and sexual selection. I. Theory. *Evolution* 38:709–719.

—. 1984b. On the measurement of natural and sexual selection. II. Applications. *Evolution* 38:720–734.

Arnqvist, G. 1998. Comparative evidence for the evolution of genitalia by sexual selection. *Nature* 393:784–786.

Arnqvist, G., Edvardsson, M., Friberg, U., and Nilsson, T. 2000. Sexual conflict promotes speciation in insects. *Proceedings of the National Academy of Sciences USA* 97:10460–10464.

Arnqvist, G. and Rowe, L. 1995. Sexual conflict and arms races between the sexes: a morphological adaptation for control of mating in a female insect. *Proceedings of the Royal Society London B* 261:123–127.

Asmussen, M. A. 1983. Density-dependent selection incorporating intraspecific competition. *Genetics* 103:335–350.

Asmussen, M. A. and Basnayake, E. 1990. Frequency-dependent selection: the high potential for permanent genetic variation in the diallelic, pairwise interaction model. *Genetics* 125:215–230.

Avise, J. C. 2000. *Phylogeography*. Harvard University Press, Cambridge, MA.

Babajide, A., Farber, R., Hofacker, I. L., Inman, J., Lapedes, A. S., and Stadler, P. F. 2001. Exploring protein sequence space using knowledge-based potentials. *Journal of Theoretical Biology* 212:35–46.

Babajide, A., Hofacker, I. L., Sippl, M. J., and Stadler, P. F. 1997. Neutral networks in protein space: a computational study based on knowledge-based potentials of mean force. *Folding and Design* 2:261–269.

Bak, P. 1996. *How nature works*. Copernicus, New York.

Bakker, T. C. M. and Pomiankowski, A. 1995. The genetic basis of female mate preference. *Journal of Evolutionary Biology* 8:129–171.

Balkau, B. J. and Feldman, M. W. 1973. Selection for migration modification. *Genetics* 74:171–174.

Barbash, D. A., Siino, D. F., Tarone, A. M., and Roote, J. 2003. A rapidly evolving MYB-related protein causes species isolation in *Drosophila*. *Proceedings of the National Academy of Sciences USA* 100:5302–5307.

Barnett, L. 1997. Tangled webs: evolutionary dynamics on fitness landscapes with neutrality. Master's thesis, University of Sussex.

Barraclough, T. G., Birky, C. W., and Burt, A. 2003. Diversification in sexual and asexual organisms. *Evolution* 57:2166–2172.

Barraclough, T. G., Harvey, P. H., and Nee, S. 1995. Sexual selection and taxonomic diversity in passerine birds. *Proceedings of the Royal Society London B* 259:211–215.

Barraclough, T. G. and Vogler, A. P. 2000. Detecting the geographic pattern of speciation from species-level phylogenies. *American Naturalist* 43:419–434.

Barton, N. H. 1979. Gene flow past a cline. *Heredity* 43:333–339.

—. 1989a. The divergence of a polygenic system subject to stabilizing selection, mutation and drift. *Genetical Research* 54:59–77.

—. 1989b. Founder effect speciation. In D. Otte and J. A. Endler, editors, *Speciation and its consequences*, pages 229–256. Sinauer, Sunderland, MA.

—. 1992. On the spread of a new gene combination in the third phase of Wright's shifting-balance. *Evolution* 46:551–557.

—. 1993. The probability of fixation of a favoured allele in a subdivided population. *Genetical Research* 62:149–157.

—. 1999. Clines in polygenic traits. *Genetical Research* 74:223–236.

Barton, N. H. and Bengtsson, B. O. 1986. The barrier to genetic exchange between hybridizing populations. *Heredity* 56:357–376.

Barton, N. H. and Charlesworth, B. 1984. Genetic revolutions, founder effects, and speciation. *Annual Review of Ecology and Systematics* 15:133–164.

Barton, N. H. and Gale, K. 1993. Genetic analysis of hybrid zones. In R. G. Harrison, editor, *Hybrid zones and the evolutionary process*, pages 13–45. Oxford University Press, New York.

Barton, N. H. and Rouhani, S. 1987. The frequency of shifts between alternative states. *Journal of Theoretical Biology* 125:397–418.

—. 1991. The probability of fixation of a new karyotype in a continuous population. *Genetical Research* 45:499–517.

—. 1993. Adaptation and the shifting balance. *Genetical Research* 61:57–74.

Barton, N. H. and Shpak, M. 2000. The stability of symmetric solutions to polygenic models. *Theoretical Population Biology* 57:249–263.

Barton, N. H. and Turelli, M. 1987. Adaptive landscapes, genetic distance and the evolution of quantitative characters. *Genetical Research* 49:157–173.

—. 1991. Natural and sexual selection on many loci. *Genetics* 127:229–255.

Bastolla, U., Roman, H. E., and Vendruscolo, M. 1999. Neutral evolution of model proteins: diffusion in sequence space and overdispersion. *Journal of Theoretical Biology* 200:49–64.

Bateson, W. 1909. Heredity and variation in modern lights. In A. C. Seward, editor, *Darwin and modern science*, pages 85–101. Cambridge University Press, Cambridge.

Bazykin, A. D. 1965. On the possibility of sympatric species formation (in Russian). *Byulleten' Moskovskogo Obshchestva Ispytateley Prirody. Otdel Biologicheskiy* 70:161–165.

—. 1969. Hypothetical mechanism of speciation. *Evolution* 23:685–687.

Beiles, A., Heth, G., and Nevo, E. 1984. Origin and evolution of assortative mating in actively speciating mole rats. *Theoretical Population Biology* 26:265–270.

Bell, G. 1997. *Selection. The mechanism of evolution.* Chapman and Hall, New York.

Bengtsson, B. O. 1985. The flow of genes through a genetic barrier. In J. J. Greenwood, P. H. Harvey, and M. Slatkin, editors, *Evolution Essays in Honor of John Maynard Smith*, pages 31–42. Cambridge University Press, Cambridge.

Bengtsson, B. O. and Christiansen, F. B. 1983. A two-locus mutation selection model and some of its evolutionary implications. *Theoretical Population Biology* 24:59–77.

Benkman, C. W. 1996. Are the ratios of bill crossing morphs in crossbills the result of frequency-dependent selection? *Evolutionary Ecology* 108:119–126.

Bennett, J. H. 1954. On the theory of random mating. *Annals of Eugenics* 18:311–317.

Bergstrom, C. T. and Real, L. A. 2000. Towards a theory of mutual mate choice: lessons from two-sided matching. *Evolutionary Ecology Research* 2:493–508.

Berlocher, S. H. and Feder, J. L. 2002. Sympatric speciation in phytophagous insects: Moving beyond controversy? *Annual Review of Entomology* 47:773–815.

Birky, C. and Walsh, J. 1988. Effects of linkage on rates of molecular evolution. *Proceedings of the National Academy of Sciences USA* 85:6414–6418.

Bodmer, W. F. 1965. Differential fertility in population genetics models. *Genetics* 51:411–424.

Bollobas, B. 2001. *Random graphs*. Cambridge University Press, New York.

Bolnick, D. and Doebeli, M. 2003. Sexual dimorphism and adaptive speciation: Two sides of the same ecological coin. *Evolution* 57:2433–2449.

Bornberd-Bauer, E. 1997. How are model protein structures distributed in sequences space? *Biophysical Journal* 73:2393–2403.

—. 2002. Randomness, structural uniqueness, modularity and neutral evolution in sequence space of model proteins. *Zeitschrift für Physikalische Chemie -International Journal of Research in Physical Chemistry and Chemical Physics* 216:139–154.

Bornberd-Bauer, E. and Chan, H. S. 1999. Modeling evolutionary landscapes: mutational stability, topology, and superfunnels in sequence space. *Proceedings of the National Academy of Sciences USA* 96:10689–10694.

Bowen, B. W., Bass, A. L., Rocha, L. A., Grant, W. S., and Robertson, D. R. 2001. Phylogeography of the trumpetfishes (Aulostomus): ring species complex on a global scale. *Evolution* 55:1029–1039.

Bradshaw, H. D., Wilbert, S. M., Otto, K. G., and Schemske, D. W. 1995. Genetic mapping of floral traits associated with reproductive isolation in monkeyflowers (*Mimulus*). *Nature* 376:762–765.

Bramson, M., Cox, J. T., and Durrett, R. 1996. Spatial models for species area curves. *Annals of Probability* 24:1727–1751.

Brett, C. E. and Baird, G. 1995. Coordinated stasis and evolutionary ecology of Silurian to Middle Devonian faunas in the Appalachian Basin. In R. Anstey and D. H. Erwin, editors, *Speciation in the fossil record*, pages 285–315. Columbia University Press, New York.

Brett, C. E., Ivany, L., and Schopf, K. 1996. Coordinated stasis: an overview. *Palaeogeography, Palaeoclimatology, Palaeoecology* 127:1–20.

Briggs, J. C. 1999a. Coincident biogeographic patterns: Indo-West Pacific Ocean. *Evolution* 53:326–335.

—. 1999b. Extinction and replacement in the Indo-West Pacific Ocean. *Journal of Biogeography* 26:777–783.

Bro, A. 1995. *The Oxford Companion to Philosophy*. Oxford University Press, Oxford.

Brooks, D. R. and McLennan, D. A. 1991. *Phylogeny, ecology, and behavior:*

A research program in comparative biology. University of Chicago Press, Chicago.

Brown, J. L. J. 1957. Centrifugal speciation. *Quarterly Review of Biology* 32:247–277.

Brown, J. L. J. and Wilson, E. O. 1956. Character displacement. *Systematic Zoology* 5:49–64.

Brown, J. M., LeebensMack, J. H., Thompson, J. N., Pellmyr, O., and Harrison, R. G. 1997. Phylogeography and host association in a pollinating seed parasite *Greya politella* (Lepidoptera: Prodoxidae). *Molecular Ecology* 6:215–224.

Bullini, L. 1994. Origin and evolution of animal hybrid species. *Trends in Ecology and Evolution* 9:422–426.

Bulmer, M. G. 1974. Density-dependent selection and character displacement. *American Naturalist* 108:45–58.

—. 1980. *Mathematical theory of quantitative genetics.* Oxford University Press, New York.

Burch, C. L. and Chao, L. 1999. Evolution by small steps and rugged landscapes in the RNA virus phi 6. *Genetics* 151:921–927.

Bürger, R. 2000. *The mathematical theory of selection, recombination, and mutation.* Wiley, Chichester.

—. 2002a. Additive genetic variation under intraspecific competition and stabilizing selection: a two-locus study. *Theoretical Population Biology* 61:197–213.

—. 2002b. On a genetic model of intraspecific competition and stabilizing selection. *American Naturalist* 160:661–682.

Bürger, R. and Gimelfarb, A. 2002. Fluctuating environments and the role of mutation in maintaining quantitative genetic variation. *Genetical Research* 80:31–46.

Burtin, Y. D. 1977. On the probability of connectedness of a random subgraph of the n-cube. *Problemy peredachi informatsii* 13:90–95.

Bush, G. 1969. Sympatric host race formation and speciation in frugivorous flies of the genus *Rhagoletis* (Diptera, Tephritidae). *Evolution* 23:237–251.

—. 1975. Modes of animal speciation. *Annual Review of Ecology and Systematics* 6:339–364.

Butlin, R. 1987. Speciation by reinforcement. *Trends in Ecology and Evolution* 2:8–13.

Butlin, R. K. 1995. Reinforcement: an idea evolving. *Trends in Ecology and Evolution* 10:432–434.

Cabot, E. L., Davis, A. W., Johnson, N. A., and Wu, C.-I. 1994. Genetics of reproductive isolation in the *Drosophila simulans* clade: complex epistasis underlying hybrid male sterility. *Genetics* 137:175–189.

Cain, M. L., Andreasen, V., and Howard, D. J. 1999. Reinforcing selection is effective under a relatively broad set of conditions in a mosaic hybrid zone. *Evolution* 53:1343–1353.

Caisse, M. and Antonovics, J. 1978. Evolution of reproductive isolation in clinal populations. *Heredity* 40:371–384.

Campbell, I. A. 1985. Random-walks on a closed-loop and spin-glass relaxation. *Journal de Physique Lettres* 46:1159–1162.

Campbell, I. A., Flesselles, J. M., Jullien, R., and Botet, R. 1987. Random walks on a hypercube and spin glass relaxation. *Journal of Physics. C. Solid State Physics* 20:L47–L51.

Campos, P. R. A., Adami, C., and Wilke, C. O. 2002. Optimal adaptive performance and delocalization in NK fitness landscapes. *Physica A* 304:495–506.

Carroll, R. L. 1988. *Vertebrate paleontology and evolution*. Freeman, New York.

—. 2000. Towards a new evolutionary synthesis. *Trends in Ecology and Evolution* 15:27–32.

Carson, H. L. 1968. The population flush and its genetic consequences. In R. C. Lewontin, editor, *Population biology and evolution*, pages 123–137. Syracuse University Press, Syracuse, NY.

—. 1982. Evolution of *Drosophila* on the newer Hawaiian volcanos. *Heredity* 48:3–25.

—. 2002. Female choice in *Drosophila*: evidence from Hawaii and implications for evolutionary biology. *Genetica* 116:383–393.

Carson, H. L. and Clague, D. A. 1995. Geology and biogeography of the Hawaiian islands. In W. L. Wagner and V. A. Funk, editors, *Hawaiian biogeography: Evolution on a hot spot archipelago*, pages 14–29. Smithsonian Institution Press, Washington, DC.

Carson, H. L. and Templeton, A. R. 1984. Genetic revolutions in relation to speciation phenomena: the founding of new populations. *Annual Review of Ecology and Systematics* 15:97–131.

Cavalli-Sforza, L. L. and Feldman, M. W. 1981. *Cultural transmission and evolution: a quantitative approach*. Princeton University Press, Princeton, NJ.

Chapman, T. and Partridge, L. 1996. Female fitness in *Drosophila melanogaster*: an interaction between the effect of nutrition and of encounter rate with males. *Proceedings of the Royal Society London B* 263:755–759.

Charlesworth, B. 1990. Speciation. In D. E. G. Briggs and P. R. Crowther, editors, *Paleobiology. A synthesis*, pages 100–106. Blackwell Scientific Publications, Oxford.

—. 1997. Is founder-flush speciation defensible? *American Naturalist* 149:600–603.

Charlesworth, B. and Rouhani, S. 1988. The probability of peak shifts in a founder population. II. An additive polygenic trait. *Evolution* 42:1129–1145.

Charlesworth, B. and Smith, D. B. 1982. A computer model for speciation by founder effect. *Genetical Research* 39:227–236.

Cherry, J. L. 2003. Selection in a subdivided population with dominance of local frequency dependence. *Genetics* 163:1511–1518.

Chesser, R. T. and Zink, R. M. 1994. Modes of speciation in birds: a test of Lynch's method. *Evolution* 48:490–497.

Chown, S. L. 1997. Speciation and rarity: separating cause from consequence. In W. E. Kunin and K. J. Gaston, editors, *The biology of rarity*, pages 91–109. Chapman & Hall, London.

Christensen, K., de Collobiano, S. A., Hall, M., and Jensen, H. J. 2002. Tangled nature: A model of evolutionary ecology. *Journal of Theoretical Biology* 216:73–84.

Christiansen, F. B. 1988. Frequency dependence and competition. *Philosophical Transactions of the Royal Society London B* 319:587–600.

—. 2000. *Population genetics of multiple loci*. Wiley, Chichester, NY.

Christiansen, F. B. and Frydenberg, O. 1977. Selection-mutation balance for two nonallelic recessive producing an inferior double homozygote. *American Journal of Human Genetics* 29:195–207.

Christiansen, F. B. and Loeschcke, V. 1980. Evolution of intraspecific exploitative competition. I. One-locus theory for small additive gene effects. *Theoretical Population Biology* 18:297–313.

Church, S. A. P. and Taylor, D. R. 2002. The evolution of reproductive isolation in spatially structured populations. *Evolution* 56:1859–1862.

Claridge, M. F., Dawah, H. A., and Wilson, M. R. 1997. *Species: The units of biodiversity*. Chapman & Hall, London.

Clarke, B. C. 1979. The evolution of genetic diversity. *Proceedings of the Royal Society London B* 205:453–474.

Cockerham, C. C., Burrows, P. M., Young, S. S., and Prout, T. 1972. Frequency-dependent selection in randomly mating population. *American Naturalist* 106:493–515.

Cohan, F. M. 1995. Does recombination constrain neutral divergence among bacterial taxa? *Evolution* 49:164–175.

Cook, J. M. and Rasplus, J.-Y. 2003. Mutualists with attitude: coevolving fig wasps and figs. *Trends in Ecology and Evolution* 18:241–248.

Cooper, V. S. and Lenski, R. E. 2000. The population genetics of ecological specialization in evolving *Escherichia coli* populations. *Nature* 407:736–739.

Coyne, J. A. 1984. Genetic basis of male sterility in hybrids between two closely related species of *Drosophila*. *Proceedings of the National Academy of Sciences USA* 81:4444–4447.

—. 1992. Genetics and speciation. *Nature* 355:511–515.

Coyne, J. A., Barton, N. H., and Turelli, M. 1997. A critique of Sewall Wright's shifting balance theory of evolution. *Evolution* 51:643–671.

—. 2000. Is Wright's shifting balance process important in evolution? *Evolution* 54:306–317.

Coyne, J. A. and Orr, H. A. 1989. Patterns of speciation in *Drosophila*. *Evolution* 43:362–381.

—. 1997. "Patterns of speciation in *Drosophila*" revisited. *Evolution* 51:295–303.

—. 1998. The evolutionary genetics of speciation. *Philosophical Transactions of the Royal Society London B* 353:287–305.

Crosby, J. L. 1970. The evolution of genetic discontinuity: computer models of the selection of barriers to interbreeding between subspecies. *Heredity* 25:253–297.

Crow, J. F., Engels, W. R., and Denniston, C. 1990. Phase three of Wright's shifting-balance theory. *Evolution* 44:233–247.

Crow, J. F. and Felsenstein, J. 1968. The effect of assortative mating on the genetic composition of a population. *Eugenics Quarterly* 15:85–97.

Crow, J. F. and Kimura, M. 1970. *An introduction to population genetic theory.* Harper and Row, New York.

Darwin, C. 1859. *The origin of species by means of natural selection, or the preservation of favoured races in the struggle for life.* Modern Library, New York.

Datta, A., Hebdrix, M., Lipsitch, M., and Jinks-Robertson, S. 1997. Dual role for DNA sequence identity and the mismatch repair system in the regulation of mitotic crossing-over in yeast. *Proceedings of the National Academy of Sciences USA* 94:9757–9762.

Davis, A. and Wu, C.-I. 1996. The broom of the sorcerer's apprentice: the fine structure of a chromosomal region causing reproductive isolation between two sibling species of *Drosophila. Genetics* 143:1287–1298.

de Almeida, R. M. C., Lemke, N., and Campbell, I. A. 2000. Stretched exponential relaxation on the hypercube and fractal phase space. *European Physical Journal* 18:513–518.

—. 2001. Stretched exponential relaxation on the hypercube and the glass transition. *Journal of Magnetism and Magnetic Materials* 226-230:1296–1297.

de Queiroz, K. 1998. The general lineage concept of species, species criteria, and the process of speciation: a conceptual unification and terminological recommendations. In D. J. Howard and S. H. Berlocher, editors, *Endless forms: Species and speciation*, pages 57–75. Oxford University Press, New York.

Dempster, E. R. 1955. Maintenance of genetic heterogeneity. *Cold Spring Harbor Symposia on Quantitative Biology* 20:25–32.

Demuth, J. P. and Rieseberg, L. H. 2001. Ring species and speciation. In *Encyclopedia of life sciences.* Macmillan, Oxford.

Derrida, B. and Peliti, L. 1991. Evolution in flat landscapes. *Bulletin of Mathematical Biology* 53:255–282.

DeSalle, R. 1995. Molecular approaches to biogeographic analysis of Hawaiian Drosophilidae. In W. L. Wagner and V. A. Funk, editors, *Hawaiian biogeography: Evolution on a hot spot archipelago*, pages 72–89. Smithsonian Institution Press, Washington, DC.

Diaconis, P., Graham, R. L., and Morrison, J. A. 1990. Asymptotic analysis of a random walk on a hypercube with many dimensions. *Random Structures and Algorithms* 1:51–72.

Diamond, J. 1986. Evolution of ecological segregation in the New Guinea mountane avifauna. In J. Diamond and T. J. Case, editors, *Community ecology*, pages 98–125. Harper and Row, Publishers, New York.

Dickinson, H. and Antonovics, J. 1973. Theoretical considerations of sympatric divergence. *American Naturalist* 107:256–274.

Dieckmann, U. 1997. Can adaptive dynamics invade. *Trends in Ecology and Evolution* 12:128–131.

Dieckmann, U. and Doebeli, M. 1999. On the origin of species by sympatric speciation. *Nature* 400:354–357.

Diehl, S. R. and Bush, G. L. 1989. The role of habitat preference in adaptation and speciation. In D. Otte and J. A. Endler, editors, *Speciation and its consequences*, pages 345–365. Sinauer, Sunderland, MA.

Dobzhansky, T., Ayala, F. J., Stebbins, G. L., and Valentine, J. W. 1977. *Evolution*. W.H. Freeman, San Francisco.

Dobzhansky, T. G. 1936. Studies on hybrid sterility. II. Localization of sterility factors in *Drosophila Pseudoobscura* hybrids. *Genetics* 21:113–135.

—. 1937. *Genetics and the origin of species*. Columbia University Press, New York.

—. 1940. Speciation as a stage in evolutionary divergence. *American Naturalist* 74:312–321.

Doebeli, M. 1996. A quantitative genetic competition model for sympatric speciation. *Journal of Evolutionary Biology* 9:893–909.

Doebeli, M. and Dieckmann, U. 2003. Speciation along environmental gradient. *Nature* 421:259–264.

Doi, M., Matsuda, M., Tomaru, M., Matsubayashi, H., and Oguma, Y. 2001. A locus for female discrimination behavior causing sexual isolation in Drosophila. *Proceedings of the National Academy of Sciences USA* 98:6714–6719.

Drossel, B. and McKane, A. 2000. Competitive speciation in quantitative genetic models. *Journal of Theoretical Biology* 204:467–478.

Duncan, K. E., Ferguson, N., and Istock, C. A. 1995. Fitness of a conjugative plasmid and its host bacteria in soil microcosms. *Molecular Biology and Evolution* 12:1012–1021.

Durrett, R. and Levin, S. 1996. Spatial models for species-area curves. *Journal of Theoretical Biology* 179:119–127.

Edmands, S. 1999. Heterosis and outbreeding depression in interpopulation crosses spanning a wide range of divergence. *Evolution* 53:1757–1768.

—. 2002. Does parental divergence predict reproductive compatibility? *Trends in Ecology and Evolution* 17:520–527.

Ehrlich, P. R. and Wilson, E. O. 1991. Biodiversity studies: science and policy. *Science* 253:758–762.

Eigen, M., McCaskill, J., and Schuster, P. 1989. The molecular quasispecies. *Advances in Chemical Physics* 75:149–263.

Eldredge, N. 1971. The allopatric model and phylogeny in Paleozoic invertebrates. *Evolution* 25:156–167.

—. 1985. *Unfinished synthesis. Biological hierarchies and modern evolutionary thought*. Oxford University Press, New York.

—. 1995. *Reinventing Darwin. The great debate at the high table of evolutionary theory*. Wiley, New York.

—. 2003. The sloshing bucket: how the physical realm controls evolution. In J. Crutchfield and P. Schuster, editors, *Towards a comprehensive dynamics of evolution: Exploring the interplay of selection, neutrality, accident, and function*, pages 3–32. Oxford University Press, New York.

Eldredge, N. and Gould, S. J. 1972. Punctuated equilibria: an alternative to phyletic gradualism. In T. J. Schopf, editor, *Models in paleobiology*, pages 82–115. Freeman, Cooper, San Francisco.

Endler, J. A. 1977. *Geographic variation, speciation and clines*. Princeton University Press, Princeton.

—. 1978. A predator's view of animal color patterns. *Evolutionary Biology* 11:319–364.

—. 1980. Natural selection on color patterns in *Poecilia reticulata*. *Evolution* 34:76–91.

—. 1986. *Natural selection in the wild*. Princeton University Press, Princeton.

—. 1988. Frequency-dependent predation, crypsis, and aposematic coloration. *Philosophical Transactions of the Royal Society London B* 319:505–523.

Engels, F. 1940. *Dialectics of nature*. International Publishers, New York.

Erdös, P. and Spencer, J. 1979. Evolution of the n-cube. *Computers and Math. with Applications* 5:33–40.

Erwin, D. H. 1994. Early introduction of major morphological innovations. *Acta Palaentol. Polonica* 38:281–294.

Eshel, I. 1983. Evolutionary and continuous stability. *Journal of Theoretical Biology* 103:99–111.

—. 1997. Continuous stability and evolutionary convergence. *Journal of Theoretical Biology* 185:333–343.

Everitt, B. S. 1993. *Cluster analysis*. Arnold, London.

Ewens, W. J. 1979. *Mathematical population genetics*. Springer-Verlag, Berlin.

Fear, K. K. and Price, T. 1998. The adaptive surface in ecology. *Oikos* 82:440–448.

Feder, J. L. 1998. The apple maggot fly, *Rhagoletis pomonella*: flies in the face of conventional wisdom about speciation? In D. J. Howard and S. H. Berlocher, editors, *Endless forms: Species and speciation*, pages 130–144. Oxford University Press, New York.

Feder, J. L., Opp, S. B., Wlazlo, B., Reynolds, K., and Go, W. 1994. Host fidelity is an effective premating barrier between sympatric races of the apple maggot fly. *Proceedings of the National Academy of Sciences USA* 91:7990–7994.

Feldman, M. W., Franklin, I., and Thomson, G. J. 1974. Selection in complex genetic systems. I. The symmetric equilibria of the three-locus symmetric viability model. *Genetics* 76:135–162.

Feldman, M. W. and Liberman, U. 1984. A symmetric two-locus fertility model.

Genetics 109:229–253.

Feller, W. 1957. *An introduction to probability theory and its application. Vol. I.* Wiley, New York.

Felsenstein, J. 1976. The theoretical population genetics of variable selection and migration. *Annual Review of Genetics* 10:253–280.

——. 1981. Skepticism towards Santa Rosalia, or why are there so few kinds of animals? *Evolution* 35:124–138.

Filchak, K. E., Roethele, J. B., and Feder, J. L. 2000. Natural selection and sympatric divergene in the apple maggot fly *Rhagoletis pomonella*. *Nature* 407:739–742.

Finjord, J. 1996. Sex and self-organization on rugged landscapes. *International Journal of Modern Physics C* 7:705–715.

Finkel, S. E. and Kolter, R. 1999. Evolution of microbial diversity during prolonged starvation. *Proceedings of the National Academy of Sciences USA* 96:4023–4027.

Fisher, R. A. 1930. *The genetical theory of natural selection.* Oxford University Press, Oxford.

——. 1937. The wave of advance of advantageous genes. *Annals of Eugenics* 7:355–369.

——. 1950. Gene frequencies in a cline determined by selection and diffusion. *Biometrics* 6:353–361.

Fishman, L. and Willis, J. H. 2001. Evidence for Dobzhansky-Muller incompatibilities contributing to the sterility of hybrids between *Mimulus guttatus* and *M. nasutus*. *Evolution* 55:1932–1942.

Florin, A.-B. and Ödeen, A. 2002. Laboratory environments are not conducive for allopatric speciation. *Journal of Evolutionary Biology* 15:10–19.

Fontana, W. and Schuster, P. 1998a. Computer model of evolutionary optimization. *Biophysical Chemistry* 26:123–147.

——. 1998b. Continuity in evolution: on the nature of transitions. *Science* 280:1451–1455.

Fontana, W., Stadler, P. F., Tarazona, P., Weinberger, E. O., and Schuster, P. 1993. RNA folding and combinatory landscapes. *Physical Review E* 47:2083–2099.

Foote, M. 1992. Paleozoic record of morphological diversity in blastozoan echinoderms. *Proceedings of the National Academy of Sciences USA* 89:7325–7329.

——. 1999. Morphological diversity in the evolutionary radiation of Paleozoic and post-Paleozoic crinoids. *Paleobiology* 25:1–115 Suppl.

Franke, E. S., Styan, C. A., and Babcock, R. 2002. Sexual conflict and polyspermy under sperm-limited conditions: in situ evidence from field simulations with the free-spawning marine echinoid *Evechinus chloroticus*. *American Naturalist* 160:485–496.

Freeberg, T. M. 1998. The cultural transmission of courtship patterns in cowbirds, *Molothrus ater*. *Animal Behavior* 56:1063–1073.

Freidlin, M. I. and Wentzell, A. D. 1998. *Random perturbations of dynamical systems*. Springer-Verlag, New York.

Friesen, V. L., Montevecchi, W. A., Baker, A. J., Barrett, R. T., and Davidson, W. 1996. Population differentiation and evolution in the common guillemot Uria aalge. *Molecular Ecology* 5:793–805.

Fry, J. D. 2003. Multilocus models of sympatric speciation: Bush versus Rice versus Felsenstein. *Evolution* 57:1735–1746.

Fryer, G. 2001. On the age and origin of the species flock of haplochromine cichlid fishes of Lake Victoria. *Proceedings of the Royal Society London B* 268:1147–1152.

Futuyma, D. J. 1987. On the role of species in anagenesis. *American Naturalist* 130:465–473.

——. 1998. *Evolutionary biology*. Sinauer, Sunderlands, MA.

Futuyma, D. J. and Moreno, G. 1988. The evolution of ecological specialization. *Annual Review of Ecology and Systematics* 19:207–233.

Gage, M. J. G., Parker, G. A., Nylin, S., and Wiklund, C. 2002. Sexual selection and speciation in mammals, butterflies and spiders. *Proceedings of the Royal Society London B* 269:2309–2316.

Gantmacher, F. R. 1959. *Applications of the theory of matrices*. Interscience, New York.

García-Pelayo, R. and Stadler, P. F. 1997. Correlation length, isotropy and meta-stable states. *Physica D* 107:240–254.

Gardiner, C. W. 1983. *Handbook of stochastic methods*. Springer-Verlag, Berlin.

Gaston, K. J. 1996. Species-range-size distributions: patterns, mechanisms and implications. *Trends in Ecology and Evolution* 11:197–201.

——. 1998. Species-range size distributions: products of speciation, extinction and transformation. *Philosophical Transactions of the Royal Society London B* 353:219–230.

Gaston, K. J., Blackburn, T. M., and Gregory, R. D. 1997. Abundance-range size relationships of breeding and wintering birds in Britain: a comparative analysis. *Ecography* 20:569–579.

Gavrilets, S. 1996. On phase three of the shifting-balance theory. *Evolution* 50:1034–1041.

——. 1997a. Coevolutionary chase in exploiter-victim systems with polygenic characters. *Journal of Theoretical Biology* 186:527–534.

——. 1997b. Evolution and speciation on holey adaptive landscapes. *Trends in Ecology and Evolution* 12:307–312.

——. 1997c. Hybrid zones with epistatic selection of Dobzhansky type. *Evolution* 51:1027–1035.

——. 1997d. Single locus clines. *Evolution* 51:979–983.

——. 1998. One-locus two-allele models with maternal (parental) selection. *Genetics* 149:1147–1152.

——. 1999a. A dynamical theory of speciation on holey adaptive landscapes. *American Naturalist* 154:1–22.

——. 1999b. Dynamics of clade diversification on the morphological hypercube. *Proceedings of the Royal Society London B* 266:817–824.

—. 2000a. Rapid evolution of reproductive isolation driven by sexual conflict. *Nature* 403:886–889.

—. 2000b. Waiting time to parapatric speciation. *Proceedings of the Royal Society London B* 267:2483–2492.

—. 2003a. Evolution and speciation in a hyperspace: the roles of neutrality, selection, mutation and random drift. In J. Crutchfield and P. Schuster, editors, *Towards a comprehensive dynamics of evolution: Exploring the interplay of selection, neutrality, accident, and function*, pages 135–162. Oxford University Press, New York.

—. 2003b. Models of speciation: what have we learned in 40 years? *Evolution* 57:2197–2215.

—. 2004. Speciation in metapopulations. In I. Hanski and O. Gaggiotti, editors, *Ecology, genetics and evolution of metapopulations*, pages 275–303. Elsevier, Amsterdam.

Gavrilets, S., Acton, R., and Gravner, J. 2000a. Dynamics of speciation and diversification in a metapopulation. *Evolution* 54:1493–1501.

Gavrilets, S., Arnqvist, G., and Friberg, U. 2001. The evolution of female mate choice by sexual conflict. *Proceedings of the Royal Society London B* 268:531–539.

Gavrilets, S. and Boake, C. R. B. 1998. On the evolution of premating isolation after a founder event. *American Naturalist* 152:706–716.

Gavrilets, S. and Cruzan, M. B. 1998. Neutral gene flow across single locus clines. *Evolution* 52:1277–1284.

Gavrilets, S. and Gibson, N. 2002. Fixation probabilities in a spatially heterogeneous environment. *Population Ecology* 44:51–58.

Gavrilets, S. and Gravner, J. 1997. Percolation on the fitness hypercube and the evolution of reproductive isolation. *Journal of Theoretical Biology* 184:51–64.

Gavrilets, S. and Hastings, A. 1994. A quantitative genetic model for selection on developmental noise. *Evolution* 48:1478–1486.

—. 1995a. Dynamics of polygenic variability under stabilizing selection, recombination, and drift. *Genetical Research* 65:63–74.

—. 1995b. Intermittency and transient chaos from simple frequency-dependent selection. *Proceedings of the Royal Society London B* 261:233–238.

—. 1996. Founder effect speciation: a theoretical reassessment. *American Naturalist* 147:466–491.

—. 1998. Coevolutionary chase in two-species systems with applications to mimicry. *Journal of Theoretical Biology* 191:415–427.

Gavrilets, S., Li, H., and Vose, M. D. 1998. Rapid parapatric speciation on holey adaptive landscapes. *Proceedings of the Royal Society London B* 265:1483–1489.

—. 2000b. Patterns of parapatric speciation. *Evolution* 54:1126–1134.

Gavrilets, S. and Waxman, D. 2002. Sympatric speciation by sexual conflict. *Proceedings of the National Academy of Sciences USA* 99:10533–10538.

Geritz, S. A. H., Gyllenberg, M., Jacobs, F. J. A., and Parvinen, K. 2002.

Invasion dynamics and attractor inheritance. *Journal of Mathematical Biology* 44:548–560.

Geritz, S. A. H. and Kisdi, E. 2000. Adaptive dynamics in diploid, sexual populations and the evolution of reproductive isolation. *Proceedings of the Royal Society London B* 267:1671–1678.

Geritz, S. A. H., Kisdi, E., Meszéna, G., and Metz, J. A. J. 1998. Evolutionary singular strategies and the adaptive growth and branching of the evolutionary tree. *Evolutionary Ecology* 12:35–57.

Geritz, S. A. H., Metz, J. A. J., Kisdi, E., and Meszéna, G. 1997. Dynamics of adaptation and evolutionary branching. *Physical Review Letters* 78:2024–2027.

Gibbons, J. R. H. 1979. A model for sympatric speciation in *Megarhyssa* (Hymenoptera: Ichneumonidae): competitive speciation. *American Naturalist* 114:719–741.

Gigord, L. D. B., Macnair, M. R., and Smithson, A. 2001. Negative frequency-dependent selection maintains a dramatic flower color polymorphism in the rewardless orchid *Dactylorhiza sambucina* (L.) Soo. *Proceedings of the National Academy of Sciences USA* 98:6253–6255.

Gillespie, J. H. 1991. *The causes of molecular evolution*. Oxford University Press, Oxford.

Gleason, J. M. and Ritchie, M. G. 1998. Evolution of courtship song and reproductive isolation in the *Drosophila willistoni* complex: do sexual signals diverge the most quickly? *Evolution* 52:1493–1500.

Glendinning, P. 1994. *Stability, instability, and chaos*. Cambridge University Press, Cambridge.

Gonzalez, B. M. 1923. Experimental studies on the duration of life. VIII. The influence upon duration of life of certain mutant genes of *Drosophila melanogaster*. *American Naturalist* 57:289–325.

Gordon, R. 1994. Evolution escapes rugged landscapes by gene or genome doubling: the blessing of higher dimensionality. *Computers and Chemistry* 18:325–441.

Gould, S. J. 2002. *The structure of evolutionary theory*. Harvard University Press, Boston.

Govindarajan, S. and Goldstein, R. A. 1997. The foldability landscape of model proteins. *Biopolymers* 42:427–438.

Gradshteyn, I. S. and Ryzhik, I. M. 1994. *Tables of integrals, series, and products*. Academic Press, San Diego.

Grant, B. R. and Grant, P. R. 1996. Cultural inheritance of song and its role in the evolution of Darwin's finches. *Evolution* 50:2471–2487.

—. 1998. Hybridization and speciation in Darwin's finches: the role of sexual imprinting on a culturally transmitted trait. In D. J. Howard and S. H. Berlocher, editors, *Endless forms: Species and speciation*, pages 404–422. Oxford University Press, New York.

Grant, P. R. 1999. *Ecology and evolution of Darwin's finches*. Princeton University Press, Princeton.

Grant, P. R. and Grant, B. R. 1997. Genetics and the origin of bird species. *Proceedings of the National Academy of Sciences USA* 94:7768–7775.

Grant, V. 1949. Pollinating system as isolating mechanisms in angiosperms. *Evolution* 3:82–97.

—. 1971. *Plant speciation*. Columbia University Press, New York.

—. 1991. *The evolutionary process: A critical study of evolutionary theory*. Columbia University Press, New York.

Grasman, J. and van Herwaarden, O. 1999. *Asymptotic methods for the Fokker-Plank equation and exit problem in applications*. Springer-Verlag, Berlin.

Greene, E., Lyon, B. E., Muehter, V. R., Ratcliffe, L., Oliver, S. J., and Boag, P. T. 2000. Disruptive sexual selection for plumage coloration in a passerine bird. *Nature* 407:1000–1003.

Griffiths, A. J. F., Miller, J. H., Suzuki, D. T., Lewontin, R. C., and Gelbart, W. M. 1996. *An introduction to genetic analysis*. W. H. Freeman, New York.

Grimmett, G. 1989. *Percolation*. Springer-Verlag, Berlin.

Grüner, W., Giegerich, R., Strothmann, D., Reidys, C., Weber, J., Hofacker, I. L., Stadler, P. F., and Schuster, P. 1996a. Analysis of RNA sequence structure maps by exhaustive enumeration: neutral networks. *Monatshefte fur Chemie* 127:355–374.

—. 1996b. Structure of neutral networks and shape space covering. *Monatshefte fur Chemie* 127:375–389.

Hadeler, K. P. and Liberman, U. 1975. Selection models with fertility differences. *Journal of Mathematical Biology* 2:19–32.

Haldane, J. B. S. 1922. Sex ratio and unisexual sterility in animal hybrids. *Journal of Genetics* 12:101–109.

—. 1927. A mathematical theory of natural and artificial selection. V. Selection and mutation. *Proceedings of the Cambridge Philosophical Society* 23:838–844.

—. 1931. A mathematical theory of natural and artificial selection. VIII. Metastable populations. *Proceedings of the Cambridge Philosophical Society* 27:137–142.

—. 1948. The theory of a cline. *Journal of Genetics* 48:277–284.

—. 1959. Natural selection. In P. R. Bell, editor, *Darwin's biological work: Some aspects reconsidered*, pages 101–149. Wiley, New York.

Haldane, J. B. S. and Jayakar, S. D. 1965. Selection for a single pair of allelomorphs with complete replacement. *Journal of Genetics* 59:171–177.

Hammond, P. M. 1995. The current magnitude of biodiversity. In V. H. Heywood, editor, *Global Biodiversity Assessment*, pages 113–138. Cambridge University Press, Cambridge.

Hamrick, J. L. and Godt, M. J. W. 1996. Effects of life history traits on genetic diversity in plant species. *Philosophical Transactions of the Royal Society London B* 351:1291–1298.

Harris, T. E. 1964. *The theory of branching processes*. Rand Corporation, Santa Monica, CA.

Harrison, R. G. 1990. Hybrid zones: windows on evolutionary process. *Oxford Surveys on Evolutionary Biology* 7:69–128.

—. 1991. Molecular changes at speciation. *Annual Review of Ecology and Systematics* 22:281–308.

—. 1998. Linking evolutionary pattern and process: the relevance of species concepts for the study of evolution. In D. J. Howard and S. H. Berlocher, editors, *Endless forms: Species and speciation*, pages 19–31. Oxford University Press, New York.

Hartl, D. L. and Clark, A. G. 1997. *Principles of population genetics*. Sinauer Associates, Sunderland, MA.

Hartl, D. L., Dykhuizen, D. E., and Dean, A. M. 1985. Limits of adaptation: the evolution of selective neutrality. *Genetics* 111:655–674.

Hartl, D. L. and Taubes, C. H. 1996. Compensatory nearly neutral mutations: selection without adaptation. *Journal of Theoretical Biology* 182:303–309.

Harvey, I. and Thompson, A. 1997. Through the labyrinth evolution finds a way: a silicon ridge. *Lecture Notes in Computer Science* 1259:406–422.

Hastings, A. 1981. Stable cycling in discrete time genetic models. *Proceedings of the National Academy of Sciences USA* 78:7224–7225.

—. 1989. Deterministic multilocus population genetics: an overview. In A. Hastings, editor, *Some mathematical questions in biology: Models in population biology. Lectures on mathematics in the life sciences. Vol. 20*, pages 27–54. American Mathematical Society, Providence, RI.

Hawthorn, D. J. and Via, S. 2001. Genetic linkage of ecological specialization and reproductive isolation in pea aphids. *Nature* 412:904–907.

Hedrick, P. W. 1986. Genetic polymorphism in heterogeneous environments: a decade later. *Annual Review of Ecology and Systematics* 17:536–566.

Hedrick, P. W., Givenan, M. E., and Ewing, E. P. 1976. Genetic polymorphism in heterogeneous environments. *Annual Review of Ecology and Systematics* 7:1–32.

Hegel, G. W. F. 1975. *Hegel's Logic: Being part one of the encyclopaedia of the philosophical sciences (1830)*. Clarendon Press, Oxford.

Henry, C. S. 1985. Sibling species, call differences, and speciation in green lacewings (Neuroptera: Chrysopidae: *Chrysoperla*). *Evolution* 39:965–984.

Higashi, M., Takimoto, G., and Yamamura, N. 1999. Sympatric speciation by sexual selection. *Nature* 402:523–526.

Higgs, P. G. 1994. Error thresholds and stationary mutant distributions in multilocus diploid genetics models. *Genetical Research* 63:63–78.

—. 1998. Compensatory neutral mutations and the evolution of RNA. *Genetica* 102/103:91–101.

Higgs, P. G. and Derrida, B. 1991. Stochastic models for species formation in evolving populations. *Journal of Physics A: Mathematical and General,* 24:L985–L991.

—. 1992. Genetic distance and species formation in evolving populations. *Journal of Molecular Evolution* 35:454–465.

Hill, W. and Caballero, A. 1992. Artificial selection experiments. *Annual Review of Ecology and Systematics* 23:287–310.

Hoekstra, H., Hoekstra, J. M., Berrigan, D., Vignieri, S. N., Hoang, A., Hill, C. E., Beerli, P., and Kingsolver, J. G. 2001. Strength and tempo of directional selection in the wild. *Proceedings of the National Academy of Sciences USA* 98:9157–9160.

Hofbauer, J. and Sigmund, K. 1988. *The theory of evolution and dynamical systems*. Cambridge University Press, Cambridge.

—. 1998. *Evolutionary games and population dynamics*. Cambridge University Press, Cambridge.

Holland, B. and Rice, W. R. 1998. Chase-away sexual selection: antagonistic seduction versus resistance. *Evolution* 52:1–7.

Holmes, M. H. 1995. *Introduction to perturbation methods*. Springer-Verlag, New York.

Hori, M. 1993. Frequency-dependent natural selection in the handedness of scale-eating cichlid fish. *Science* 260:216–219.

Hostert, E. E. 1997. Reinforcement: a new perspective on an old controversy. *Evolution* 51:697–702.

Howard, D. J. 1993. Reinforcement: origin, dynamics, and fate of an evolutionary hypothesis. In R. G. Harrison, editor, *Hybrid zones and the evolutionary process*, pages 46–69. Oxford University Press, New York.

—. 1999. Conspecific sperm and pollen precedence and speciation. *Annual Review of Ecology and Systematics* 30:109–132.

Howard, D. J. and Berlocher, S. H. 1998. *Endless forms: Species and speciation*. Oxford University Press, New York.

Howard, D. J., Reece, M., Gregory, P. G., Chu, J., and Cain, M. L. 1998. The evolution of barriers to fertilization between closely related organisms. In D. J. Howard and S. H. Berlocher, editors, *Endless forms: Species and speciation*, pages 279–288. Oxford University Press, New York.

Hubbell, S. P. 2001. *The unified neutral theory of biodiversity and biogeography*. Princeton University Press, Princeton.

Huber, B. A. and González, A. P. 2001. Female genital dimorphism in a spider (Araneae: Pholcidae). *Journal of Zoology* 255:301–304.

Huges, B. D. 1996. *Random walks and random environments. Volume 2. Random environments*. Clarendon Press, Oxford.

Huynen, M. A. 1996. Exploring phenotype space through neutral evolution. *Journal of Molecular Evolution* 43:165–169.

Huynen, M. A. and Hogeweg, P. 1994. Pattern generation in molecular evolution: Exploitation of the variation in RNA landscapes. *Journal of Molecular Evolution* 39:71–79.

Huynen, M. A., Stadler, P. F., and Fontana, W. 1996. Smoothness within ruggedness: the role of neutrality in adaptation. *Proceedings of the National Academy of Sciences USA* 93:397–401.

Ihara, Y., Aoki, K., and Feldman, M. W. 2003. Runaway sexual selection with paternal transmission of the male trait and gene-culture determination of the female preference. *Theoretical Population Biology* 63:53–62.

Ihara, Y. and Feldman, M. W. 2003. Evolution of disassortative and assortative mating preference based on imprinting. *Theoretical Population Biology* 64:193–200.

Iizuka, M. and Takefu, M. 1996. Average time until fixation of mutants with compensatory fitness interactions. *Genes and Genetic Systems* 71:167–173.

Iliadi, K., Iliadi, N., Rashkovetsky, E., Minkov, I., Nevo, E., and Korol, A. 2001. Sexual and reproductive behavior of *Drosophila melanogaster* from a microclimatically interslope differentiated population of "Evolution Canyon" (Mount Carmel, Israel). *Proceedings of the Royal Society London B* 268:2365–2374.

Innan, H. and Stephan, W. 2001. Selection intensity against deleterious mutations in RNA secondary structures and rate of compensatory nucleotide substitutions. *Genetics* 159:389–399.

Irwin, D. E., Irwin, J. H., and Price, T. 2001. Ring species as bridges between microevolution and speciation. *Genetica* 112:223–243.

Iwasa, I. and Pomiankowski, A. 1995. Continual change in mate preferences. *Nature* 377:420–422.

Iwasa, I., Pomiankowski, A., and Nee, S. 1991. The evolution of costly mate preferences. II. The "handicap" principle. *Evolution* 45:1431–1442.

Jackson, J. B. C. and Cheetham, A. H. 1999. Tempo and mode of speciation in the sea. *Trends in Ecology and Evolution* 14:72–77.

Johannesson, K., Rolan-Alvarez, E., and Ekendahl, A. 1995. Incipient reproductive isolation between two sympatric morphs of the intertidal snail *Littorina saxatilis*. *Evolution* 49:1180–1190.

Johnson, N. A. and Porter, A. H. 2000. Rapid speciation via parallel, directional selection on regulatory genetic pathways. *Journal of Theoretical Biology* 205:527–542.

Johnson, P. A., Hoppensteadt, F. C., Smith, J. J., and Bush, G. L. 1996. Conditions for sympatric speciation: a diploid model incorporating habitat fidelity and non-habitat assortative mating. *Evolutionary Ecology* 10:187–205.

Jones, A. G., Moore, G. I., Kvarnemo, C., Walker, D., and Avise, J. C. 2003. Sympatric speciation as a consequence of male pregnancy in seahorses. *Proceedings of the National Academy of Sciences USA* 100:6598–6603.

Jordan, D. S. 1905. The origin of species through isolation. *Science* 22:545–562.

Kacser, H. and Burnes, J. A. 1973. The control of flux. *Symp. Soc. Exp. Biol.* 32:65–104.

—. 1981. The molecular basis of dominance. *Genetics* 97:639–666.

Kaneko, K. and Yomo, T. 2000. Sympatric speciation: compliance with phenotype diversification from a single genotype. *Proceedings of the Royal Society London B* 267:2367–2373.

Kaneshiro, K. Y. 1980. Sexual isolation, speciation and the direction of evolution. *Evolution* 34:437–444.

Karlin, S. 1982. Classification of selection-migration structures and conditions for a protected polymorphism. *Evolutionary Biology* 14:61–204.

Karlin, S. and Feldman, M. W. 1970. Linkage and selection: two-locus symmetric viability model. *Theoretical Population Biology* 1:39–71.

Karlin, S. and McGregor, J. 1972. Application of method of small parameters in multi-niche population genetics models. *Theoretical Population Biology* 3:180–209.

—. 1974. Towards a theory of the evolution of modifier genes. *Theoretical Population Biology* 5:59–103.

Karlin, S. and Taylor, H. M. 1975. *A first course in stochastic processes.* Academic Press, San Diego.

Kauffman, S. A. 1993. *The origins of order.* Oxford University Press, Oxford.

Kauffman, S. A. and Levin, S. 1987. Towards a general theory of adaptive walks on rugged landscapes. *Journal of Theoretical Biology* 128:11–45.

Kawata, M. 2002. Invasion of empty niches and subsequent sympatric speciation. *Proceedings of the Royal Society London B* 269:55–63.

Kawata, M. and Yoshimura, J. 2000. Speciation by sexual selection in hybridizing populations without viability selection. *Evolutionary Ecology Research* 2:897–909.

Kawecki, T. J. 1996. Sympatric speciation driven by beneficial mutations. *Proceedings of the Royal Society London B* 263:1515–1520.

—. 1997. Sympatric speciation via habitat specialization driven by deleterious mutations. *Evolution* 51:1751–1763.

Kelly, J. K. and Noor, M. A. F. 1996. Speciation by reinforcement: a model derived from studies of *Drosophila. Genetics* 143:1485–1497.

Kiester, A. R., Lande, R., and Schemske, D. W. 1984. Models of coevolution and speciation in plants and their pollinators. *American Naturalist* 124:220–243.

Kimura, M. 1964. Diffusion models in population genetics. *Journal of Applied Probability* 1:177–232.

—. 1969. The number of heterozygous nucleotide sites maintained in a finite population due to steady flux of mutations. *Genetics* 61:893–903.

—. 1983. *The neutral theory of molecular evolution.* Cambridge University Press, New York.

—. 1985. The role of compensatory neutral mutations in molecular evolution. *Journal of Genetics* 64:7–19.

Kimura, M. and Crow, J. F. 1964. The theory of genetic loads. In *Proceedings of the XI International Congress of Genetics*, pages 495–506. The Hague, Netherlands.

Kimura, M. and Maruyama, T. 1966. The mutational load with epistatic gene interactions in fitness. *Genetics* 54:1337–1351.

Kimura, M. and Ohta, T. 1969. The average number of generation until fixation of a mutant gene in a finite population. *Genetics* 61:763–771.

King, M. 1993. *Species evolution. The role of chromosome change.* Cambridge University Press, Cambridge.

Kingsolver, J. G., Hoekstra, H. E., Hoekstra, J. M., Berrigan, D., Vignieri, S. N., Hill, C. E., Hoang, A., Gibert, P., and Beerli, P. 2001. The strength of phenotypic selection in natural populations. *American Naturalist* 157:245–261.

Kirkpatrick, M. 1982. Sexual selection and the evolution of female mate choice. *Evolution* 36:1–12.

—. 1987. Sexual selection by female choice in polygynous animals. *Annual Review of Ecology and Systematics* 18:43–70.

—. 2000. Reinforcement and divergence under assortative mating. *Proceedings of the Royal Society London B* 267:1649–1655.

—. 2001. Reinforcement during ecological speciation. *Proceedings of the Royal Society London B* 268:1259–1263.

Kirkpatrick, M. and Barton, N. H. 1997. Evolution of a species' range. *American Naturalist* 150:1–23.

Kirkpatrick, M. and Dugatkin, L. A. 1994. Sexual selection and the evolutionary effects of copying mate choice. *Behavioral Ecology and Sociobiology* 34:443–449.

Kirkpatrick, M. and Ravigné, V. 2002. Speciation by natural and sexual selection: models and experiments. *American Naturalist* 159:S22–S35.

Kirkpatrick, M. and Servedio, M. R. 1999. The reinforcement of mating preferences on an island. *Genetics* 151:865–884.

Kisdi, E. 1999. Evolutionary branching under asymmetric competition. *Journal of Theoretical Biology* 197:149–162.

Kisdi, E. and Geritz, S. A. H. 1999. Adaptive dynamics in allele space: evolution of genetic polymorphism by small mutations in a heterogeneous environment. *Evolution* 53:993–1008.

Kliman, R. M., Andolfatto, P., Coyne, J. A., Depaulis, F., Kreitman, M., Berry, A. J., McCarter, J., Wakeley, J., and Hey, J. 2000. The population genetics of the origin and divergence of the *Drosophila simulans* complex species. *Genetics* 156:1913–1931.

Koenig, W. D. and Ashley, M. V. 2003. Is pollen limited? The answer is blowin' in the wind. *Trends in Ecology and Evolution* 18:157–159.

Kolář, M. and Slanina, F. 2003. Selective advantage of topological disorder in biological evolution. *European Physical Journal B* 31:379–3848.

Kolmogorov, A., Petrovsky, I., and Piscounoff, N. 1937. Etude de l'equation de la diffusion avec croissance de la quantile de matiere et son applications a un problem biologique. *Bull. State Univ. Moscow. Section A. Math. Mech.* 1:1–25.

Kondrashov, A. S. 1983a. Multilocus model of sympatric speciation. I. One character. *Theoretical Population Biology* 24:121–135.

—. 1983b. Multilocus model of sympatric speciation. II. Two characters. *Theoretical Population Biology* 24:136–144.

—. 1986. Multilocus model of sympatric speciation. III. Computer simulation. *Theoretical Population Biology* 29:1–15.

—. 1987. Deleterious mutations and the evolution of sexual reproduction. *Na-*

ture 336:435–440.

—. 1992. The third phase of Wright's shifting-balance: a simple analysis of the extreme case. *Evolution* 46:1972–1975.

—. 2003. Accumulation of Dobzhansky-Muller incompatibilities within a spatially structured population. *Evolution* 57:151–153.

Kondrashov, A. S. and Kondrashov, F. A. 1999. Interactions among quantitative traits in the course of sympatric speciation. *Nature* 400:351–354.

Kondrashov, A. S. and Mina, S. I. 1986. Sympatric speciation: when is it possible? *Biological Journal of the Linnean Society* 27:201–223.

Kondrashov, A. S. and Shpak, M. 1998. On the origin of species by means of assortative mating. *Proceedings of the Royal Society London B* 265:2273–2278.

Kondrashov, A. S., Sunyaev, S., and Kondrashov, F. A. 2002. Dobzhansky-Muller incompatibilities in protein evolution. *Proceedings of the National Academy of Sciences USA* 99:14878–14883.

Kondrashov, A. S. and Yampolsky, L. Y. 1996. High genetic variability under the balance between symmetric mutation and fluctuating stabilizing selection. *Genetical Research* 68:157–164.

Kondrashov, A. S., Yampolsky, L. Y., and Shabalina, S. A. 1998. On the origin of species by means of natural selection. In D. J. Howard and S. H. Berlocher, editors, *Endless forms: Species and speciation*, pages 90–98. Oxford University Press, New York.

Koopman, K. F. 1950. Natural selection for reproductive isolation between *Drosophila pseudoobscura* and *Drosophila persimilis*. *Evolution* 4:135–148.

Kornfield, I. and Smith, P. F. 2000. African cichlid fishes: model systems for evolutionary biology. *Annual Review of Ecology and Systematics* 31:163–196.

Korol, A., Kirzhner, V. M., Ronin, E., and Nevo, E. 1996. Cyclical environmental changes as a factor maintaining genetic polymorphism. 2. Diploid selection for an additive trait. *Evolution* 50:1432–1441.

Korol, A., Rashkovetsky, E., Iliadi, K., Michalak, P., Ronin, E., and Nevo, E. 2000. Nonrandom mating in *Drosophila melanogaster* laboratory populations derived from closely adjacent ecologically contrasting slopes at "Evolution canyon". *Proceedings of the National Academy of Sciences USA* 97:12637–12642.

Kyriacou, C. P. 2002. Single gene mutations in *Drosophila*: what can they tell us about the evolution of sexual behavior? *Genetica* 116:197–203.

Laland, K. N. 1994. Sexual selection with a culturally transmitted mating preference. *Theoretical Population Biology* 45:1–15.

Lande, R. 1976. Natural selection and random genetic drift in phenotypic evolution. *Evolution* 30:314–334.

—. 1979. Effective deme size during long-term evolution estimated from rates of chromosomal rearrangement. *Evolution* 33:234–251.

—. 1980. Genetic variation and phenotypic evolution during allopatric speciation. *American Naturalist* 116:463–479.

—. 1981. Models of speciation by sexual selection on polygenic characters. *Proceedings of the National Academy of Sciences USA* 78:3721–3725.

—. 1982. Rapid origin of sexual isolation and character divergence in a cline. *Evolution* 36:213–223.

—. 1985a. Expected time for random genetic drift of a population between stable phenotypic states. *Proceedings of the National Academy of Sciences USA* 82:7641–7645.

—. 1985b. The fixation of chromosomal rearrangements in a subdivided population with local extinction and colonization. *Heredity* 54:323–332.

—. 1986. The dynamics of peak shifts and the pattern of morphological evolution. *Paleobiology* 12:343–354.

Lande, R. and Arnold, S. J. 1983. The measurement of selection on correlated characters. *Evolution* 37:1210–1226.

Lande, R. and Kirkpatrick, M. 1988. Ecological speciation by sexual selection. *Journal of Theoretical Biology* 133:85–98.

Lande, R., Seehausen, O., and van Alphen, J. J. M. 2001. Mechanisms of rapid sympatric speciation by sex reversal and sexual selection in cichlid fish. *Genetica* 112-113:435–443.

Lau, K. F. and Dill, K. A. 1989. A lattice statistical mechanics model of the conformational and sequence spaces of proteins. *Macromolecules* 22:3986–3997.

—. 1990. Theory for protein mutability and biogenesis. *Proceedings of the National Academy of Sciences USA* 87:638–642.

Lemke, N. and Campbell, I. A. 1996. Random walks in a closed space. *Physica A* 230:554–562.

Lessios, H. A. 1998. The first stage of speciation as seen in organisms separated by the Isthnus of Panama. In D. J. Howard and S. H. Berlocher, editors, *Endless forms: Species and speciation*, pages 186–201. Oxford University Press, New York.

Levene, H. 1953. Genetic equilibrium when more than one ecological niche is available. *American Naturalist* 87:331–333.

Levin, D. A. 2000. *The origin, expansion, and demise of plant species*. Oxford University Press, New York.

Lewontin, R., Kirk, D., and Crow, J. 1966. Selective mating, assortative mating, and inbreeding: definitions and implications. *Eugenics Quarterly* 15:141–143.

Lewontin, R. R. 1974. *The genetic basis of evolutionary change*. Columbia University Press, New York.

Li, W.-H. 1976. Distribution of nucleotide differences between two randomly chosen cistrons in a subdivided population: the finite island model. *Theoretical Population Biology* 10:303–308.

—. 1997. *Molecular evolution*. Sinauer Associates, Sunderland, MA.

Liberman, U. and Feldman, M. W. 1985. A symmetric two locus model with viability and fertility selection. *Journal of Mathematical Biology* 22:31–

60.
Liou, L. W. and Price, T. D. 1994. Speciation by reinforcement of premating isolation. *Evolution* 48:1451–1459.

Lipman, D. J. and Wilbur, W. J. 1991. Modeling neutral and selective evolution of protein folding. *Proceedings of the Royal Society London B* 245:7–11.

Littlejohn, M. J. 1965. Premating isolation in the *Hyla ewingi* complex (anura: Hylidae). *Evolution* 19:234–243.

Loeschcke, V. and Christiansen, F. B. 1984. Evolution and intraspecific exploitative competition. II. A two-locus model for additive genetic effects. *Theoretical Population Biology* 268:228–264.

Losos, J. B. 1998. Ecological and evolutionary determinants of the species-area relationship in Carribean anoline lizards. In P. Grant, editor, *Evolution on Islands*, pages 210–224. Oxford University Press, Oxford.

Losos, J. B. and Glor, R. E. 2003. Phylogenetic comparative methods and the geography of speciation. *Trends in Ecology and Evolution* 18:220–227.

Losos, J. B., Jackman, T. R., Larson, A., de Queiroz, K., and Rodriguez-Schettino, L. 1998. Contingency and determinism in replicated adaptive radiations of island lizards. *Science* 279:2115–2118.

Losos, J. B., Schoener, T. W., Warheit, K. I., and Creer, D. 2001. Experimental studies of adaptive differentiation in Bahamian *Anolis lizards*. *Genetica* 112:399–415.

Losos, J. B., Warheit, K. I., and Schoener, T. W. 1997. Adaptive differentiation following experimental island colonization in *Anolis* lizards. *Nature* 387:70–73.

Ludwig, W. 1950. Zur theorie der konkurrenz. die annidation (einnischung) als funfter evolutionsfaktor. *Zoologischer Anzeiger* 145 (Ergänzungsband):516–537.

Lupia, R. 1999. Discordant morphological disparity and taxonomic diversity during the cretaceous angiosperm radiation: North American pollen record. *Paleobiology* 25:1–28.

Lynch, J. D. 1989. The gauge of speciation: on the frequencies of modes of speciation. In D. Otte and A. Endler, J., editors, *Speciation and its consequences*, pages 527–553. Sinauer, Sunderland, MA.

Lynch, M. 2002. Gene duplication and evolution. *Science* 297:945–947.

Lynch, M. and Conery, J. S. 2000. The evolutionary fate and consequences of duplicate genes. *Science* 290:1151–1155.

Lynch, M. and Force, A. G. 2000. The origin of interspecific genomic incompatibility via gene duplication. *American Naturalist* 156:590–605.

Lynch, M. and Walsh, B. 1998. *Genetics and analysis of quantitative characters*. Sinauer, Sunderland, MA.

Machado, C. A., Kliman, R. M., Markert, J. A., and Hey, J. 2002. Inferring the history of speciation from multiple DNA sequence data: the case of *Drosophila pseudoobscura* and close relatives. *Molecular Biology and Evolution* 19:472–488.

Macnair, M. R. and Christie, P. 1983. Reproductive isolation as a pleiotropic effect of copper tolerance in *Mimulus guttatus*. *Heredity* 50:295–302.

Macnair, M. R. and Gardner, M. 1998. The evolution of edaphic endemics. In D. J. Howard and S. H. Berlocher, editors, *Endless forms: species and speciation*, pages 157–171. Oxford University Press, New York.

Majewski, J. and Cohan, F. M. 1998. The effect of mismatch repair and heteroduplex formation on sexual isolation in *Bacillus*. *Genetics* 148:13–18.

——. 1999. DNA sequence similarity requirements for interspecific recombination in *Bacillus*. *Genetics* 153:1525–1533.

Mallet, J. 1995. A species definition for the modern synthesis. *Trends in Ecology and Evolution* 10:294–299.

Mani, G. S. and Clarke, B. C. C. 1990. Mutational order: a major stochastic process in evolution. *Proceedings of the Royal Society London B* 240:29–37.

Manzo, F. and Peliti, L. 1994. Geographic speciation in the Derrida-Higgs model of species formation. *Journal of Physics A: Mathematical and General* 27:7079–7086.

Markert, J. A., Arnegard, M. E., Danley, P. D., and Kocher, T. D. 1999. Biogeography and population genetics of the Lake Malawi cichlid *Melanochromis auratus*: habitat transience, philopatry and speciation. *Molecular Ecology* 8:1013–1026.

Martin, O. and Hosken, D. 2003. The evolution of reproductive isolation through sexual conflict. *Nature* 423:979–982.

Martinez, M. A., Pezo, V., Marlière, P., and Wain-Hobson, S. 1996. Exploring the functional robustness of an enzyme by *in vitro* evolution. *EMBO Journal* 15:1203–1210.

Matessi, C. and Cori, R. 1972. Models of population genetics of Batesian mimicry. *Theoretical Population Biology* 3:41–68.

Matessi, C., Gimelfarb, A., and Gavrilets, S. 2001. Long term buildup of reproductive isolation promoted by disruptive selection: how far does it go? *Selection* 2:41–64.

Mather, K. 1947. Species crosses in *antirrhinum*. I. Genetic isolation of the species *Majus, Glutinosum* and *Orontium*. *Heredity* 1:175–186.

——. 1955. Polymorphism as an outcome of disruptive selection. *Evolution* 9:52–61.

May, R. M. 1990. How many species? *Philosophical Transactions of the Royal Society London B* 330:293–304.

Maynard Smith, J. 1962a. Disruptive selection, polymorphism and sympatric speciation. *Nature* 195:60–62.

——. 1962b. The limitations of molecular evolution. In I. J. Good, editor, *The scientist speculates: An anthology of partly-baked ideas*, pages 252–256. Basic Books, New York.

——. 1966. Sympatric speciation. *American Naturalist* 100:637–650.

——. 1970. Natural selection and the concept of a protein space. *Nature* 225:563–564.

——. 1982. *Evolution and the theory of games*. Cambridge University Press,

Cambridge.

—. 1983. The genetics of stasis and punctuation. *Annual Review of Genetics* 17:11–25.

Maynard Smith, J. and Szathmáry, E. 1995. *The major transitions in evolution.* W. H. Freeman & Company/Spectrum, Oxford.

Mayr, E. 1942. *Systematics and the origin of species.* Columbia University Press, New York.

—. 1947. Ecological factors in speciation. *Evolution* 1:263–288.

—. 1954. Change of genetic environment and evolution. In C. Barigozzi, editor, *Evolution as a process*, pages 1–19. Liss, New York.

—. 1963. *Animal species and evolution.* Belknap Press, Cambridge, MA.

—. 1982. Processes of speciation in animals. In C. Barigozzi, editor, *Mechanisms of speciation*, pages 1–19. Liss, New York.

Mayr, E. and Diamond, J. 2001. *The Birds of Northern Melanesia: Speciation, Ecology, and Biogeography.* Oxford University Press, New York.

McCune, A. R. 1997. How fast is speciation? Molecular, geological, and phylogenetic evidence from adaptive radiations of fishes. In T. J. Givnish and K. Sytsma, editors, *Molecular evolution and adaptive radiation*, pages 585–610. Cambridge University Press, Cambridge.

McKinnon, J. S. and Rundle, H. D. 2002. Speciation in nature: the threespine stickleback model systems. *Trends in Ecology and Evolution* 17:480–488.

McNeily, T. and Antonovics, J. 1968. Evolution in closely adjacent populations. IV. Barriers to gene flow. *Heredity* 23:205–218.

Metz, J. A. J., Geritz, S. A. H., Meszéna, G., Jacobs, F. J. A., and van Heerwaarden, J. S. 1996. Adaptive dynamics, a geometrical study of the consequences of nearly faithful reproduction. In S. J. van Strien and S. M. Verduyn Lunel, editors, *Stochastic and spatial structures of dynamical systems*, pages 183–231. North Holland, Amsterdam.

Michalak, P., Minkov, I., Helin, A., Lerman, D. N., Bettencourt, B. R., Feder, M. E., Korol, A. B., and Nevo, E. 2001. Genetic evidence for adaptation-driven incipient speciation of *Drosophila melanogaster* along a microclimatic contrast in "Evolution Canyon", Israel. *Proceedings of the National Academy of Sciences USA* 98:13191–13200.

Michalakis, I. and Slatkin, M. 1996. Interaction of selection and recombination in the fixation of negative-epistatic genes. *Genetical Research* 67:257–269.

Mitra, S., Landel, H., and Pruett-Jones, S. 1996. Species richness covaries with mating system in birds. *Auk* 113:544–551.

Mitton, J. B. 1997. *Selection in natural populations.* Oxford University Press, Oxford.

Montalvo, A. M. and Ellstrand, N. C. 2001. Nonlocal transplantation and outbreeding depression in the subshrub *Lotus scoparius (Fabaceae)*. *American Journal of Botany* 88:258–269.

Moore, F. B. G., Rozen, D. E., and Lenski, R. E. 2000. Pervasive compensatory adaptation in *Escherichia coli*. *Proceedings of the Royal Society London B* 267:515–522.

Moore, F. B. G. and Tonsor, S. J. 1994. A simulation of Wright shifting-balance process: migration and the 3 phases. *Evolution* 48:69–80.

Moore, W. S. 1979. A single locus mass-action model of assortative mating, with comments on the process of speciation. *Heredity* 42:173–186.

—. 1981. Assortative mating genes selected along a gradient. *Heredity* 46:191–195.

Morrow, E., Pitcher, T. E., and Arnqvist, G. 2003. No evidence that sexual selection is an "engine of speciation" in birds. *Ecology Letters* 6:228–234.

Muller, H. J. 1940. Bearing of the *Drosophila* work on systematics. In J. Huxley, editor, *The new systematics*, pages 185–268. Oxford University Press, Oxford.

—. 1942. Isolating mechanisms, evolution and temperature. *Biological Symposia* 6:71–125.

Murray, J. D. 1989. *Mathematical biology*. Springer-Verlag, Berlin.

Nagylaki, T. 1975. Conditions for the existence of clines. *Genetics* 80:595–615.

—. 1992. *Introduction to theoretical population genetics*. Springer-Verlag, Berlin.

Naisbit, R. E., Jiggins, C. D., and Mallet, J. 2001. Disruptive sexual selection against hybrids contributes to speciation between *Heliconius cydno* and *Heliconius melpomene*. *Proceedings of the Royal Society London B* 268:1–6.

Nakajima, A. and Aoki, K. 2002. On Richerson and Boyd's model of cultural evolution by sexual selection. *Theoretical Population Biology* 61:73–81.

Navaro, A. and Barton, N. H. 2003a. Accumulating postzygotic isolation genes in parapatry: a new twist on chromosomal speciation. *Evolution* 57:447–459.

—. 2003b. Chromosomal speciation and molecular divergence: accelerated evolution in rearranged chromosomes. *Science* 300:321–324.

Naveira, H. F. and Masida, X. R. 1998. The genetic of hybrid male sterility in *Drosophila*. In D. J. Howard and S. H. Berlocher, editors, *Endless forms: Species and speciation*, pages 330–338. Oxford University Press, New York.

Nee, S., Harvey, P. H., and May, R. M. 1991. Lifting the veil on abundance patterns. *Proceedings of the Royal Society London B* 243:161–163.

Nei, M. 1972. Genetic distance between populations. *American Naturalist* 106:283–292.

—. 1976. Mathematical models of speciation and genetic distance. In S. Karlin and E. Nevo, editors, *Population genetics and ecology*, pages 723–768. Academic Press, New York.

—. 1987. *Molecular evolutionary genetics*. Columbia University Press, New York.

Nei, M. and Graur, D. 1984. Extent of protein polymorphism and the neutral mutation theory. *Evolutionary Biology* 17:73–118.

Nei, M., Maruyama, T., and Wu, C.-I. 1983. Models of evolution of reproductive isolation. *Genetics* 103:557–579.

Nei, M. and Roychoudhury, A. K. 1973. Probability of fixation of nonfunctional genes at duplicate loci. *American Naturalist* 107:362–372.

Nevo, E. 1997. Evolution in action across phylogeny caused by microclimatic stresses at "Evolution canyon". *Theoretical Population Biology* 52:231–243.

———. 2001. *Adaptive radiation of blind subterranean mole rats.* Backhuys Publishers, Leiden.

Newman, C. M., Cohen, J. E., and Kipnis, C. 1985. Neo-Darwinian evolution implies punctuated equilibrium. *Nature* 315:400–401.

Newman, M. E. J. and Engelhardt, R. 1998. Effects of selective neutrality on the evolution of molecular species. *Proceedings of the Royal Society London B* 265:1333–1338.

Noest, A. J. 1997. Instability of the sexual continuum. *Proceedings of the Royal Society London B* 264:1389–1393.

Noor, M. A. F. 1999. Reinforcement and other consequences of sympatry. *Heredity* 83:503–508.

Noor, M. A. F., Grams, K. L., Bertucci, L. A., and Reiland, J. 2001. Chromosomal inversions and the persistence of species. *Proceedings of the National Academy of Sciences USA* 98:12084–12088.

Norris, J. R. 1997. *Markov chains.* Cambridge University Press, Cambridge.

Nosil, P., Crespi, B. J., and Sandoval, C. P. 2003. Reproductive isolation driven by the combined effects of ecological adaptation and reinforcement. *Proceedings of the Royal Society London B* 270:1911–1918.

Nuismer, S. L., Thompson, J. N., and Gomulkiewicz, R. 1999. Gene flow and geographically structured coevolution. *Proceedings of the Royal Society London B* 266:605–609.

———. 2000. Coevolutionary clines across selection mosaics. *Evolution* 54:1102–1115.

Ödeen, A. and Florin, A.-B. 2000. Effective population size may limit the power of laboratory experiments to demonstrate sympatric and parapatric speciation. *Proceedings of the Royal Society London B* 267:601–606.

———. 2002. Sexual selection and peripatric speciation: the Kaneshiro model revisited. *Journal of Evolutionary Biology* 15:301–306.

O'Donald, P. 1960. Assortative mating in a population in which two alleles are segregating. *Heredity* 15:389–396.

———. 1963. Sexual selection for dominant and recessive genes. *Heredity* 18:451–457.

Ogden, R. and Thorpe, R. S. 2002. Molecular evidence for ecological speciation in tropical habitats. *Proceedings of the National Academy of Sciences USA* 99:13612–13615.

Ohta, T. 1972. Population size and rate of evolution. *Molecular Evolution* 1:305–314.

———. 1992. The nearly neutral theory of molecular evolution. *Annual Review of Ecology and Systematics* 23:263–286.

———. 1995. Synonymous and nonsynonymous substitutions in mammalian genes and the nearly neutral theory. *Journal of Molecular Evolution* 40:56–63.

——. 1997. The meaning of near-neutrality at coding and non-coding regions. *Gene* 205:261–267.

——. 1998. Evolution by nearly-neutral mutations. *Genetica* 102/103:83–90.

Ohta, T. and Kimura, M. 1973. A model of mutation appropriate to estimate the number of electrophoretically detectable alleles in a finite population. *Genetical Research* 22:201–204.

Omholt, S. W., Plahte, E., Oyehaug, L., and Xiang, K. F. 2000. Gene regulatory networks generating the phenomena of additivity, dominance and epistasis. *Genetics* 155:969–980.

Orr, H. A. 1995. The population genetics of speciation: the evolution of hybrid incompatibilities. *Genetics* 139:1803–1813.

——. 1997. Dobzhansky, Bateson, and the genetics of speciation. *Genetics* 144:1331–1335.

——. 1998. The population genetics of adaptation: the distribution of factors fixed during adaptive evolution. *Evolution* 52:935–949.

——. 2001. The genetic of species differences. *Trends in Ecology and Evolution* 16:343–350.

Orr, H. A. and Orr, L. H. 1996. Waiting for speciation: the effect of population subdivision on the waiting time to speciation. *Evolution* 50:1742–1749.

Orr, H. A. and Presgraves, D. C. 2000. Speciation by postzygotic isolation: forces, genes and molecules. *BioEssays* 22:1085–1094.

Orr, H. A. and Turelli, M. 2001. The evolution of postzygotic isolation: accumulating Dobzhansky-Muller incompatibilities. *Evolution* 55:1085–1094.

Pál, C. and Miklós, I. 1999. Epigenetic inheritance, genetic assimilation and speciation. *Journal of Theoretical Biology* 200:19–37.

Palumbi, S. R. 1992. Marine speciation on a small planet. *Trends in Ecology and Evolution* 7:114–118.

——. 1998. Species formation and the evolution of gamete recognition loci. In D. J. Howard and S. H. Berlocher, editors, *Endless forms: Species and speciation*, pages 271–278. Oxford University Press, New York.

——. 1999. All males are not created equal: fertility differences depend on gamete recognition polymorphisms in sea urchins. *Proceedings of the National Academy of Sciences USA* 967:12632–12637.

Panhuis, T. M., Butlin, R., Zuk, M., and Tregenza, T. 2001. Sexual selection and speciation. *Trends in Ecology and Evolution* 16:364–371.

Parker, G. A. 1979. Sexual selection and sexual conflict. In M. S. Blum and N. A. Blum, editors, *Sexual selection and reproductive competition in insects*, pages 123–166. Academic Press, New York.

Parker, G. A. and Partridge, L. 1998. Sexual conflict and speciation. *Philosophical Transactions of the Royal Society London B* 353:261–274.

Patterson, H. E. H. 1985. The recognition concept of species. In E. S. Vrba, editor, *Species and speciation*, pages 21–29. Transvaal Museum Monograph, Pretoria, South Africa.

Peck, S. L., Ellner, S. P., and Gould, F. 1998. A spatially explicit stochastic model demonstrates the feasibility of Wright's shifting balance theory. *Evolution* 52:1834–1839.

Peliti, L. and Bastolla, U. 1994. Collective adaptation in a statistical model of an evolving population. *Comptes Rendus de l'Academie des Sciences Series III: Sciences de la Vie – Life Sciences* 317:371–374.

Phillips, P. C. 1993. Peak shifts and polymorphism during phase three of Wright's shifting-balance process. *Evolution* 47:1733–1743.

—. 1996. Waiting for a compensatory mutation: phase zero of the shifting-balance process. *Genetical Research* 67:271–283.

Phillips, P. C. and Johnson, N. A. 1998. The population genetics of synthetic lethals. *Genetics* 150:449–458.

Piálek, J. and Barton, N. H. 1997. The spread of advantageous allele across a barrier: the effects of random drift and selection against heterozygotes. *Genetics* 145:493–504.

Pimm, S. L. 1979. Sympatric speciation: a simulation study. *Biological Journal of the Linnean Society* 11:131–139.

Podos, J. 2001. Correlated evolution of morphology and vocal signal structure in Darwin's finches. *Nature* 409:185–188.

Pomiankowski, A. 1987. The costs of choice in sexual selection. *Journal of Theoretical Biology* 128:195–218.

Presgraves, D. C. 2002. Patterns of postzygotic isolation in lepidoptera. *Evolution* 56:1168–1183.

Presgraves, D. C., Balagolapan, L., Abmayr, S. M., and Orr, H. A. 2003. Adaptive evolution drives divergence of a hybrid incompatibility gene between two species of *Drosophila*. *Nature* 423:715–719.

Price, T. and Bouvier, M. M. 2002. The evolution of F_1 postzygotic incompatibilities in birds. *Evolution* 56:2083–2089.

Price, T., Turelli, M., and Slatkin, M. 1993. Peak shift by correlated response to selection. *Evolution* 47:280–290.

Prokopy, R. J., Diehl, S. R., and Cooley, S. S. 1988. Behavioral evidence for host races in *Rhagoletis pomonella* flies. *Oecologia* 76:138–147.

Provine, W. B. 1986. *Sewall Wright and evolutionary biology*. University of Chicago Press, Chicago.

—. 1989. Founder effects and genetic revolutions in microevolution and speciation: a historical perspective. In L. V. Gidding, K. Y. Kaneshiro, and W. W. Anderson, editors, *Genetics, speciation and the founder principle*, pages 43–76. Oxford University Press, Oxford.

Raup, D. M. 1991. A kill curve for phanerozoic marine species. *Paleobiology* 17:37–48.

Reidys, C. M. 1997. Random induced subgraphs of generalized n-cubes. *Advances in Applied Mathematics* 19:360–377.

Reidys, C. M., Forst, C. V., and Schuster, P. 2001. Replication and mutation on neutral networks. *Bulletin of Mathematical Biology* 63:57–94.

Reidys, C. M. and Stadler, P. F. 2001. Neutrality in fitness landscapes. *Applied Mathematics and Computation* 117:321–350.

—. 2002. Combinatorial landscapes. *SIAM Review* 44:3–54.

Reidys, C. M., Stadler, P. F., and Schuster, P. 1997. Generic properties of combinatory maps: neutral networks of RNA secondary structures. *Bulletin of Mathematical Biology* 59:339–397.

Ribera, I., Barraclough, T., and Vogler, A. 2001. The effect of habitat type on speciation rates and range movements in aquatic beetles: inferences from species-level phylogenies. *Molecular Ecology* 10:721–735.

Rice, W. R. 1984. Disruptive selection on habitat preferences and the evolution of reproductive isolation. *Evolution* 38:1251–1260.

—. 1985. Disruptive selection on habitat preferences and the evolution of reproductive isolation: an exploratory experiment. *Evolution* 39:645–656.

—. 1996. Sexually antagonistic male adaptation triggered by experimental arrest of female evolution. *Nature* 381:232–234.

Rice, W. R. and Hostert, E. E. 1993. Laboratory experiments on speciation: what have we learned in 40 years? *Evolution* 47:1637–1653.

Rice, W. R. and Salt, G. 1988. Speciation via disruptive selection on habitat preference: experimental evidence. *American Naturalist* 131:911–917.

—. 1990. The evolution of reproductive isolation as a correlated character under sympatric conditions: experimental evidence. *Evolution* 44:1140–1152.

Richerson, P. and Boyd, R. 1989. The role of evolved predispositions in cultural evolution. *Ethology and Sociobiology* 10:195–219.

Ridley, M. 1993. *Evolution*. Blackwell Scientific Publications, Boston.

Riechert, S. E. 1993. Investigation of potential gene flow limitation of behavioral adaptation in an aridland spider. *Behavioral Ecology and Sociobiology* 32:355–363.

Riechert, S. E., Singer, F. D., and Jones, T. C. 2001. High gene flow levels lead to gamete wastage in a desert spider system. *Genetica* 112-113:297–319.

Rieseberg, L. H. 1995. The role of hybridization in evolution: old wine in new skins. *American Journal of Botany* 82:944–953.

—. 1999. Hybrid origin of plant species. *Annual Review of Ecology and Systematics* 28:359–389.

—. 2001. Chromosomal arrangements and speciation. *Trends in Ecology and Evolution* 16:351–358.

Ritchie, M. G. and Phillips, S. D. F. 1998. The genetics of sexual isolation. In D. J. Howard and S. H. Berlocher, editors, *Endless forms: Species and speciation*, pages 291–308. Oxford University Press, New York.

Rosenzweig, M. L. 1978. Competitive speciation. *Biological Journal of the Linnean Society* 10:275–289.

Roberts, M. S. and Cohan, F. M. 1993. The effect of DNA sequence divergence on sexual isolation in *Bacillus*. *Genetics* 134:401–408.

Rosenzweig, M. L. 1995. *Species diversity in space and time*. Cambridge University Press, Cambridge.

Rost, B. 1997. Protein structures sustain evolutionary drift. *Folding & Design* 2:S19–S24.

Roughgarden, J. 1972. Evolution of niche width. *American Naturalist* 106:683–

718.

—. 1979. *Theory of population genetics and evolutionary ecology: An introduction*. Macmillan, New York.

Rouhani, S. and Barton, N. H. 1987. The probability of peak shifts in a founder population. *Journal of Theoretical Biology* 126:51–62.

Ryan, M. J. and Wilczynski, W. 1988. Coevolution of sender and receiver: effects on local mate preference in cricket frogs. *Science* 240:1786–1788.

Sadedin, S. and Littlejohn, M. J. 2003. A spatially explicit individual-based model of reinforcement in hybrid zones. *Evolution* 57:962–970.

Sanderson, N. 1989. Can gene flow prevent reinforcement? *Evolution* 43:1223–1235.

Sasa, M. M., Chippindale, P. T., and Johnson, N. A. 1998. Patterns of postzygotic isolation in frogs. *Evolution* 52:1811–1820.

Sawyer, S. 1977a. Asymptotic properties of the equilibrium probability of identity in a geographically structured population. *Advances in Applied Probability* 9:268–282.

—. 1977b. Rates of consolidation in a selectively neutral migration model. *Annals of Probability* 5:486–493.

—. 1979. A limit theorem for patch size in a selectively-neutral migration model. *Journal of Applied Probability* 16:482–495.

Sawyer, S. and Hartl, D. 1981. On the evolution of behavioral reproductive isolation: the Wallace effect. *Theoretical Population Biology* 19:261–273.

Schluter, D. 1994. Experimental evidence that competition promotes divergence in adaptive radiation. *Science* 266:798–801.

—. 1998. Ecological causes of speciation. In D. J. Howard and S. H. Berlocher, editors, *Endless forms: Species and speciation*, pages 114–129. Oxford University Press, New York.

—. 2000. *The ecology of adaptive radiation*. Oxford University Press, Oxford.

Schmalhausen, I. I. 1949. *Factors of evolution: the theory of stabilizing selection*. University of Chicago Press, Chicago.

Schmidt, P. S. and Rand, D. M. 1999. Intertidal microhabitat and selection at Mpi: Interlocus contrasts in the northern acorn barnacle, *Semibalanus balanoides*. *Evolution* 53:135–146.

Schultes, E. A. and Bartel, D. P. 2000. One sequence, two ribozymes: implications for the emergence of new ribozyme folds. *Science* 289:448–452.

Schuster, P. 1995. How to search for RNA structures. theoretical concepts in evolutionary biotechnology. *Journal of Biotechnology* 41:239–257.

Schuster, P., Fontana, W., Stadler, P. F., and Hofacker, I. L. 1994. From sequences to shapes and back: a case study in RNA secondary structures. *Proceedings of the Royal Society London B* 255:279–284.

Schuster, P. and Swetina, J. 1982. Stationary mutant distributions and evolutionary optimization. *Bulletin of Mathematical Biology* 50:635–660.

Seehausen, O., van Alphen, J. J. M., and Lande, R. 1999. Color polymorphism and sex ratio distortion in a cichlid fish as an incipient stage in sympatric speciation by sexual selection. *Ecology Letters* 2:367–378.

Seger, J. 1985a. Intraspecific resource competition as a cause of sympatric speciation. In P. J. Greenwood, P. H. Harveyand, and M. Slatkin, editors, *Evolution*, pages 43–53. Cambridge University Press, Cambridge.

—. 1985b. Unifying genetic models for the evolution of female choice. *Evolution* 39:1185–1193.

—. 1988. Dynamics of some simple host-parasite models with more than two genotypes in each species. *Philosophical Transactions of the Royal Society London B* 319:541–555.

Sepkoski, J. J. 1998. Rates of speciation in the fossil record. *Philosophical Transactions of the Royal Society London B* 353:315–326.

Servedio, M. R. 2000. Reinforcement and the genetic of nonrandom mating. *Evolution* 54:21–29.

Servedio, M. R. and Kirkpatrick, M. 1997. The effects of gene flow on reinforcement. *Evolution* 51:1764–1772.

Shaw, K. L. 1998. Species and the diversity of natural groups. In D. J. Howard and S. H. Berlocher, editors, *Endless forms: Species and speciation*, pages 44–56. Oxford University Press, New York.

Sherrington, D. and Kirkpatrick, S. 1975. Solvable model of a spin-glass. *Physical Review Letters* 35:1792–1795.

Shpak, M. and Kondrashov, A. S. 1999. Applicability of the hypergeometric phenotypic model to haploid and diploid populations. *Evolution* 53:600–604.

Simpson, G. G. 1953. *The major features of evolution*. Columbia University Press, New York.

Singh, R. S. 1990. Patterns of species divergence and genetic theories of speciation. In K. Wöhrmann and C. K. Jain, editors, *Population biology: Ecological and evolutionary viewpoints*, pages 231–265. Springer-Verlag, Berlin.

Slatkin, M. 1972. On treating the chromosomes as the unit of selection. *Genetics* 72:157–168.

—. 1976. The rate of spread of an advantageous allele in a subdivided population. In S. Karlin and E. Nevo, editors, *Population genetics and ecology*, pages 767–780. Academic Press, New York.

—. 1978. Spatial patterns in the distributions of polygenic characters. *Journal of Theoretical Biology* 70:213–228.

—. 1979. Frequency- and density-dependent selection on a quantitative character. *Genetics* 93:755–771.

—. 1981. Fixation probabilities and fixation times in a subdivided population. *Evolution* 35:477–488.

—. 1982. Pleiotropy and parapatric speciation. *Evolution* 36:263–270.

—. 1987. The average number of sites separating DNA sequences drawn from a subdivided population. *Theoretical Population Biology* 32:42–49.

—. 1996. In defense of founder-flush theories of speciation. *American Naturalist* 147:493–505.

Smith, H. M. 1955. The perspective of species. *Turtox News* 33:74–77.

—. 1965. More evolutionary terms. *Systematic Zoology* 14:57–58.

—. 1969. Parapatry: sympatry or allopatry? *Systematic Zoology* 18:254–255.

Smith, T. B. 1993. Disruptive selection and the genetic basis of bill size polymorphism in the African finch *Pyrenestes*. *Nature* 363:618–620.

Smith, T. B. and Skúlason, S. 1996. Evolutionary significance of resource polymorphisms in fishes, amphibians, and birds. *Annual Review of Ecology and Systematics* 27:111–133.

Smith, T. B., Wayne, R. K., Girman, D. J., and Bruford, M. W. 1997. A role for ecotones in generating rainforest biodiversity. *Science* 276:1855–1857.

Sokal, R. R. and Rohlf, F. J. 1995. *Biometry*. Freeman and Company, New York.

Soshnikov, A. and Sudakov, B. 2003. On the largest eigenvalue of a random subgraph of the hypercube. *Communications in Mathematical Physics* 239:53–63.

Sota, T., Kusumoto, F., and Kubota, K. 2000. Consequences of hybridization between Ohomopterus insulicola and O-arrowianus (*Coleoptera, Carabidae*) in a segmented river basin: parallel formation of hybrid swarms. *Biological Journal of the Linnean Society* 71:297–313.

Spencer, H. G., McArdle, B. H., and Lambert, D. M. 1982. A theoretical investigation of speciation by reinforcement. *American Naturalist* 128:241–262.

Spirito, F. 1987. The reduction of gene exchange due to a prezygotic isolating mechanisms with monogenic inheritance. *Theoretical Population Biology* 32:216–239.

Spirito, F., Rossi, C., and Rizzoni, M. 1983. The reduction of gene flow due to partial sterility of heterozygotes for a chromosome mutation. I. Studies on a "neutral" gene not linked to the chromosome mutation in a two population model. *Evolution* 37:785–797.

Spirito, F. and Sampogna, F. 1995. Conditions for the maintenance of prezygotic and zygotic isolation in a continent-island model. *Theoretical Population Biology* 48:235–250.

Stadler, P. F. and Schnabl, W. 1992. The landscape of the traveling salesman problem. *Physical Letters A* 161:337–344.

Stam, P. 1983. The evolution of reproductive isolation in closely adjacent populations through differential flowering time. *Heredity* 50:105–118.

Stanley, S. M. 1986. Population size, extinction, and speciation: the fission effect in Neogene Bivalvi. *Paleobiology* 12:89–110.

—. 1990. Adaptive radiation and macroevolution. In P. D. Taylor and G. P. Larwood, editors, *Major evolutionary radiations*, pages 1–15. Clarendon Press, Oxford.

Stephan, W. 1996. The rate of compensatory evolution. *Genetics* 144:419–426.

Stockley, P. 1997. Sexual conflict resulting from adaptations to sperm competition. *Trends in Ecology and Evolution* 12:154–159.

Stork, N. E. 1993. How many species are there? *Biodiversity and Conservation* 35:321–337.

Stratton, D. A. and Bennington, C. C. 1998. Fine-grained spatial and temporal variation in selection does not maintain genetic variation in *Erigeron annuus*. *Evolution* 52:678–691.

Streelman, J. T. and Danley, P. D. 2003. The stages of vertebrate evolutionary radiation. *Trends in Ecology and Evolution* 18:126–131.

Strobeck, C. 1974. Sufficient conditions for polymorphism with N niches and M mating groups. *American Naturalist* 108:152–156.

—. 1987. Average number of nucleotide differences in a sample from a single subpopulation: a test for population subdivision. *Genetics* 117:149–153.

Stutt, A. D. and Siva-Jothy, M. T. 2001. Traumatic insemination and sexual conflict in the bed bug *Cimex lectulariu*. *Proceedings of the National Academy of Sciences USA* 98:5683–5687.

Sugihara, G. 1980. How do species divide resources? *American Naturalist* 116:770–787.

Sved, J. A. 1981a. A two-sex polygenic model for the evolution of premating isolation. I. Deterministic theory for natural populations. *Genetics* 97:197–215.

—. 1981b. A two-sex polygenic model for the evolution of premating isolation. II. Computer simulation of experimental selection procedures. *Genetics* 97:217–235.

Swanson, W. J., Aquadro, C. F., and Vacquier, V. D. 2001. Polymorphism in abalone fertilization proteins is consistent with the neutral evolution of the egg's receptor for lysin (VERL) and positive Darwinian selection of sperm lysin. *Molecular Biology and Evolution* 18:376–383.

Tachida, H. and Iizuka, M. 1991. Fixation probability in spatially changing environments. *Genetical Research* 58:243–251.

Takeshi, M. 1993. Species abundance patterns and community structure. *Advances in Ecological Research* 24:111–186.

Takimoto, G., Higashi, M., and Yamamura, N. 2000. A deterministic genetic model for sympatric speciation by sexual selection. *Evolution* 54:1870–1881.

Tauber, C. A. and Tauber, M. J. 1977. Sympatric speciation based on allelic changes at 3 loci - evidence from natural populations in 2 habitats. *Science* 197:1298–1299.

Tauber, E., Roe, E., Costa, R., Hennessy, J. M., and Kyriacou, C. P. 2003. Temporal mating isolation driven by behavioral gene in *Drosophila*. *Current Biology* 13:140–145.

Templeton, A. R. 1980. The theory of speciation via the founder principle. *Genetics* 94:1011–1038.

—. 1989. The meaning of species and speciation: a genetic perspective. In D. Otte and J. A. Endler, editors, *Speciation and its consequences*, pages 3–27. Sinauer, Sunderland, MA.

—. 1994. The role of molecular genetics in speciation studies. In B. Schierwater, B. Streit, G. P. Wagner, and R. DeSalle, editors, *Molecular ecology and evolution: approaches and applications*, pages 455–477. Birkhaüser, Basel.

—. 1996. Experimental evidence for the genetic transilience model of speciation. *Evolution* 50:909–915.

—. 1998. Species and speciation: geography, population structure, ecology, and gene trees. In D. J. Howard and S. H. Berlocher, editors, *Endless forms: Species and speciation*, pages 32–43. Oxford University Press, New York.

Thompson, A. and Layzell, P. 2000. Evolution of robustness in an electrics design. In J. Miller, P. Thompson, P. Thomson, and T. Fogarty, editors, *Proceedings of the Third International Conference on Evolvable Systems: From Biology to Hardware*, volume 18 of LNCS, pages 218–228. Springer-Verlag, Berlin.

Thompson, J. N. 1994. *The coevolutionary process*. University of Chicago Press, Chicago.

—. 1999a. The evolution of species interactions. *Science* 284:2116–2118.

—. 1999b. Specific hypotheses on the geographic mosaic of coevolution. *American Naturalist* 153:S1–S14.

Thompson, V. 1986. Synthetic lethals: a critical review. *Evolutionary Theory* 8:1–13.

Tilley, S. G., Verrell, P. A., and Arnold, S. J. 1990. Correspondence between sexual isolation and allozyme differentiation: a test in the salamander *Desmognathus ochrophaeus*. *Proceedings of the National Academy of Sciences USA* 87:2715–2719.

Timoféeff-Ressovsky, N. W. and Svirezhev, Y. M. 1966. On adaptive polymorphism in populations of *Adalia bipunctata* (in Russian). *Problemy Kibernetiki* 16:137–146.

Ting, C.-T., Takahashi, A., and Wu, C.-I. 2001. Incipient speciation by sexual isolation in *Drosophila*: concurrent evolution at multiple loci. *Proceedings of the National Academy of Sciences USA* 98:6709–6713.

Ting, C.-T., Tsaur, S.-C., Wu, M.-L., and Wu, C.-I. 1998. A rapidly evolving homeobox at the site of a hybrid sterility gene. *Science* 282:1501–1504.

Tregenza, T. and Wedell, N. 2000. Genetic compatibility, mate choice and patterns of parentage. *Molecular Ecology* 9:1013–1027.

Turelli, M. and Barton, N. H. 1994. Genetic and statistical analyses of strong selection on polygenic traits: what, me normal? *Genetics* 138:913–941.

Turelli, M., Barton, N. H., and Coyne, J. A. 2001. Theory and speciation. *Trends in Ecology and Evolution* 16:330–343.

Turner, G. F. and Burrows, M. T. 1995. A model of sympatric speciation by sexual selection. *Proceedings of the Royal Society London B* 260:287–292.

Udovic, D. 1980. Frequency-dependent selection, disruptive selection, and the evolution of reproductive isolation. *American Naturalist* 116:621–641.

Vacquier, V. D. 1998. Evolution of gamete recognition proteins. *Science* 281:1995–1998.

Vacquier, V. D. and Lee, Y.-H. 1993. Abalone sperm lysin: unusual mode of evolution of a gamete recognition protein. *Zygote* 1:181–196.

Vacquier, V. D. and Moy, G. W. 1997. The fucose sulfate polymer of egg jelly binds to sperm REJ and is the inducer of the sea urchin sperm acrosome reaction. *Developmental Biology* 192:125–135.

Valentine, J. W. 1980. Determinants of diversity in higher taxonomic categories. *Paleobiology* 6:444–450.

van Dijk, P. and Bijlsma, R. 1994. Simulations of flowering time displacement between two cytotypes that form inviable hybrids. *Heredity* 72:522–535.

van Dooren, T. J. M. 1999. The evolutionary ecology of dominance-recessivity. *Journal of Theoretical Biology* 198:519–532.

—. 2003. Adaptive dynamics for Mendelian genetics. In U. Dieckmann, , and J. A. J. Metz, editors, *Elements of Adaptive Dynamics*. Oxford University Press, Oxford (in press).

van Doorn, G. S., Luttikhuizen, P. C., and Weissing, F. J. 2001. Sexual selection at the protein level drives the extraordinary divergence of sex-related genes during sympatric speciation. *Proceedings of the Royal Society London B* 268:2155–2161.

van Doorn, G. S. and Weissing, F. J. 2001. Ecological versus sexual selection models of sympatric speciation: a synthesis. *Selection* 2:17–40.

van Nimwegen, E. and Crutchfield, J. P. 2000. Optimizing epochal evolutionary search: population-size independent theory. *Computer Methods in Applied Mechanics and Engineering* 186:171–194.

—. 2001. Optimizing epochal evolutionary search: Population-size dependent theory. *Machine Learning* 45:77–114.

van Nimwegen, E., Crutchfield, J. P., and Huynen, M. 1999. Neutral evolution of mutational robustness. *Proceedings of the National Academy of Sciences USA* 96:9716–9720.

van Nimwegen, E., Crutchfield, J. P., and Mitchell, M. 1997a. Finite populations induce metastability in evolutionary search. *Physics Letters A* 229:144–150.

—. 1997b. Statistical dynamics of the royal road genetic algorithm. *Theoretical Computer Science* 229:41–102.

Van Tienderen, P. H. 1991. Evolution of generalists and specialists in spatially heterogeneous environments. *Evolution* 45:1317–1331.

Van Valen, L. 1973. A new evolutionary law. *Evolutionary Theory* 1:1–30.

Via, S. 2001. Sympatric speciation in animals: the ugly duckling grows up. *Trends in Ecology and Evolution* 16:381–390.

Vrba, E. S. 1985. Environment and evolution: alternative causes of the temporal distribution of evolutionary events. *South African Journal of Science* 81:229–236.

Vulić, M., Dionisio, F., Taddei, F., and Radman, M. 1997. Molecular keys to speciation: DNA polymorphism and the control of genetic exchange in enterobacteria. *Proceedings of the National Academy of Sciences USA* 94:9763–9767.

Waddington, C. H. 1957. *The strategy of genes*. Macmillan, New York.

Wade, M. J. and Goodnight, C. J. 1998. The theories of Fisher and Wright in the context of metapopulations: when nature does many small experiments. *Evolution* 52:1537–1553.

—. 2000. The ongoing synthesis: a reply to Coyne at al. *Evolution* 54:317–324.

Wade, M. J., Patterson, H., Chang, N. W., and Johnson, N. A. 1994. Postcopulatory, prezygotic isolation in flour beetles. *Heredity* 72:163–167.

Wagner, A. 1996. Redundant gene functions and natural selection. *Journal of Evolutionary Biology* 12:1–16.

Wagner, G. P. and Altenberg, L. 1996. Complex adaptation and the evolution of evolvability. *Evolution* 50:967–976.

Wagner, G. P., Booth, G., and Bagheri-Chaichian, H. 1997. A population genetic theory of canalization. *Evolution* 51:329–347.

Wagner, P. J. 1995. Testing evolutionary constraint hypothesis with early Paleozoic gastropods. *Paleobiology* 21:248–272.

Wagner, P. J. and Erwin, D. H. 1995. Phylogenetic patterns as tests of speciation models. In D. H. Erwin and R. L. Anstey, editors, *New approaches to speciation in the fossil record*, pages 87–122. Columbia University Press, New York.

Wake, D. B. 1997. Incipient species formation in salamanders of the *Ensatina* complex. *Proceedings of the National Academy of Sciences USA* 94:7761–7767.

Walsh, J. B. 1982. Rate of accumulation of reproductive isolation by chromosome rearrangements. *American Naturalist* 120:510–532.

Ward, R. D., Skibinski, D. O. F., and Woodwark, M. 1992. Protein heterozygosity, structure, and taxonomic differentiation. *Evolutionary Biology* 26:73–159.

Waxman, D. and Gavrilets, S. 2004. 20 questions on adaptive dynamics. *Journal of Evolutionary Biology* .

Weber, K. E. 1996. Large genetic change at small fitness cost in large populations of *Drosophila melanogaster* selected for wind tunnel flight: rethinking fitness surfaces. *Genetics* 144:205–213.

Welch, J. J. 2004. Rates of adaptation in complex genetic systems. Ph.D. thesis, University of Sussex.

Werth, C. R. and Windham, M. D. 1991. A model for divergent, allopatric speciation of polyploid pteridophytes resulting from silencing of duplicate-gene expression. *American Naturalist* 137:515–526.

White, M. J. D. 1978. *Modes of speciation*. Freeman, San Francisco.

Whitlock, M. C. 1995. Variance induced peak shift. *Evolution* 49:252–259.

Whitlock, M. C., Phillips, P. C., Moore, F. B.-M., and Tonsor, S. J. 1995. Multiple fitness peaks and epistasis. *Annual Reviews of Ecology and Systematics* 26:601–629.

Wiese, L. and Wiese, W. 1977. On speciation by evolution of gametic incompatibility: a model case in *Chlamydomonas*. *American Naturalist* 111:733–742.

Wilke, C. O. 2001a. Adaptive evolution on neutral landscapes. *Bulletin of Mathematical Biology* 63:715–730.

—. 2001b. Selection for fitness versus selection for robustness in RNA secondary structure folding. *Evolution* 55:2412–2420.

Wills, C. J. 1977. A mechanism for rapid allopatric speciation. *American Naturalist* 111:603–605.

Wilson, D. S. and Turelli, M. 1986. Stable underdominance and the evolutionary invasion of empty niches. *American Naturalist* 127:835–850.

Wolf, J. B., Brodie III, E. D., and Wade, M. J. 2000. *Epistasis and the evolutionary process*. Oxford University Press, New York.

Woodcock, G. and Higgs, P. G. 1996. Population evolution on a multiplicative single-peak fitness landscape. *Journal of Theoretical Biology* 179:61–73.

Wright, S. 1921. Systems of mating. III. Assortative mating based on somatic resemblance. *Genetics* 6:144–161.

—. 1931. Evolution in Mendelian populations. *Genetics* 16:97–159.

—. 1932. The roles of mutation, inbreeding, crossbreeding and selection in evolution. In D. F. Jones, editor, *Proceedings of the Sixth International Congress on Genetics*, volume 1, pages 356–366. Austin, Texas.

—. 1935. Evolution in populations in approximate equilibrium. *Journal of Genetics* 30:257–266.

—. 1941. On the probability of fixation of reciprocal translocations. *American Naturalist* 75:513–522.

—. 1969. *Evolution and the genetics of populations. Vol. 2. The theory of gene frequencies*. University of Chicago Press, Chicago.

—. 1982. The shifting balance theory and macroevolution. *Annual Review of Genetics* 16:1–19.

—. 1988. Surfaces of selective value revisited. *American Naturalist* 131:115–123.

Wu, C.-I. 1985. A stochastic simulation study of speciation by sexual selection. *Evolution* 39:66–82.

—. 2001. The genic view of the process of speciation. *Journal of Evolutionary Biology* 14:851–865.

Wu, C.-I. and Hollocher, H. 1998. Subtle is nature: the genetics of species differentiation and speciation. In D. J. Howard and S. H. Berlocher, editors, *Endless forms: Species and speciation*, pages 339–351. Oxford University Press, New York.

Wu, C.-I., Jonhson, N. A., and Palopoli, M. F. 1996. Haldane's rule and its legacy: why are there so many sterile males? *Trends in Ecology and Evolution* 11:281–284.

Wu, C.-I. and Palopoli, M. F. 1994. Genetics of postmating reproductive isolation in animals. *Annual Review of Genetics* 27:283–308.

Zawadzki, P., Roberts, M. S., and Cohan, F. M. 1995. The log-linear relationship between sexual isolation and sequence divergence in *Bacillus* transformation is robust. *Genetics* 140:917–932.

Zhivotovsky, L. A. and Gavrilets, S. 1992. Quantitative variability and multilocus polymorphism under epistatic selection. *Theoretical Population Biology* 42:254–283.

Zipf, G. K. 1949. *Human behavior and the principle of least effort*. Adison-Wesley, Reading, MA.

Index

abalone, 284

abiotic and biotic factors: and fitness, 40; and fitness landscapes, 400; and natural selection, 10, 120; and punctuated equilibrium, 409; spatially heterogeneity of, 235; spatially heterogeneous selection, induced by, 277, 401; and speciation, 190, 409, 416

absolute preference, in mating, 321, 323, 384

accumulation of genetic incompatibilities, 158-74

AD. *See* adaptive dynamics

adaptation (*see also* local adaptation), 21, 76, 103, 113, 333, 418; to macrohabitat, 413; to microhabitat, 413; in a neighborhood of fitness peak, 38; and shifting balance, 69

adaptive diversification: in *Anolis* lizards, 408; in Darwin's finches, 368

adaptive dynamics: description, 265-76; limitations, 274-77; and matching-based nonrandom mating, 362

adaptive evolution, 35

adaptive landscapes. *See* fitness landscapes

adaptive radiation, 228, 406-8, 417

adaptive speciation, 234

adaptive topographies. *See* fitness landscapes

adaptive walks, 50, 83

additive fitness regime, 37

additive quantitative character: disruptive selection on, 260, 261, 307, 313, 321, 325, 352-56, 364-68, 374-84, 393-96; fitness landscapes for, 41-46; in growing populations, 73-74; and nonrandom mating, 301-4, 321-27, 352-56, 364-68, 374-87, 391, 393-

96; and spatially heterogeneous selection, 243-48, 263-65, 393-96; stabilizing selection on, 43, 91-93, 243-48, 258, 260-62, 264, 271, 273, 319, 325, 345, 360, 375, 377, 381, 385, 386, 393, 395

additive quantitative traits. *See* additive quantitative character

additive selection, in Levene model, 237

African finch, 237

allele frequencies, changes in: by migration, 242; by migration and selection, 252, by mutation, 49, 57, 179; by nonrandom mating, 279, 294, 328; by selection, 23, 49, 55, 57, 179, 184, 236; by random drift and selection, 24, 58, 78-80

alleles: advantageous, 64, 78, 134, 136, 184, 187, 221-25, 226, 241, 335-37, 369; deleterious, 38, 64, 78, 91, 127-29, 139-41, 187, 336, 341, 388; incompatible, 121, 135, 141-42, 151, 223-25, 229; neutral, 55-56, 132-34, 137, 139, 140, 147-48, 153, 193, 222, 387; underdominant, 56, 68, 71-73, 153

allopatric speciation: in Bateson-Dobzhansky-Muller model, 131-37; definition, 11-14; and divergent degeneration of duplicated genes, 154-55; in the four-locus model, 157-58; law of distribution and, 12; mechanisms of, 401; in multidimensional Bateson-Dobzhansky-Muller model, 174-84; in one- and two-locus, multiallele models, 149-51; and sexual conflict, 370-74; and stochastic peak shifts, 53-66, 71-75; by sexual selection, 321-24, 386, 387; in the three-locus model,

MONOGRAPHS IN POPULATION BIOLOGY
Edited by Simon A. Levin and Henry S. Horn

Titles available in the series (by monograph number)

9 780691 119830